An Introduction to the Mathematics of Breathers

Basic Theory and Models

An Introduction to the Mathematics of Breathers

Basic Theory and Models

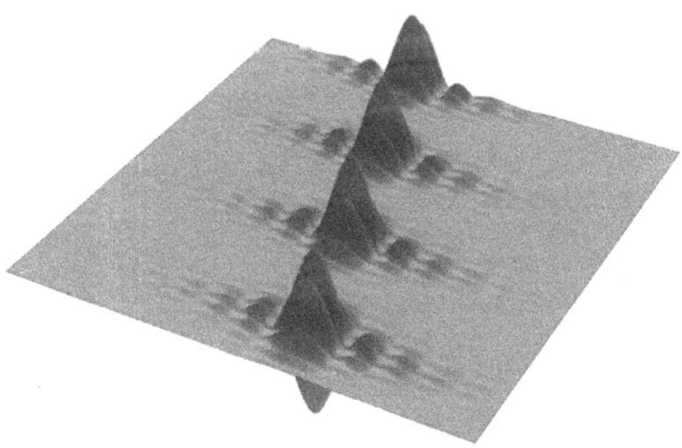

Miguel Á Alejo
University of Córdoba, Spain

Claudio Muñoz
University of Chile, Chile

World Scientific

NEW JERSEY • LONDON • SINGAPORE • BEIJING • SHANGHAI • HONG KONG • TAIPEI • CHENNAI • TOKYO

Published by

World Scientific Publishing Co. Pte. Ltd.
5 Toh Tuck Link, Singapore 596224
USA office: 27 Warren Street, Suite 401-402, Hackensack, NJ 07601
UK office: 57 Shelton Street, Covent Garden, London WC2H 9HE

British Library Cataloguing-in-Publication Data
A catalogue record for this book is available from the British Library.

AN INTRODUCTION TO THE MATHEMATICS OF BREATHERS
Basic Theory and Models

ISBN 978-981-98-3230-9 (hardcover)
ISBN 978-981-98-3231-6 (ebook for institutions)
ISBN 978-981-98-3232-3 (ebook for individuals)

For any available supplementary material, please visit
https://www.worldscientific.com/worldscibooks/10.1142/14785#t=suppl

Preface

This book is, as its title suggests, an introduction to the mathematics of continuous breathers. It serves as the starting point for a theory that has developed in many directions over the past twenty years and is now under intense scrutiny and active research. Accordingly, what is presented here are the first steps-the initial results needed to understand the main goals and challenges in the study of breathers.

Breathers are fascinating objects, with striking mathematical and visual representations that highlight the beauty of these phenomena. Geometrically, they can be elegantly characterized as curves with zero mean, while physically they represent dynamical, periodic motions in long-lived scalar field models. They are ubiquitous in almost every dispersive model with integrability properties, yet they also appear in nonintegrable equations. Consequently, for almost every proposed classification in breather theory, there exists a counterexample that does not satisfy the given definition. In this sense, breathers are notoriously difficult to classify.

There are several approaches to studying breathers, and here we present the most significant ones developed in recent years, with emphasis on rigorous proofs and results that have led to a limited but coherent understanding of these solutions. One may study breathers through integrable techniques such as the inverse scattering transform, or by using Bäcklund transformations to deduce specific properties. Alternatively, nonintegrable techniques can establish results applicable to breathers beyond integrable models. In any case, one feature remains constant: breathers capture highly intricate dynamics in the underlying theory while retaining a remarkable simplicity in the resolution of the problems they address.

The preparation of these notes began many years ago, in 2014. The task was not easy, as much of the theory was still evolving. More than a decade later, it is gratifying to see that numerous important results by many researchers have since emerged. This volume is therefore only the first installment in a planned series of three books on continuous rigorous breather theory. The two forthcoming volumes will focus, respectively, on the nonlinear Schrödinger model in full detail (not covered here), and on deeper topics and open questions. We expect both to be available in the near future, although the amount of material has grown enormously in recent years.

The opening chapter (Chapter 1) sets the stage by defining breathers, presenting their geometric and physical characterizations, and situating them within the broader context of dispersive equations. It introduces the main equations where breathers arise and outlines the challenges in their classification. The second chapter (Chapter 2) turns to the analytical framework, developing the functional setting and core estimates required for a precise mathematical treatment, and discussing how integrability and nonintegrability influence the behavior of solutions. The third chapter (Chapter 3) focuses on presenting the most important breather solutions. The following chapter (Chapter 4) explores the modified Korteweg-de Vries (mKdV) breather, probably the simplest breather in the literature. Foundational tools and results are introduced to the reader, and the most basic properties are proved. The most important proved property is the stability of these solutions. Chapter 5 then details how one can deal with mKdV breathers using the integrability theory, specially when using weighted estimates. Chapter 6 presents recent results related to existence of multi-breathers and the proposed cubic Whitham model, where a hypothesized breather is proposed. Chapter 7 deals now with numerical aspects related to the stability of breathers. Chapter 8 is devoted to the study of the Gardner breather, a close relative to the mKdV breather, which is very important when dealing with general nonlinearities. Chapter 9 now deals with the sine-Gordon. Chapter 10 describes recent results on the uniqueness of these breathers, and the parabolic case. Chapter 11 describes the zero-mean periodic in space breathers, where proposed stability results are reviewed, with a final section dealing with the only known examples of nonzero mean breathers. Finally, Chapter 12 deals with the Gardner and mKdV hierarchies, their breathers and their variational characterization. As mentioned before, we have chosen to write a different monograph dealing with the nonlinear Schrödinger breathers, and many deep topics such as nonexistence of breathers, asymptotic stability versus nonexistence, and higher dimensional breathers.

Balancing rigor with accessibility, this text serves mathematical, physical, and computational interests. By blending geometric intuition, analysis, and explicit constructions, it highlights the complexity and elegance of breathers, offering a coherent overview and inspiration for future research.

Miguel Á. Alejo and Claudio Muñoz

Contents

List of Figures

List of Tables

List of Acronyms

A	Akhmediev
BT	Bäcklund Transformation
BO	Benjamin-Ono
eKdV	extended Korteweg-de Vries
gKdV	generalized Korteweg-de Vries
IBVP	initial boundary value problem
ILW	Intermediate Long-Wave
ISM	Inverse Scattering Method
IST	Inverse Scattering Theory
JEFs	Jacobi Elliptic Functions
KdV	Korteweg-de Vries
KKSH	Kevrekidis, Khare, Saxena and Herring
KM	Kuznetsov-Ma
mKdV	modified Korteweg-de Vries
NLS	nonlinear Schrödinger
NSF	National Science Foundation
NVBC	nonvanishing boundary value or condition
ODE	Ordinary Differential Equation
P	Peregrine
PDE	Partial Differential Equation
SG	sine-Gordon
SS	Sasa-Satsuma
SY	Satsuma-Yajima
1D	one dimensional

Chapter 1

Introduction

1.1 Preface

These lecture notes are focused on describing, in a rigorous but still readable form, some recent mathematical results corresponding to the theory of continuous, periodic and non periodic *breathers*, as well as presenting to the reader a series of unsolved problems concerning more complicated solutions. Loosely speaking, a breather is a solution of a dispersive partial differential equation (PDE, usually Hamiltonian), which is periodic in time, and localized in space, although the second condition can be weakened by allowing periodicity or almost periodicity in space. Breathers are beautiful mathematical solutions describing a mixed behavior, they are oscillatory and localized, and seem very related to the notion of integrability: it is very difficult to find exact breathers in nonintegrable models.

Figure 1.1 describes the evolution of one of the most important of these objects, the mKdV breather. One can clearly see its really complex nature, starting from the oscillatory behavior around its center, its periodicity modulo translations, and its solitary wave behavior. In this book, our work will be to give a complete description of this area of research, that has been expanding constantly during past years.

1.2 What is a breather

This is not an easy question to answer. The main reason is that the definition depends on the model that you want to consider. However, there are some concurrent notions in every breather solution:

(1) it has some sort of time periodicity, at least under an appropriate space translation, or phase shift.

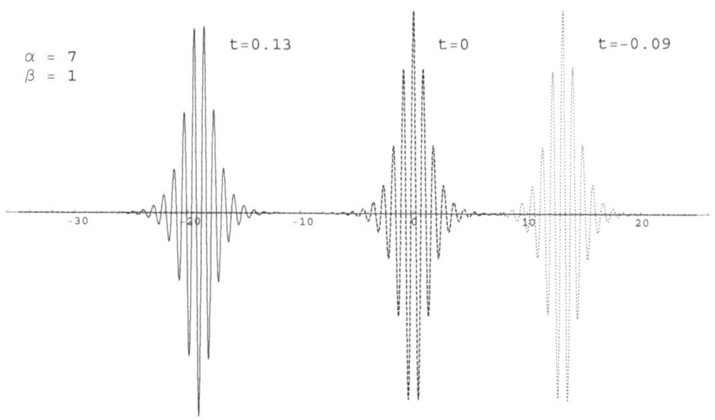

Fig. 1.1 Snapshots of an mKdV breather at three different times.

(2) Spatial decay is not mandatory, since periodic breathers are not satisfying this property.

However, in nearly all of this first part we will only consider localized breathers, in the sense that they spatially converge to zero at infinity.

1.3 What is not a breather

This is not universal, but for us, standard solitary waves will not be declared as breathers. One can give a formal argument as follows: breathers *do have two independent variables*, and solitary waves, up to the symmetries of the problem, have only one. This can be seen in Fig. 1.1: in addition to the movement in the x variable, there is an oscillatory behavior present that cannot be ruled out by a simple symmetry invariance. This makes breathers hard objects to describe.

1.4 The models

In this book we shall exemplify the above ideas with a (far from being exhaustive) account of breather solutions of the non periodic and periodic modified Korteweg-de Vries (mKdV) equation

$$u_t + (u_{xx} + u^3)_x = 0, \qquad u(t,x) \in \mathbb{R},\ t \in \mathbb{R},\ x \in \mathbb{R} \text{ or } \mathbb{T}_x := \mathbb{R}/L\mathbb{Z}; \qquad (1.4.1)$$

the Gardner equation

$$w_t + (w_{xx} + \mu w^2 + w^3)_x = 0, \quad \mu > 0, \qquad w(t,x) \in \mathbb{R},\ (t,x) \in \mathbb{R}^2; \qquad (1.4.2)$$

and the sine-Gordon (SG) equation

$$u_{tt} - u_{xx} + \sin u = 0, \quad (u, u_t) = (u, u_t)(t, x) \in \mathbb{R}^2, \ (t, x) \in \mathbb{R}^2. \tag{1.4.3}$$

The above equations are well-known *completely integrable* models [GKM68, AbCl91, Lam80], with infinitely many conserved quantities, and a suitable Lax-pair formulation. The Inverse Scattering Theory (IST) has been applied by many authors in order to describe the behavior of solutions in certain generality, see e.g. [Sch86, AbCl91, Lam80, PeGr97] and references therein. In particular, the evolution of *rapidly decaying* initial data can be described by purely algebraic methods. Moreover, solutions for these equations can be obtained explicitly using this theory [AbCl91, Lam80]. Some of them are shown to decompose into a very particular set of solutions, usually referred as solitons, breathers and radiation (see e.g. Schuur [Sch86] and references therein).

Solutions of (1.4.1), (1.4.2) and (1.4.3) are invariant under space and time translations. Indeed, for any $t_0, x_0 \in \mathbb{R}$, $u(t - t_0, x - x_0)$ is also a solution. Furthermore, an additional feature of the SG equation is the invariance under Lorentz transformations: given any $v \in (-1, 1)$, then

$$u(\gamma(t - vx), \gamma(x - vt)), \quad \gamma := (1 - v^2)^{-1/2},$$

is a new solution of (1.4.3).

In addition to the above mentioned special solutions of (1.4.1), (1.4.2) and (1.4.3), and as we mentioned before, there exists another large family of explicit and oscillatory solutions, known in the physical and mathematical literature as the *breather* solution, and which is a periodic in time, spatially localized real function. Although there is no universal definition for a breather, we will adopt the following convention, that will match the mKdV, Gardner and SG cases.

1.5 A brief historical account

For the mKdV equation, the breather was discovered by Wadati [Wad73]; and for the Gardner equation, by [PeGr97, Ale12c], using the IST. These solutions have become a canonical example of complexity in nonlinear integrable systems [Lam80, AbCl91]. Moreover, their surprising mixed behavior, combining oscillatory and soliton character, has focused the attention of many researchers since thirty years ago [Sch86, Den93]. From the physical point of view, breather solutions seem to be relevant to localization-type phenomena in optics, condensed matter physics and biological processes [Aub97]. They also play an important role in the modeling of freak and rogue waves events on surface gravity waves and also of internal waves in the stratified ocean, in Josephson junctions and even in nonlinear optics. See [DyTr99, GGK12, AKF10, KSB98] for a representative set of these examples.

The previous models appear in many applications. For instance, the mKdV equation in the whole real line is an integrable and canonical nonlinear dispersive equation, originally describing shallow water wave dynamics [Whi74], and therefore appearing as a good approximation of different physical problems. A few examples are the motion of the curvature of some geometric fluxes [GoPe91, NSW92, NaWa93], vortex patches, ferromagnetic vortices [Wex99], traffic models [KoSa95], [Naga99], anharmonic lattices [Ono92], hyperbolic surfaces [Schi95], among others.

Chapter 2

Dispersive type equations

2.1 Introduction: wave packets, dispersion

The purpose of this chapter is to introduce to the reader the main equations that will be considered in these notes, together their most important properties. Most of these equations are of *dispersive* nature, a concept that we will explain in a moment. Usually, dispersive models appear as consequence of nature phenomena, and in general they share important physical properties. On the other hand, these properties must be used at large when studying the behavior of solutions.

In what follows, we will review the mean features of the following dispersive models:

(1) the Korteweg-de Vries and related equations;
(2) the nonlinear Schrödinger equations;
(3) some scalar field equations, including the well-known sine-Gordon equation.

For a complete description, the reader may consult the monograph by Linares and Ponce [LiPo14], Cazenave [Caz03], Klein and Saut [KS15], and Yang [Yan10]. Here we will state only the main properties needed in our setting. Before introducing the notion of dispersive equation, we need some simple but very important concepts.

Definition 2.1.1. Given a vector $k \in \mathbb{R}^d$, $d \geq 1$, called *wave vector*, and a real-valued scalar $w \in \mathbb{R}$, usually referred as a *frequency*, we say that the function

$$\mathbb{R} \times \mathbb{R}^d \ni (t, x) \longmapsto u_{k,w}(t, x) := e^{i(k\cdot x - wt)} \in \mathbb{C} \tag{2.1.1}$$

is a *wave packet* with wave vector k and frequency w.

Wave packets are useful for several reasons, particularly in Physics. In our case, they will be key elements to reveal the exact dispersive nature of the equation.

Before introducing the equations that we will study, we need some simple classification rules. Let $u = u(t, x)$ be a sufficiently smooth function, for x in some spatial domain and $t \in \mathbb{R}$. In this book we will consider two different kinds of equations: the linear equations, written in an abstract form as

$$\partial_t u + Lu = 0, \tag{2.1.2}$$

where L is a linear operator, most of the time a differential operator, and nonlinear equations

$$\partial_t u + Lu + N(u) = 0, \tag{2.1.3}$$

where N is a truly nonlinear operator in nature. Examples of linear equations are the transport equation in one dimension,

$$\partial_t u + \partial_x u = 0, \tag{2.1.4}$$

the **Airy equation** in $\mathbb{R}_t \times \mathbb{R}_x$,

$$\partial_t u + \partial_x^3 u = 0, \tag{2.1.5}$$

among others. On the other hand, if $\Delta = \sum_{j=1}^d \partial_{x_j}^2$ represents the Laplacian in $\mathbb{R}^d$,

$$\partial_t u - i\Delta u - i|u|^2 u = 0,$$

is a nonlinear equation as in (2.1.3) for a function $u(t, x) \in \mathbb{C}$ because of the nonlinear term $i|u|^2 u$. In fact it is currently known as the *cubic NLS* equation.

Definition 2.1.2. Given a linear equation as in (2.1.2) defined in $\mathbb{R}_t \times \mathbb{R}_x^d$, we say that it is *dispersive* if the following conditions are satisfied:

(1) There exists a real-valued frequency $w = w(k)$ such that the wave packet $u_{k,w(k)}$ defined in (2.1.1) is solution to (2.1.2).
(2) As a function of the wave vector $k \in \mathbb{R}^d$, $w = w(k)$ is differentiable and $\nabla w(k)$ varies with k.

Finally, we will say that a nonlinear equation is dispersive if its linear part is dispersive.

In physical terms, $\nabla w(k)$ is called the **velocity group** of the equation. Therefore, an equation is dispersive if different frequencies propagate at different group velocities.

Example: The Airy equation in (2.1.5) is dispersive, since $u_{k,w} = e^{i(kx-wt)}$ solves (2.1.5) provided

$$w(k) = -k^3.$$

Therefore we will have $w'(k) = -k^2 \leq 0$. The transport equation in (2.1.4) is not dispersive because one gets

$$w(k) = k, \quad w'(k) = 1,$$

which does not vary with k. Another equation that is not dispersive is the standard linear heat equation in $\mathbb{R}_t \times \mathbb{R}_x^d$

$$\partial_t u - \Delta u = 0,$$

for which a wave packet is solution if and only if $w(k)$ is complex valued, which is not allowed in view of Definition 2.1.2.

Other dispersive models of key importance in this book are the Korteweg-de Vries (KdV) equation

$$\partial_t u + \partial_x(\partial_x^2 u + u^2) = 0, \quad u = u(t,x) \in \mathbb{R}, \quad t, x \in \mathbb{R},$$

and the (*cubic*) nonlinear Schrödinger (NLS) equation,

$$i\partial_t u + \partial_x^2 u + |u|^2 u = 0, \quad u = u(t,x) \in \mathbb{C}, \quad t, x \in \mathbb{R}.$$

These are models describing the behavior of shallow water waves in different regimes (more on this in next chapters) and the evolution of atomic and sub-atomic objects. These two equations also have breathers, as we will see in forthcoming chapters. Finally, the sine-Gordon (SG) equation (1.4.3)

$$\partial_t^2 u - \partial_x^2 u + \sin u = 0,$$

is another example of very important dispersive model with breather solutions. We leave to the reader the verification of the dispersive nature of each of these models.

A second example. Maxwell's equations model the evolution of electric and magnetic fields E, B in the vacuum. They are given by

$$\text{(Maxwell)} \quad \begin{cases} \nabla \cdot E = 0, \\ \nabla \cdot B = 0, \\ \partial_t B = -\nabla \times E, \\ \partial_t E = \nabla \times B. \end{cases}$$

Here, the fields $E = E(t,x)$ and $B = B(t,x)$ take values in $\mathbb{R}^3$, for $x \in \mathbb{R}^3$ and $t \in \mathbb{R}$.

As usual, divergence and curl are defined as follows: if $E = (E_1, E_2, E_3) \in \mathbb{R}^3$ and $x = (x_1, x_2, x_3) \in \mathbb{R}^3$,

$$\nabla \cdot E = \partial_{x_1} E_1 + \partial_{x_2} E_2 + \partial_{x_3} E_3 \in \mathbb{R},$$

and for $\mathbf{e_1}, \mathbf{e_2}, \mathbf{e_3}$ the canonical basis of $\mathbb{R}^3$,

$$\begin{aligned}\nabla \times E &= \det \begin{pmatrix} \mathbf{e_1} & \mathbf{e_2} & \mathbf{e_3} \\ \partial_{x_1} & \partial_{x_2} & \partial_{x_3} \\ E_1 & E_2 & E_3 \end{pmatrix} \\ &= \left(\partial_{x_2} E_3 - \partial_{x_3} E_2\right)\mathbf{e_1} + \left(\partial_{x_3} E_1 - \partial_{x_1} E_3\right)\mathbf{e_2} \\ &\quad + \left(\partial_{x_1} E_2 - \partial_{x_2} E_1\right)\mathbf{e_3} \in \mathbb{R}^3,\end{aligned}$$

with det the usual 3×3 determinant.

Now that we have a good understanding of (Maxwell), we perform some standard tricks to realize that indeed, these are vector wave-type equations. Formally taking time derivative in the equation for $\partial_t E$, and using Maxwell twice, we obtain

$$\partial_t^2 E = \nabla \times \partial_t B = -\nabla \times (\nabla \times E) = -\nabla(\nabla \cdot E) + \Delta E = \Delta E,$$

where ΔE is the vector Laplacian, taken component-wise. We conclude that

$$\partial_t^2 E - \Delta E = 0.$$

Therefore, we can say that light in the vacuum behaves as a wave.

Let us look for plane wave solutions. Given $\mathbf{e} \in \mathbb{R}^3 \backslash \{0\}$ fixed, we set

$$E(t, x) = e^{i(k\cdot x - wt)}\mathbf{e}, \quad k \in \mathbb{R}^3, \ w \in \mathbb{R}.$$

First of all, notice that in order to satisfy $\nabla \cdot E = 0$, we need

$$k \cdot \mathbf{e} = 0,$$

something that can be ensured if we set $\mathbf{e} = \mathbf{e_1}$, the first unit vector of the canonical basis in $\mathbb{R}^3$, and $k = (0, k_2, k_3)$, namely k lies in the plane $k_2 k_3$.

Assuming this fact, let us set

$$E_0(t, x) = e^{i(k\cdot x - wt)}\mathbf{e_1}, \quad k \in \mathbb{R}^3, \ w \in \mathbb{R}.$$

We look for $w = w(k)$ such that E_0 solves the wave equation. Replacing, one gets

$$-w^2 + |k|^2 = 0, \implies w = w(k) = |k| = \sqrt{k_2^2 + k_3^2}.$$

Consequently, the frequency $w(k)$ of the plane wave varies with k, and the so-called group velocity, namely $\nabla w(k)$, is given by

$$\nabla w(k) = \left(0, \frac{k}{|k|}\right) \in \mathbb{S}^2,$$

which varies with k. So, in some sense this equation is dispersive, but at a very weak level, since the absolute value of the vector $\nabla w(k)$ is constant, and only angular variables move with k.

2.2 Invariances

Many dispersive PDEs enjoy standard symmetries, usually appearing from physical properties. In the following, we assume that $u = u(t,x)$ is a solution of a standard dispersive PDE, where $t \in \mathbb{R}$ is the time variable, and $x \in \mathbb{R}^d$ is the space variable.

If the equation has no explicit time dependence, then

$$u(t+t_0,x), \ t_0 \in \mathbb{R}, \qquad \text{(translation in time)},$$

will also be another solution of the equation, and if the equation has no explicit dependence on the variable x, it will be invariant by shifts:

$$u(t,x+x_0), \ x_0 \in \mathbb{R}^d, \qquad \text{(translation in space)},$$

is also another solution. As an example, the linear wave equation for $u = u(t,x) \in \mathbb{R}$, $t \in \mathbb{R}$, $x \in \mathbb{R}^d$,

$$\partial_t^2 u - \Delta u = 0, \tag{2.2.1}$$

satisfies the time and space invariances.

Another important symmetry is the **scaling invariance**, which reveals to be very important when working in the local well-posedness theory and some stability problems. If the equation is posed on the entire $\mathbb{R}^d$, then, thanks to the lack of compactness of the domain, a suitable scaling of the solution $u(t,x)$ is a new solution of the same dispersive equation. For example, for any $\lambda > 0$,

$$u_\lambda(t,x) := u(\lambda t, \lambda x)$$

is a new solution of the linear wave equation (2.2.1).

Through these notes we will see different examples of equations with particular scaling symmetries. Sometimes an equation may not have a scaling invariance; for instance, the Klein-Gordon equation:

$$\partial_t^2 u - \partial_x^2 u + u = 0,$$

has no scaling invariance. However, it shares many similarities with the standard wave equation (2.2.1) that makes its understanding not too far from it.

2.3 The Fourier transform

Now we recall some standard techniques appearing in dispersive equations. For more details, the reader may consult the book by Linares and Ponce [LiPo14].

First of all, we consider the Fourier transform in $\mathbb{R}^d$. Recall that L^p, $1 \le p < +\infty$ corresponds to the Lebesgue space with the standard Lebesgue measure:

$$L^p := \left\{ f : \mathbb{R}^d \to \mathbb{R} \ : \ f \text{ is measurable and } \int_{\mathbb{R}^d} |f|^p < +\infty \right\},$$

and

$$L^\infty := \left\{ f : \mathbb{R}^d \to \mathbb{R} \ : \ f \text{ is measurable and } \operatorname{esssup}_{\mathbb{R}^d} |f| < +\infty \right\},$$

where esssup denotes the essential supremum (i.e., taken up to a zero measure set). The extension to the complex-valued case is direct: both real and imaginary parts must be in L^p.

Definition 2.3.1 (Fourier transform). Given $f \in L^1(\mathbb{R}^d)$, we define its Fourier transform by

$$\mathcal{F}[f](\xi) = \hat{f}(\xi) := c_d \int_{\mathbb{R}^d} e^{-ix\cdot\xi} f(x)dx, \quad c_d := (2\pi)^{-d/2}.$$

It is clear that $\hat{f} \in L^\infty(\mathbb{R}^d)$, provided $f \in L^1(\mathbb{R}^d)$. Moreover, $\hat{f}$ is continuous and decays at infinity. If $f \in L^1(\mathbb{R}^d) \cap L^2(\mathbb{R}^d)$, the Fourier transform can be extended to $L^2(\mathbb{R}^d)$, in a form to be specified later. Additionally, the following identities are satisfied:

(1) If $\alpha \in \mathbb{N}^d$ is a muti-index and $x^\beta f \in L^1$ for all $|\beta| \le |\alpha|$, then $\partial_\xi^\alpha \hat{f}$ is well-defined and

$$\partial_\xi^\alpha \hat{f}(\xi) = \mathcal{F}((-ix)^\alpha f)(\xi). \tag{2.3.1}$$

(2) If $\beta \in \mathbb{N}^d$ is a muti-index and $\partial_x^\beta f$ is in L_x^1,

$$\mathcal{F}(\partial_x^\beta f)(\xi) = (i\xi)^\beta \hat{f}(\xi). \tag{2.3.2}$$

(3) If $f, g \in L^1$, then its **convolution**

$$(f \star g)(x) := \int f(x-y)g(y)dy = \int f(y)g(x-y)dy,$$

is in L^1 and one has

$$\mathcal{F}(f \star g)(\xi) = (2\pi)^{d/2} \hat{f}(\xi)\hat{g}(\xi).$$

(4) Given the Gaussian $f(x) := \frac{1}{(4\pi a)^{d/2}} e^{-|x|^2/4a}$, $a > 0$, one has

$$\mathcal{F}[f](\xi) = \frac{1}{(2a)^{d/2}} e^{-|\xi|^2/(4a)}.$$

One can also define the inverse Fourier transform $\check{f}$ as

$$\check{f}(\xi) := c_d \int e^{ix\cdot\xi} f(x)dx,$$

for f in $L^1(\mathbb{R}^d)$. One can check that if f **and** $\hat{f}$ **are in** $L^1(\mathbb{R}^d)$**, then** $f = \check{\hat{f}}$. Consequently, we denote

$$\mathcal{F}^{-1}(f) := \check{f}$$

in this case. The **Parseval's identity** states that if $f, g \in L^1(\mathbb{R}^d)$, then

$$\int \hat{f}(\xi)g(\xi)d\xi = \int f(\xi)\hat{g}(\xi)d\xi.$$

This identity is quite useful because if $f \in L^1 \cap L^2$, and formally choosing $g = \overline{\hat{f}}$ we will get

$$\int |\hat{f}(\xi)|^2 d\xi = \int f^2(x)dx,$$

known as the **Plancherel's identity**. As a consequence, one can define by density $\hat{f}$ for f in L^2, and the Fourier transform is an isometry in L^2.

2.4 Tempered distributions

First of all, we need the notion of Schwartz's class. Before defining it, recall that for $f : \mathbb{R}^d \to \mathbb{R}$ real-valued, $\partial_x^\alpha f(x)$ denotes the partial derivative of f with respect to a multi-index $\alpha = (\alpha_1, \ldots, \lambda_d) \in \mathbb{N}^d$, such that

$$\partial_x^\alpha f(x) := \frac{\partial^{|\alpha|} f(x)}{\partial x_1^{\alpha_1} \partial x_2^{\alpha_2} \cdots \partial x_d^{\alpha_d}}, \qquad |\alpha| := \alpha_1 + \cdots + \alpha_d.$$

Additionally,

$$x^\alpha := x_1^{\alpha_1} x_2^{\alpha_2} \cdots x_d^{\alpha_d} \quad \text{and} \quad |x| = \left(x_1^2 + x_2^2 + \cdots + x_d^2\right)^{1/2}.$$

Given α, β multi-indexes, and $f \in C^\infty(\mathbb{R}^d)$, we introduce the semi-norms

$$p_{\alpha,\beta}(f) := \sup_{x\in\mathbb{R}^d} |x^\alpha \partial_x^\beta f(x)|. \tag{2.4.1}$$

Here $|\cdot|$ is the classical absolute value in $\mathbb{R}$. Notice that $f \in C^\infty$ does not imply $p_{\alpha,\beta}(f)$ finite.

Definition 2.4.1 (Schwartz's class). We denote by $\mathcal{S}(\mathbb{R}^d)$ the space of smooth functions $f \in C^\infty(\mathbb{R}^d)$ which are rapidly decaying in space, namely for each $m \in \mathbb{N}$ and any muti-index $\alpha \in \mathbb{N}^d$, there is a constant $C_{\alpha,m}$, independent of $x \in \mathbb{R}^d$, such that

$$\sup_{x\in\mathbb{R}^d} |x|^m |\partial_x^\alpha f(x)| \le C_{\alpha,m} < +\infty.$$

Examples of functions in the Schwartz class are exponentials like $e^{-|x|^2}$, and compactly supported smooth functions in $C_0^\infty(\mathbb{R}^d)$. However, $1/(1+|x|^2)$ in $\mathbb{R}^d$ does not belong to $\mathcal{S}(\mathbb{R}^d)$.

The standard topology on $\mathcal{S}(\mathbb{R}^d)$ is induced by the family of semi-norms $p_{\alpha,\beta}$ in (2.4.1): we will have a separable vector space and $f_n \to 0$ in $\mathcal{S}(\mathbb{R}^d)$ if and only if $p_{\alpha,\beta}(f_n) \to 0$ for any α, β multi-indexes.

Also, notice that properties (2.3.1)-(2.3.2) are naturally transferred into functions in $S(\mathbb{R}^d)$: they ensure that the Fourier transform of a function in $\mathcal{S}(\mathbb{R}^d)$ is in $\mathcal{S}(\mathbb{R}^d)$, and similarly for the inverse Fourier transform. Consequently, $\mathcal{F}$ is an homeomorphism in $\mathcal{S}(\mathbb{R}^d)$.

The dual space of $\mathcal{S}(\mathbb{R}^d)$, namely, the vector space of real-valued linear continuous functionals in $\mathcal{S}(\mathbb{R}^d)$, is denoted as $\mathcal{S}(\mathbb{R}^d)' = \mathcal{S}'(\mathbb{R}^d)$, and it is called the space of **tempered distributions**. The topology on $\mathcal{S}'(\mathbb{R}^d)$ is natural and simple:

$$T_n \to 0 \text{ in } \mathcal{S}' \quad \text{if and only if} \quad T_n(f) \to 0 \text{ for each } f \in \mathcal{S}(\mathbb{R}^d).$$

Finally, the Fourier transform on $\mathcal{S}'$ is defined using the dual characterization on S: if $T \in \mathcal{S}'(\mathbb{R}^d)$,

$$\langle \mathcal{F}(T), f\rangle := \langle T, \mathcal{F}(f)\rangle, \qquad \text{for each } f \in \mathcal{S}(\mathbb{R}^d).$$

Consequently, $\mathcal{F} : \mathcal{S}' \to \mathcal{S}'$ is again a homeomorphism.

2.5 Sobolev spaces

An important tool to understand the behavior of dispersive PDEs is the existence of L^2-based Hilbert spaces adapted to the evolution of the equation. In these sections we will assume that the reader has some knowledge about these spaces, but in order to make things self-contained, we will briefly recall the main ingredients needed in this book.

Definition 2.5.1. Let $s \in \mathbb{R}$. We denote by $H^s = H^s(\mathbb{R}^d)$ the inhomogeneous Sobolev space of index s given by

$$H^s(\mathbb{R}^d) := \left\{ f \in \mathcal{S}'(\mathbb{R}^d) \ : \ (1+|\xi|^2)^{s/2}\hat{f}(\xi) \in L^2(\mathbb{R}^d) \right\}.$$

Similarly, we define the homogeneous Sobolev space $\dot{H}^s = \dot{H}^s(\mathbb{R}^d)$ as

$$\dot{H}^s(\mathbb{R}^d) = \left\{ f \in \mathcal{S}'(\mathbb{R}^d) \ : \ |\xi|^s \hat{f}(\xi) \in L^2(\mathbb{R}^d) \right\}.$$

Clearly $H^0 = L^2$. Note that $(1+|\xi|^2)^{s/2}\hat{f}(\xi)$ makes sense as distribution in $\mathcal{S}'$. If it is additionally in L^2, we have an H^s function ($s \geq 0$) or distribution ($s < 0$).

In this book we will usually choose $s \geq 0$. Each H^s is a Hilbert space endowed with the norm induced by an inner product:

$$\begin{aligned} \|f\|_{H^s} &:= \left\|(1+|\xi|^2)^{s/2}\hat{f}(\xi)\right\|_{L^2}, \\ \langle f, g\rangle &:= \int (1+|\xi|^2)^s \hat{f}\overline{\hat{g}}. \end{aligned} \tag{2.5.1}$$

(Sometimes the inner product is taken with real part.) Additionally, and directly from (2.5.1), the following standard properties are satisfied:

Basic properties of Hilbert spaces.

(1) If $s < s'$, then $H^{s'} \subset H^s$.
(2) If $s > \frac{d}{2}$, then H^s is included in the Hölder space $C^{0,\alpha}$, for some $\alpha \in (0,1)$.
(3) If $s > \frac{d}{2}$, H^s defines a multiplicative algebra, that is to say $f, g \in H^s$ implies $fg \in H^s$, and for some constant $C > 0$ independent of f and g,

$$\|fg\|_{H^s} \leq C(\|f\|_{L^\infty}\|g\|_{H^s} + \|f\|_{H^s}\|g\|_{L^\infty}).$$

(4) If $s = n \in \mathbb{N}$, then one has the equivalence

$$H^n(\mathbb{R}^d) = \left\{f \in L^2 \ : \ \partial_x^\alpha f \in L^2, \quad |\alpha| \leq n\right\}.$$

2.6 Well-posedness

Now that we have all the ingredients, we can speak quickly about well-posedness issues. In order to make things the clearest possible, we proceed with a case study. Let us consider the linear Schrödinger model

$$i\partial_t u + \Delta u = 0, \tag{2.6.1}$$

posed for $x \in \mathbb{R}^d$, with $u(t,x) \in \mathbb{C}$, and $t \in \mathbb{R}$.

Let us recall the notion of well-posedness in a broad sense. Given

(1) A PDE of space-time character (not necessarily dispersive) which has the abstract form

$$\partial_t u + Lu + N(u) = 0, \quad u = u(t,x),$$

with L and N linear and nonlinear parts, respectively,
(2) where we identify a "time variable" as the evolutive variable (in this case t), and

(3) an initial condition u_0 in a fixed vector space X (probably of Banach type), for which the PDE makes sense at least formally,

we say that the Cauchy problem

$$\partial_t u + Lu + N(u) = 0, \quad u(t = 0) = u_0 \in X,$$

is **well-posed in the sense of Hadamard** if the following three conditions are satisfied:

(1) There exists a unique solution u defined in an open interval I containing 0 and such that $u(t) \in X$ in this interval,
(2) If the solution has a given regularity or smoothness at some time, it is preserved by the flux afterwards; and
(3) The solution map $X \ni u_0 \mapsto u$ depends continuously on the initial data u_0.

The solution u may be part of an intricate vector space of time-dependent functions with values in X, that is $F(I; X)$. This time dependence should be continuous, but it is not always the case.

If some of these conditions are not satisfied, we will say that **the problem is ill-posed**. Sometimes, if only the third condition above fails, we say that we have **existence of solutions**. This is the standard case of quasilinear models, such as Einstein equations, for which continuity of the solution map is not necessarily true. Sometimes, one has particular notions of ill-posedness: they can be mild or truly strong, some of these cases will be studied using breathers in this book.

If the solution u of a PDE ceases to exist after some time, in a sense to be specified later, we say that the solution is not globally well-defined. On the contrary, if the solution u can be extended for any possible time t with values in X, we will say that it is globally well-defined.

Example. Let us consider (2.6.1) with initial data in $\mathcal{S}(\mathbb{R}^d; \mathbb{C})$ (that is, complex-valued data)

$$i\partial_t u + \Delta u = 0, \qquad u(t = 0) = u_0 \in \mathcal{S}(\mathbb{R}^d_x).$$

Assuming that the solution is in the Schwartz class, and formally taking spatial Fourier Transform, we get

$$i\partial_t \hat{u} - |\xi|^2 \hat{u} = 0, \qquad \hat{u}(t = 0) = \hat{u}_0 \in \mathcal{S}(\mathbb{R}^d_\xi).$$

This is an ODE in the variable t, with ξ fixed. Its solution is

$$\hat{u}(t, \xi) = e^{-i|\xi|^2 t} \hat{u}_0(\xi),$$

which is usually denoted as

$$u(t) = e^{it\Delta} t u_0.$$

The dispersive character of the model is seen in the fact that the Schrödinger kernel in Fourier $e^{-i|\xi|^2 t}$ is of oscillatory type.

Notice that for each time $t \in \mathbb{R}$ one has that $e^{-i|\xi|^2 t}\hat{u}_0(\xi)$ is well-defined in $\mathcal{S}(\mathbb{R}^d_\xi)$, thus $u(t, x)$ is in the Schwartz class. On the other hand, using the inverse Fourier transform,

$$u(t, x) = c_d \int e^{ix\cdot\xi} e^{-i|\xi|^2 t}\hat{u}_0(\xi) d\xi.$$

Therefore, a solution in the Schwartz class exists and it is unique.

Even better, any semi-norm $p_{\alpha,\beta}$ of $u(t, x)$ is still finite and only dependent on u_0. Consequently, the continuity in the Schwartz class is also preserved:

$$\begin{aligned}
&\left| x^\alpha \partial_x^\beta \int e^{ix\cdot\xi} e^{-i|\xi|^2 t}\hat{u}_0(\xi) d\xi \right| \\
&\lesssim \left| x^\alpha \int e^{ix\cdot\xi} e^{-i|\xi|^2 t}(i\xi)^\beta \hat{u}_0(\xi) d\xi \right| \\
&\lesssim \left| \int \partial_\xi^\alpha (e^{ix\cdot\xi}) e^{-i|\xi|^2 t}(i\xi)^\beta \hat{u}_0(\xi) d\xi \right| \\
&\lesssim \left| \int \langle\xi\rangle^{-m} e^{ix\cdot\xi} \langle\xi\rangle^m \partial_\xi^\alpha [e^{-i|\xi|^2 t}(i\xi)^\beta \hat{u}_0(\xi)] d\xi \right| \lesssim p_{\alpha,\beta+m}(u_0).
\end{aligned}$$

Recall that $\langle\xi\rangle := \sqrt{1 + |\xi|^2}$ is the *japanese bracket*, and m is any large number. Since the function

$$\langle\xi\rangle^m \partial_\xi^\alpha \left(e^{-i|\xi|^2 t}(i\xi)^\beta \hat{u}_0(\xi)\right)$$

belong to the Schwartz class, we conclude that $u_{n,0} \to 0$ in $\mathcal{S}$ implies $u_n(t) \to 0$ in $\mathcal{S}$, for any time t.

Therefore, the equation is well-posed. Using complex integral techniques, one can find that

$$u(t, x) = \frac{1}{(4\pi t)^{d/2}} \int e^{\frac{i|x-y|^2}{4\pi t}} u_0(y) dy.$$

Another example. Let us consider the Heat equation *backwards in time*. That is,

$$\partial_t u + \Delta u = 0, \qquad u(t = 0) = u_0 \in \mathcal{S}(\mathbb{R}^d_x),$$

that is equivalent to solving the heat equation changing $u(t, x)$ by $u(-t, x)$. Using again the Fourier transform,

$$\hat{u}(t, \xi) = e^{|\xi|^2 t}\hat{u}_0(\xi),$$

and

$$u(t,x) = c_d \int e^{ix\cdot\xi} e^{|\xi|^2 t} \hat{u}_0(\xi) d\xi.$$

However, even if $\hat{u}_0$ is Schwartz, this last expression may not be defined for times large enough, (or simply for any time $t > 0$).

Therefore, the solution may not exists in the Schwartz class, and the equation is ill-posed. Compare with the heat equation, that regularizes solutions for $t > 0$.

However, something can be said for instance if u_0 has compact support in Fourier space, namely $\operatorname{supp} \hat{u}_0 \subseteq B(0, M)$ for some $M > 0$. In this case

$$|u(t,x)| = \left| \int e^{ix\cdot\xi} e^{|\xi|^2 t} \hat{u}_0(\xi) d\xi \right| \lesssim e^{M^2 t}.$$

Hence, there is effectively an instability (possibly an exponential growth of the solution), but it is well-defined for any time. In this sense, a unique solution continues to exist, with same regularity as before, but bounds grow in time.

2.7 Critical spaces

Very often, a very good criterion to study well-posedness is given by the **critical space** measured in terms of a particular invariance of the equation. For example, the Korteweg de Vries (KdV) equation [KdV95]

$$\partial_t u + \partial_x(\partial_x^2 u + u^2) = 0, \quad u = u(t,x) \in \mathbb{R}, \quad (t,x) \in \mathbb{R}^2, \tag{2.7.1}$$

has H^{-1} as **critical space** with respect to the natural scaling of the equation: if $u(t,x)$ is a solution, then for any $\lambda \in \mathbb{R}$ the modification $\lambda^2 u(\lambda^3 t, \lambda x)$ is also a solution to KdV. In particular, the H^{-1} norm remains invariant under this transformation, check this out. The space H^{-1} is particularly important because it is a natural obstruction to well-posedness, although sometimes well-posedness is lost even at higher regularity.

Another example is given by the *cubic* NLS:

$$i\partial_t u + \partial_x^2 u + |u|^2 u = 0, \quad u = u(t,x) \in \mathbb{C}, \quad t, x \in \mathbb{R},$$

for which one has the Galilean transformation

$$u(t,x) \longmapsto u(t, x - vt) e^{ixv/2 - iv^2 t/4}, \quad v \in \mathbb{R},$$

that leaves the equation invariant. Consequently, the L^2 norm is invariant under this transformation, posing an important obstruction to well-posedness, even if the scaling symmetry for NLS

$$u(t,x) \longmapsto cu(c^2 t, cx), \quad c \in \mathbb{R},$$

leaves the $H^{-1/2}$ norm invariant.

2.8 Focusing versus defocusing

The type of nonlinearity in a dispersive model may have important consequences in the long time behavior of solutions. One of the simplest aspects to have in mind is to determine its defocusing or focusing nature.

In order to exemplify this property, notice the following properties associated to ODEs:

(1) Linear ODE: global solution.

$$\dot{u} = u \implies u = u_0 e^t.$$

(2) Nonlinear focusing:

$$\dot{u} = u^2 \implies u = \frac{1}{\frac{1}{u_0} - t}.$$

Notice that $u_0 > 0$ implies blow-up for $t > 0$.

$$\dot{u} = -u^2 \implies u = \frac{1}{\frac{1}{u_0} + t}$$

Here $u_0 > 0$ implies global existence and decay for $t > 0$.

This simple example reveals that the sign in the nonlinearity leads to important outcomes. What is behind this is the natural defocusing nature of waves (they tend to scatter), but sometimes nonlinearities tend to concentrate waves in a strong fashion, forming solitary waves or even blow up solutions, in which case one is under a focusing nature of the nonlinearity.

In the case of dispersive models, the generalized KdV model

$$\partial_t u + \partial_x(\partial_x^2 u + u^p) = 0, \quad p = 2, 3, 4, \dots \tag{2.8.1}$$

is focusing if p is odd. In the case p even, changing $u \mapsto -u$ changes the sign in the nonlinearity, and we do not have a clear notion of focusing-defocusing. However,

$$\partial_t u + \partial_x(\partial_x^2 u - u^p) = 0, \quad p = 3, 5, 7, \dots$$

is clearly defocusing.

As a standard rule of thumb, defocusing models tends to scatter, and almost every time solutions are global (there are counterexamples in higher dimensions). Focusing models usually have solitary waves (Section 2.10) and blow up solutions. Defocusing models tend to have kink-like solutions (Section 2.10).

2.9 Conservation laws

Given a PDE with a time-variable direction, a conservation law is a (usually physical) quantity that is preserved along the flow of the equation.

For instance, as we have seen before, the generalized KdV (gKdV) model (2.8.1)

$$\partial_t u + \partial_x(\partial_x^2 u + u^p) = 0, \quad p = 2, 3, 4, \ldots$$

has formally conserved its L^2 norm or mass:

$$\int u^2(t,x)dx = \int u^2(0,x)dx.$$

This can be easily checked by assuming that the solution is smooth and sufficiently decreasing to zero in space, to then make rigorous by a density argument:

$$\begin{aligned}\frac{d}{dt}\int u^2(t,x)dx &= 2\int uu_t(t,x)dx \\ &= -2\int u\partial_x(\partial_x^2 u + u^p)(t,x)dx \\ &= 2\int \partial_x u(\partial_x^2 u + u^p)(t,x)dx = 0.\end{aligned}$$

Therefore, the mass is conserved along the flow.

Many dispersive models have several conserved quantities, and some special ones have infinitely many conserved quantities, such as integrable models. Conserved quantities are key for local and global well-posedness theory. In our case, we will need them to prove stability and instability results for breathers.

2.10 Solitons, solitary waves and kinks

Solitons and more generally, solitary waves, are very important objects when studying dispersive models. They are the building blocks of every solution of a dispersive model of focusing type.

Usually, an equation without solitons should completely disperse its energy, in the sense that it should behave as the linear counterpart. However, if the equation has solitons, they should appear as part of the evolution, properly speaking.

Usually, solitons appear because of a delicate balance between the nonlinearity and the dispersion. In general, models need a sort of focusing nonlinearity, competing against the linear part that disperses. Physical models where solitons naturally appear can be found in detail in Dauxois-Peyrard [DaPe06].

What is more interesting is the fact that these profiles are often regarded as *global minimizers* of a constrained functional in a suitable energy (H^1) topology. In this chapter, we will describe in some detail the main properties of solitons, considering particular models of high interest and importance.

For example, mKdV (1.4.1) has solitons of the form

$$u(t,x) = Q_c(x - ct), \quad Q_c(s) := \sqrt{c}Q(\sqrt{c}s), \quad c > 0, \tag{2.10.1}$$

with profile

$$Q(s) := \frac{\sqrt{2}}{\cosh(s)} = 2\sqrt{2}\partial_s[\arctan(e^s)].$$

Using the invariance of the equation, we have

$$u(t,x) := Q_c(x - ct - x_0)$$

is a full family of solitons, satisfying

$$Q_c'' - c\,Q_c + Q_c^3 = 0, \quad Q_c > 0, \quad Q_c \in H^1(\mathbb{R}) \quad \text{(mKdV)}. \tag{2.10.2}$$

In the same way, Gardner (1.4.2) has solitons, a sort of generalization of mKdV solitons (2.10.1), namely given by

$$Q_{c,\mu}(s) := \frac{c}{\frac{\mu}{3} + \sqrt{\frac{\mu^2}{9} + \frac{c}{2}}\cosh\left(\sqrt{c}s\right)}.$$

The reader can also check that $Q_{c,\mu} > 0$ satisfies the nonlinear elliptic equation

$$Q_{c,\mu}'' - c\,Q_{c,\mu} + \mu Q_{c,\mu}^2 + Q_{c,\mu}^3 = 0, \quad Q_{c,\mu} > 0, \quad Q_{c,\mu} \in H^1(\mathbb{R}) \quad \text{(Gardner)},$$

and that $\lim_{\mu\to 0} Q_{c,\mu} = Q_c$. See Section 8.2 for further details.

Definition 2.10.1 (Solitary wave). A solitary wave is a solution of a nonlinear dispersive equation that consists of a fixed profile traveling at a certain speed along the x axis.

For example, (2.10.1) is a solitary wave with profile Q and speed c. Naturally, solitary waves may collide. Depending on the final output after collision, a solitary wave can be considered as a soliton or not.

Definition 2.10.2 (Elasticity and solitons). We say that two nonlinear objects interact elastically if their energy and mass remain unchanged after the collision. A soliton is a nonlinear traveling wave that interacts elastically with other solitons.

Once again, (2.10.1) and (2.10.2) are examples of solitons. Similarly, a *multi-soliton* is an explicit solution describing the interaction of several solitons, discovered by Hirota [Hir72].

2.10.1 *Stability*

Additionally, solitary waves and solitons are often regarded as *global minimizers* of a constrained functional in the H^1-topology. To be more precise, it is well-known that the standard conservation laws in the case of mKdV at the H^1-level are the *mass*

$$M[u](t) := \frac{1}{2}\int_{\mathbb{R}} u^2(t,x)dx = M[u](0),$$

and *energy*

$$E[u](t) := \frac{1}{2}\int_{\mathbb{R}} u_x^2(t,x)dx - \frac{1}{4}\int_{\mathbb{R}} u^4(t,x)dx = E[u](0).$$

In the case of the Gardner equation (1.4.2), the mass has the same definition as above, but the energy is given by

$$E_\mu[w](t) := \frac{1}{2}\int_{\mathbb{R}} w_x^2 - \frac{\mu}{3}\int_{\mathbb{R}} w^3 - \frac{1}{4}\int_{\mathbb{R}} w^4 = E_\mu[w](0), \tag{2.10.3}$$

which is H^1-subcritical. (For mKdV and Gardner, the Cauchy problem is globally well-posed at such a level of regularity or even better, see e.g. Kenig-Ponce-Vega [KPV93].)

Using these conserved quantities, the variational structure can be characterized as follows: there exists a suitable *Lyapunov functional*, *invariant in time* and such that the soliton Q_c or $Q_{c,\mu}$ is a corresponding *extremal point*. Moreover, it is a global minimizer under fixed mass. For the mKdV case, this functional is given by [Ben72]

$$\mathcal{H}[u](t) = E[u](t) + c\,M[u](t), \tag{2.10.4}$$

where $c > 0$ is the scaling of the solitary wave. Indeed, a simple computation shows that for any $z(t) \in H^1(\mathbb{R})$ small,

$$\mathcal{H}[Q_c + z](t) = \mathcal{H}[Q_c] + \int_{\mathbb{R}} z(Q_c'' - cQ_c + Q_c^3) + O(\|z(t)\|_{H^1}^2).$$

The zero order term above is independent of time, while the first order term in z is zero from (2.10.2), proving the critical character of Q_c.

2.10.2 *Kinks*

Kinks are topological solutions to dispersive models [Vac06, DaPe06]. These represent nonlinear modes that carry out a certain charge or momentum due to their particular limits at infinity.

The sine-Gordon (SG) equation

$$\partial_t^2 u - \partial_x^2 u + \sin u = 0,$$

has kinks, which are soliton solutions as well. Indeed, given $v \in (0,1)$, $x_0 \in \mathbb{R}$ and $\gamma = (1-v^2)^{-1/2}$, SG has kinks of the form

$$u(t,x) = \varphi(\gamma(x - vt - x_0)), \quad \varphi(s) := 4 \arctan e^s.$$

It is not difficult to see that these solutions can be associated to a well-known functional defined in the $H^1 \times L^2$ topology. Their long time behavior is highly complicated because of breathers, see e.g. [AMP20].

2.11 Historical developments

If one studies perturbations of solitons, the concepts of *orbital*, and *asymptotic stability* emerge naturally. In particular, since energy and mass are conserved quantities, it is natural to expect that solitons are stable in a suitable energy space. Indeed, H^1-stability of mKdV and more general solitons and multi-solitons in other models has been considered e.g. in Benjamin [Ben72], Bona-Souganidis-Strauss [BSS87], Cazenave-Lions [CaLi82], Weinstein [Wei85, Wei86, Wei87], Maddocks-Sachs [MaSa93], Martel-Merle-Tsai [MMT02], Martel-Merle [MaMe09] and Muñoz [Mun16]. L^2-stability of KdV solitons has been proved by Merle-Vega [MeVe03]. Moreover, asymptotic stability properties for gKdV equations (2.8.1) have been studied by Pego-Weinstein [PeWe94] and Martel-Merle [MaMe01, MaMe05], among many other authors.

Kinks solutions are orbitally stable in the energy space $H^1 \times L^2$. Indeed, kinks can be viewed as relative minimizers of a suitable energy functional. For the proofs of this result for the SG case and more general equations, we refer to the works by Henry-Perez-Wreszinski [HPW82], and Grillakis-Shatah-Strauss [GSS87].

The underlying question is then the study of the corresponding *stability* of these breather solutions. A first step in that direction was already done in [AlMu13] (see also [AlMu12, AMP17, AMP17b]), where the nonlinear stability of mKdV breathers was presented. Recent studies about the stability-instability of these breather structures are [Mun14] and [BPS14].

2.12 Exercises

(2.1) Find the dispersion relation $w(k)$ and the velocity group of the wave equation

$$\partial_t^2 u - \Delta u = 0, \qquad u = u(t,x) \in \mathbb{R}, \ t \in \mathbb{R}, \ x \in \mathbb{R}^d.$$

Is it dispersive or not? And for the Laplace equation $\Delta u = 0$ in $\mathbb{R}^d$?

(2.2) Consider the following models

$$i\partial_t u + \Delta u = 0, \qquad \text{(Linear Schrödinger)},$$

and

$$i\partial_t u + \partial_x^2 u + 2\,\mathrm{Re}\, u = 0 \quad \text{(Modulationally unstable Schrödinger)}.$$

For each of these two models, use the plane wave analysis to determine whether or not these are dispersive, and if possible, determine regions of parameters where these equations are (not) dispersive.

Chapter 3

Breathers

3.1 A rigorous definition

Already in the introduction of this book we discussed the notion of breather. Here, we will present an intuitive and a rigorous definition of it. Once again, we emphasize that this definition is not universal, and some breather-like solutions in a particular model may be left outside because of the proposed definition. However, we will try to encompass all the previously known breathers in just one definition.

A breather is essentially a periodic in time and localized in space solution to a dispersive model. Just because of this naive definition, one can say that breathers must be strange objects, since dispersive equations, as its name suggests, are of dispersive nature, and periodic oscillatory behavior is hard to maintain. Rigorously,

Definition 3.1.1 (Breather). We say that $B = B(t,x)$ is an **aperiodic breather** solution for a particular one-dimensional dispersive equation if there are $T > 0$ and $L = L(T) \in \mathbb{R}$ such that, for all $t \in \mathbb{R}$ and $x \in \mathbb{R}$, one has

$$B(t+T, x) = B(t, x-L), \tag{3.1.1}$$

up to the symmetries of the equation, and moreover, the infimum among times $T > 0$ such that property (3.1.1) is satisfied for such a time T is uniformly positive. Finally, if B is also periodic in x, we say that B is a **periodic breather**.

Remark 3.1.2. In the particular case where $L = 0$, the breather is also periodic in time.

Remark 3.1.3. Note that the positive T condition ensures that solitons (and multi-solitons) are not breathers, since e.g. $Q_c(x - c(t+T)) = Q_c(x - L - ct)$ for $L := cT$ but T can be any real-valued time. In the case of NLS equations and their solitons (we will see later), Definition 3.1.1 include them because of the $U(1)$ invariance.

Even if breathers are rare objects, they appear more often than expected. Almost every integrable equation possesses breather solutions, but relating breathers just only to integrable models is a naive error: several nonintegrable models have breather solutions as well. So the question is clearly subtle and deep.

3.2 Examples in integrable models

Our first examples will concern integrable models.

3.2.1 *The mKdV breather*

Breather solutions for the *focusing* mKdV

$$\partial_t u + \partial_x(\partial_x^2 u + u^3) = 0, \quad u = u(t,x) \in \mathbb{R}, \quad (t,x) \in \mathbb{R}^2,$$

were discovered by Wadati in the seventies [Wad73] by using the Inverse Scattering Method (ISM). The explicit form of the Wadati breather is

$$B(t,x) = 2\sqrt{2}\partial_x\left(\arctan\left[\frac{\beta}{\alpha}\frac{\sin(\alpha y_1)}{\cosh(\beta y_2)}\right]\right), \tag{3.2.1}$$

where

$$y_1 = x + \delta t + x_1, \quad y_2 = x + \gamma t + x_2,$$
$$\delta := \alpha^2 - 3\beta^2, \quad \gamma := 3\alpha^2 - \beta^2,$$

and $x_1, x_2 \in \mathbb{R}$ free parameters.

For localized mKdV breathers, L from Definition 3.1.1 can be either positive, negative, or zero, and each case corresponds to a different type of breather solution, as we will see below. Indeed, one has for $\delta \neq 0$ that

$$T = \frac{2\pi}{\alpha\delta}, \quad L = \gamma T,$$

as it is not difficult to check. The case $\delta = 0$ is particularly special, because in this case the oscillatory part stays fixed in time, and the breather moves to the left with speed $-8\beta^2$, corresponding to a particular fixed profile. In the general case $\delta \neq 0$, the behavior contrasts with the definition of breather for the sine Gordon equation (Subsection 3.2.3), where without loss of generality, L can be chosen to be identically zero.

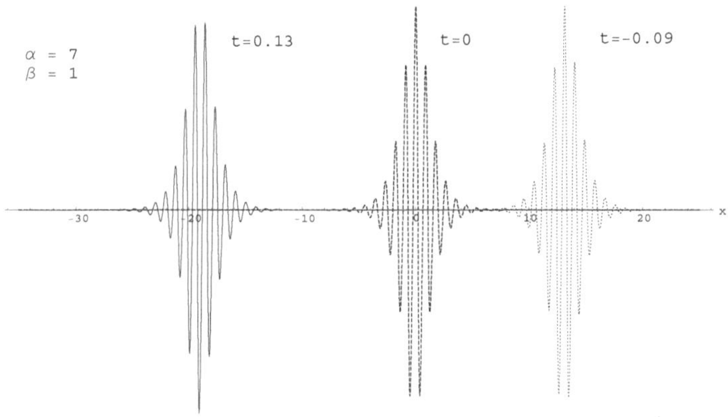

Fig. 3.1 Evolution of the mKdV breather (4.1.7) with $\alpha = 7, \beta = 1$ at instants $t = -0.09$, $t = 0$, and $t = 0.13$. Note that with the selected values of α, β, the *velocity* is given by $\gamma = 3\alpha^2 - \beta^2 = 146 > 0$ and then the breather moves to the left. (Images taken from [AlMu13].)

3.2.2 *The Gardner breather*

Breather solutions for the Gardner equation

$$\partial_t u + \partial_x(\partial_x^2 u + \mu u^2 + u^3) = 0, \quad u(t,x) \in \mathbb{R}, \quad (t,x) \in \mathbb{R}^2,$$

were discovered by Pelinovsky and Grimshaw [PeGr97] and in a clear form later by Alejo [Ale12b]. As a natural extension (or μ perturbation) of the *focusing* mKdV equation previously mentioned, it is natural to see that the corresponding breather solution as a perturbation of the mKdV breather. Indeed, its explicit form, obtained by the ISM (and also by using other integrability methods), the Gardner breather is

$$B_{\alpha,\beta,\mu}(t,x;x_1,x_2) := 2\sqrt{2}\partial_x\left[\arctan\left(\frac{G_{\alpha,\beta,\mu}(t,x)}{F_{\alpha,\beta,\mu}(t,x)}\right)\right], \tag{3.2.2}$$

with y_1 and y_2 given exactly as in the mKdV case

$$y_1 = x + \delta t + x_1, \quad y_2 = x + \gamma t + x_2,$$
$$\delta := \alpha^2 - 3\beta^2, \quad \gamma := 3\alpha^2 - \beta^2,$$

and

$$G_{\alpha,\beta,\mu}(t,x) := \frac{\beta\sqrt{\alpha^2+\beta^2}}{\alpha\sqrt{\Delta}}\sin(\alpha y_1) - \frac{\sqrt{2}\mu\beta[\cosh(\beta y_2) + \sinh(\beta y_2)]}{\Delta},$$

$$F_{\alpha,\beta,\mu}(t,x) := \cosh(\beta y_2) - \frac{\sqrt{2}\mu\beta[\alpha\cos(\alpha y_1) - \beta\sin(\alpha y_1)]}{\alpha\sqrt{\alpha^2+\beta^2}\sqrt{\Delta}},$$

and $\Delta := \alpha^2 + \beta^2 - \frac{2\mu^2}{9} > 0$. Note that when we set $\mu = 0$, we recover the above mKdV breather.

There is something deeper in the fact that the parameters δ and γ coincide in both mKdV and Gardner. In Chapter 8 we will discuss in detail this universality property.

3.2.3 *The sine-Gordon breather*

One of the most iconic breathers so far. As already introduced, the sine-Gordon (SG) model is given by

$$\partial_t^2 u - \partial_x^2 u + \sin u = 0, \quad u(t,x) \in \mathbb{R}, \quad (t,x) \in \mathbb{R}^2. \tag{3.2.3}$$

Notice that now we consider a nonlinear wave equation, with a huge impact in the literature. In Chapter 9, we will describe in detail the mathematics associated to the famous SG breather: for any $\beta \in (0,1), \beta \neq 0$,

$$B(t,x;\beta) := 4\arctan\left(\frac{\beta}{\alpha}\frac{\cos(\alpha t)}{\cosh(\beta x)}\right), \quad \alpha^2 + \beta^2 = 1. \tag{3.2.4}$$

This is just the standing breather, namely, it does not move in space. However, applying the invariances of the model, and in particular the Lorentz transform (also called Lorentz boost), one can obtain a relativistic moving breather. More details in Chapter 9.

3.2.4 *NLS breathers*

The model is the famous integrable NLS (Zakharov-Shabat [ZaSh72]), or focusing cubic nonlinear Schrödinger equation:

$$i\partial_t u + \partial_x^2 u + |u|^2 u = 0, \quad u(t,x) \in \mathbb{C}, \quad (t,x) \in \mathbb{R}^2.$$

The initial value problem for this model has been well-studied in the case of zero limit at infinity, and in the case of kink perturbations. See [Zhi87]. There are several NLS breathers, depending on the background at spatial infinite. Here we will briefly mention

(1) The Satsuma-Yajima (SY) breather (converging to zero at infinite in space),
(2) The Peregrine (P) breather (converging to the Stokes wave e^{it} at infinite in space),
(3) The Kuznetsov-Ma (KM) breather (converging to the Stokes wave e^{it} at infinite in space),

(4) The Akhmediev (A) breather (periodic in space).

All these breathers are $U(1)$ invariant, in the sense that $u \mapsto e^{i\gamma} u$ is also a solution provided u is solution, for any $\gamma \in \mathbb{R}$.

The Satsuma-Yajima (SY) breather

Let $c_1, c_2 > 0$, and $\gamma_\pm := c_2 \pm c_1$. The NLS equation with zero background has the standing, exponentially decaying breather [SaYa74]

$$B_{SY}(t,x) := \frac{2\sqrt{2}\gamma_+\gamma_- e^{ic_1^2 t}(c_1\cosh(c_2 x) + c_2 e^{i\gamma_+\gamma_- t}\cosh(c_1 x))}{\gamma_-^2\cosh(\gamma_+ x) + \gamma_+^2\cosh(\gamma_- x) + 4c_1c_2\cos(\gamma_+\gamma_- t)},$$

as solution which is a perturbation of the zero state, see Fig. 3.2.

By invariances of the equation under time-space shifts, it is possible to give a more general form involving shifts $x_1, x_2 \in \mathbb{R}$ in the t and x variables, respectively. Note that by choosing $c_1 = 1$ and $c_2 = 3$, we recover the *original breather discovered by Satsuma-Yajima* [SaYa74, Yan10]:

$$\frac{4\sqrt{2}e^{it}(\cosh(3x) + 3e^{8it}\cosh x)}{\cosh 4x + 4\cosh(2x) + 3\cos(8t)}.$$

It is important to notice that the SY breather is part of a richer family of multi-solitons, from which the SY breather is just a particular degenerate case [Yan10, SHB10]. For that reason, it is natural to expect that this breather is unstable. See Fig. 3.2 for details.

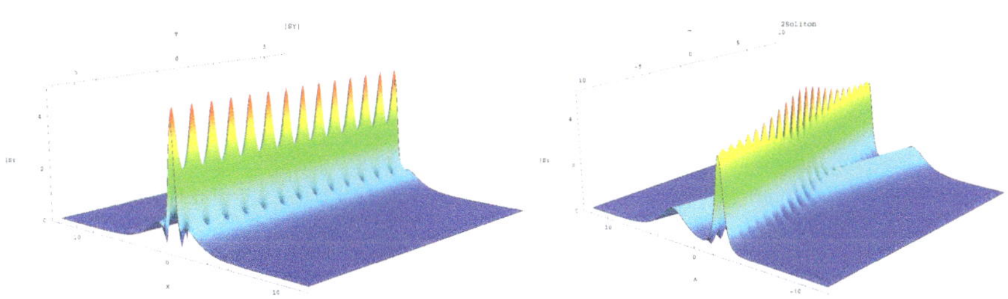

Fig. 3.2 *Left:* Absolute value of the SY breather. Note the periodic in time behavior of this solution. *Right:* Absolute value of the double soliton close to (with $\alpha_1 = 0.1$, $\alpha_2 = 0$) the SY breather. The SY breather is part of the more complex family of NLS 2-solitons, and very small perturbations of the SY breather may lead to non breather solutions. The left axis represents the x variable, and the right axis, the t variable.

The Peregrine (P) breather

Given by [Per83, Per67]

$$B_P(t,x) := e^{it}\left(1 - \frac{4(1+2it)}{1+4t^2+2x^2}\right),$$

this breather (sometimes called soliton) is a polynomially decaying (in space and time) perturbation of the **nonzero background** given by the Stokes wave e^{it}, which appears and disappears from nowhere [AAT09]. See Fig. 3.3 left for details. Despite its algebraic dependence, and therefore a lack of periodic in time behaviour, we still call it as breather because, as it was shown in [AFM18], it holds a characteristic fourth order ODE, which describe all NLS breathers. See [AFM18] for further reading.

Some interesting connections have been made between the Peregrine breather and the intensely studied subject of *rogue waves* in ocean [ZaDy06, ShGe10, AAT09, KFF10, ASA09] (see also [BPS14] for an alternative explanation to the rogue wave phenomenon).

Note that, because of time and space invariances in NLS, for any $t_0, x_0 \in \mathbb{R}$, $B_P(t-t_0, x-x_0)$ is also a Peregrine breather.

The Kuznetsov-Ma (KM) breather

Another breather in NLS that converges to the Stokes wave e^{it} at spatial infinity. It is given by [Kuz77, Ma79, AkKo86]

$$B_{KM}(t,x) := e^{it}\left[1 - \sqrt{2}\beta\frac{(\beta^2\cos(\alpha t) + i\alpha\sin(\alpha t))}{\alpha\cosh(\beta x) - \sqrt{2}\beta\cos(\alpha t)}\right],$$

$$\alpha := (8a(2a-1))^{1/2}, \quad \beta := (2(2a-1))^{1/2}, \quad a > \frac{1}{2}.$$

Notice that in the formal limit $a \downarrow \frac{1}{2}$ one recovers the Peregrine breather. See Fig. 3.3 right for details. Note that B_{KM} is a Schwartz perturbation of the Stokes wave, and therefore a smooth classical solution of NLS. It has been also observed in optical fibre experiments, see Kibler et al. [KFF12] and references therein for a complete background on the mathematical problem and its physical applications.

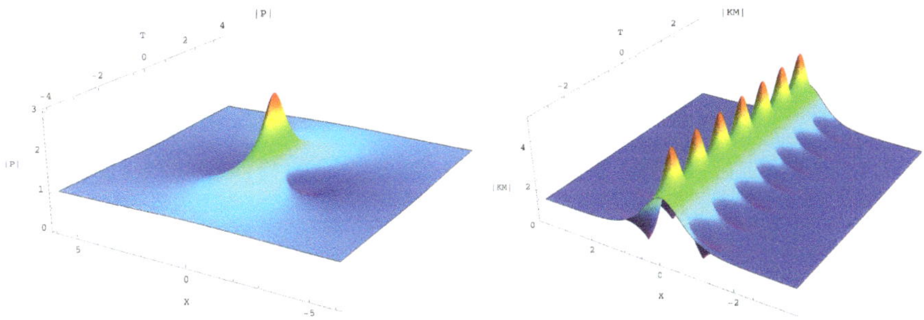

Fig. 3.3 *Left:* Absolute value of the P breather. Note the localized character in space and time. *Right:* Absolute value of the KM breather. The left axis represents the x variable, and the right one the t variable.

The Akhmediev (A) breather

Finally, we consider the $\frac{2\pi}{a}$-**periodic in space, localized in time** Akhmediev breather [AkKo86] is given by (see Fig. 3.4)

$$A(t,x) := e^{it}\left[1 + \frac{\alpha^2\cosh(\beta t) + i\beta\sinh(\beta t)}{\sqrt{2a}\cos(\alpha x) - \cosh(\beta t)}\right],$$

$$\beta = (8a(1-2a))^{1/2}, \quad \alpha = (2(1-2a))^{1/2}, \quad a \in \left(0, \frac{1}{2}\right).$$

Once again, as a grows to $\frac{1}{2}$, the Akhmediev breather converges to the Peregrine one. We anticipate that this breather is unstable [AFM18].

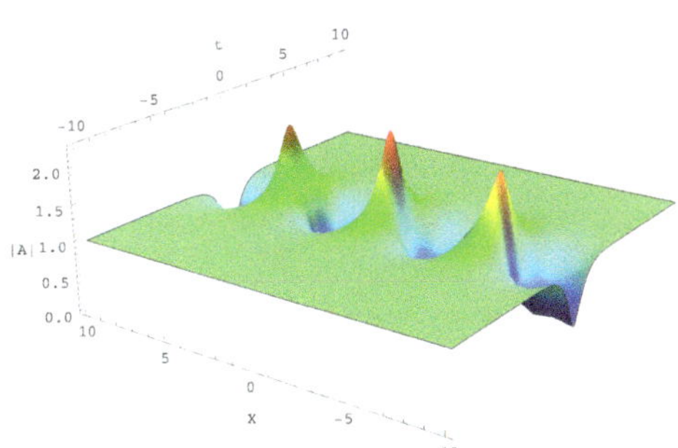

Fig. 3.4 Absolute value of the Akhmediev breather, with $a = 0.25$. Note that A is periodic in space, localized in time.

3.2.5 *The Sasa-Satsuma (SS) breather*

This is another example of a $U(1)$ invariant equation. The SS model [SaSa91] corresponds to

$$u_t + u_{xxx} + 6|u|^2 u_x + 3u(|u|^2)_x = 0,$$
$$u = u(t,x) \in \mathbb{C}, \quad t, x \in \mathbb{R},$$

a sort of complex mKdV equation. Let $\alpha, \beta > 0$ be arbitrary but fixed parameters. Following [SaSa91, eqns. (38)-(39)], and [Yan10, eqns. (3-250)-(3-252)], an exact breather solution of SS is given by the expression

$$B_{SS}(t,x) := Q_\beta(x + \gamma t + x_2)e^{i\Theta},$$
$$\Theta := \alpha(x + \delta t + x_1), \quad Q_\beta(x) := \beta Q(\beta x),$$
$$\gamma := 3\alpha^2 - \beta^2, \quad \delta := \alpha^2 - 3\beta^2.$$

Above, Q is *complex-valued*, exponentially decaying:

$$Q(x) := Q_\eta(x) := \frac{2(e^x + \eta e^{-x})}{e^{2x} + 2 + |\eta|^2 e^{-2x}}, \quad \eta := \frac{\alpha}{\alpha + i\beta}. \tag{3.2.5}$$

It is well-known that the real-valued function $|Q|$ is **single humped** when $|\eta| > 1/2$ (i.e. $|\alpha| > \frac{1}{2}\sqrt{\alpha^2+\beta^2}$), and **double humped** when $0 < |\eta| \leq 1/2$ (or $|\alpha| \leq \frac{1}{2}\sqrt{\alpha^2+\beta^2}$), see [Yan10, PeYa05]. This mixed shape is in strong contrast with the standard NLS soliton which is only single humped. See Fig. 3.5 for more details.

The previous breathers have been studied in rigour and detail in [Mun17, AFM18, AFM21], showing that, except for the Sasa-Satsuma breather, all of them are unstable.

3.3 Breathers in possibly nonintegrable models

Now we consider examples of breathers in nonintegrable systems. In a broad sense, a nonintegrable system is one that cannot be solved using known inverse scattering transform analytical techniques, requiring other methods. In comparison, while integrable systems admit exact solutions through quadratures, possibly infinitely many conserved quantities, or transformation methods, nonintegrable systems lack such complete analytical solvability.

We first start with nonlinear non integrable wave equations, specifically with the nonlinear Klein-Gordon equation. The existence of radially symmetric breather solutions for the cubic Klein-Gordon equation

$$u_{tt} - \Delta u + m^2 u = \Gamma(x)u^3, \quad \text{on } \mathbb{R} \times \mathbb{R}^3,$$

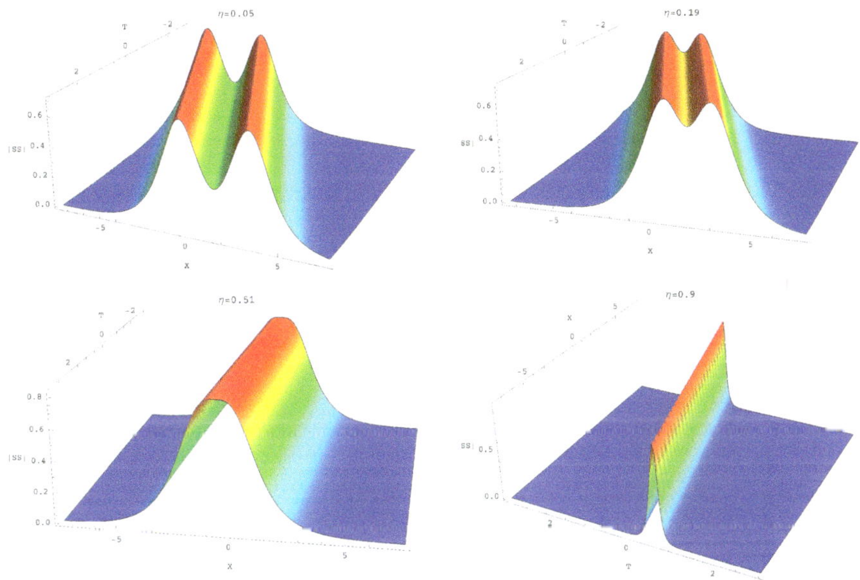

Fig. 3.5 Absolute value of the SS breather, for different values of the parameter η. *Left above:* $|B_{SS}|$ with $\eta = 0.05$; *right above:* $|B_{SS}|$ with $\eta = 0.19$; note that these are cases where the double hump is clearly devised. *Left below:* $|B_{SS}|$ with $\eta = 0.51$, and *right below:* $|B_{SS}|$ with $\eta = 0.9$. Note that for η close to 1, one recovers the NLS soliton, and for η close to zero, the breather decouples and two clearly defined humps, at equal distance for all time (of order $O(|\log \eta|)$), emerge in the dynamics.

with $\Gamma \in L^\infty_{rad}(\mathbb{R}^3) \cap C^1_{loc}(\mathbb{R}^3)$, $m > 0$ in three spatial dimensions was established in [Schei20]. These real-valued solutions are polychromatic, that is, they take the form

$$u(t,x) = u_0(x) + \sum_{k=1}^{\infty} 2\cos(\omega k t) u_k(x) = \sum_{k\in\mathbb{Z}} e^{i\omega k t} u_k(x).$$

Breathers in inhomogeneous periodic media

Here one considers the inhomogeneous wave equation, forced with a nonlinearity $f(x,u)$,

$$s(x)\partial_{tt}u - \partial_{xx}u + q(x)u = f(x,u), \quad (t,x) \in \mathbb{R}\times\mathbb{R},$$

under suitable assumptions on s, q and f. Indeed, they must satisfy periodicity conditions in space. Blank et al. [BCB11], for an $f(x,u) = r(x)u^3$, were able to present an explicit example of such inhomogeneous coefficients such that for this particular choice, the existence of spatially localized, $2\pi/\omega$-time periodic solutions (*breathers*) of finite energy is guaranteed. The proof of this existence result relies on using spatial dynamics, center manifold theory and bifurcation theory for periodic systems.

Due to the required periodicity in time of the solution they considered a polychromatic, real-valued solution ansatz

$$u(t,x) = \sum_{k\in 2\mathbb{Z}+1} u_k(x)e^{ik\omega t}, \quad \bar{u}_k(x) = u_{-k}(x), \quad \omega = \frac{2\pi}{T}.$$

Additional remarks. Scalar field equations in higher dimensions

Very few information is known about the existence of strongly localized breathers of

$$\partial_{tt}u - \Delta u = g(u), \quad \mathbb{R}\times\mathbb{R}^N, \tag{3.3.1}$$

satisfying, for instance, that $u(\cdot,t) \in L^2(\mathbb{R}^N)$ for almost all $t \in \mathbb{R}$ and $N \geq 2$. In [MaSc21] infinitely many weakly localized breathers were found for nonlinearities of the form $Q(x)|u|^{p-2}u$, where Q lies in a suitable Lebesgue space and $p > 2$ is chosen suitably depending on Q as well as the space dimension $N \geq 2$. In [PluRei16], the existence of weakly localized breathers for nonlinear wave equations of the form

$$\partial_{tt}u + \mathcal{L}u = Q(x)|u|^{p-2}u \quad \text{on } \mathbb{R}\times\mathbb{R}^N \tag{3.3.2}$$

was proved. Since they want (3.3.2) to hold in a distributional sense, they only need to assume that there is some $m \in \mathbb{N}$ and a suitable $q \in [p,\infty]$ such that $\mathcal{L} : W^{m,\infty}(K) \to L^{q'}(\mathbb{R}^N)$ is a bounded linear map for all compact $K \subset \mathbb{R}^N$. Here and in the following, q' denotes the Hölder conjugate exponent of q defined by $\frac{1}{q}+\frac{1}{q'} = 1$. As a model case one may have in mind $\mathcal{L} = -\Delta$. In contrast to [MaSc21], they can deal with nonradial Q and obtain the existence of an unbounded sequence of breathers.

In the case $N = 1$ there are nonexistence results [Den93, KMM17] dealing with small perturbations of the SG equation respectively small odd breathers (not covering the even SG breather). These results and several others will be covered in a forthcoming monograph.

The Kurth's breather solution in Vlasov's gravitation

Before introducing this special breather solution, we present some basic concepts around it. Firstly, let $f(t,x,v) \in \mathbb{R}_{\geq 0}$, be the probability density to find a particle at time $t \in \mathbb{R}$ in the position $x \in \mathbb{R}^3$ with speed $v \in \mathbb{R}^3$. Let

$$\rho(t,x) = \int_v f(t,x,v)dv$$

the density of particles in that box. The Vlasov-Poisson system is written as

$$\begin{cases} f_t + v \cdot \nabla_x f - \nabla_x V \cdot \nabla_v f = 0, & \textit{(Vlasov)} \\ \Delta_x V = 4\pi\rho, \quad \textit{(Poisson)} \end{cases} \tag{3.3.3}$$

where here V is a *gravitational* potential function. The Vlasov-Poisson system (3.3.3) is Galilean invariant, and therefore, if $(f(t,x,v), V(t,x))$ is a solution, then $(f(t,x-ct,v-c), V(t,x-ct))$ is also a solution of (3.3.3). In addition, (3.3.3) has infinitely many conserved quantities, for instance the energy

$$E[f](t) := \frac{1}{2}\int_{\mathbb{R}^6} |v|^2 f dv dx - \frac{1}{2}\int_{\mathbb{R}^3} |\nabla V|^2 dx, \tag{3.3.4}$$

and the mass of the solution

$$M[f](t) := \int_{\mathbb{R}^6} f dv dx.$$

Likewise, the total linear momentum Q and angular momentum L:

$$Q[f](t) := \int_{\mathbb{R}^6} v f dv dx, \qquad L[f](t) := \int_{\mathbb{R}^6} x \wedge f dv dx,$$

are also conserved quantities. Moreover, since the characteristic flow of the Vlasov equation (3.3.3) preserves the Lebesgue measure, all L^p norms of f are conserved. Namely

$$\int_{\mathbb{R}^6} |f|^p dv dx < \infty, \quad \forall\ 1 \le p \le \infty.$$

In 1978, Kurth [Kur78] obtained the following solution (f,ρ) of (3.3.3),

$$\rho(t,x) := \left(\frac{4\pi}{3}\right)^{-1} \phi(t)^{-3} \chi_{\{|x|<\phi(t)\}},$$

$$f(t,x,v) := \begin{cases} \frac{3}{8}\left[1 - \left(\frac{|x|^2}{\phi(t)}\right)^2 - |\phi(t)v - \phi'(t)x|^2 + |x \wedge v|^2\right]_+^{-1/2}, & |x \wedge v| < 1, \\ 0, \ \textit{otherwise}, \end{cases}$$

where ϕ is a periodic solution of the nonlinear second order ODE

$$\phi^3\phi'' + \phi = 1, \qquad \phi(0) = 1.$$

In terms of the velocity ϕ', the total energy E (3.3.4) can be simplified as follows

$$E(t) := \frac{3}{5}[(\phi'(0))^2 - 1].$$

Once the initial speed ϕ' is known, the qualitative behavior of the solution is classified in three main sets:

$$\begin{cases} \phi'(0) = 0, & \textit{static solution} \to E < 0, \\ 0 < |\phi'(0)| < 1, & \textit{periodic (in time) solution} \to E < 0, \\ |\phi'(0)| \ge 1, & \textit{dispersive solution} \to E \ge 0. \end{cases} \tag{3.3.5}$$

About the potential V, it can be split as the sum of two terms depending on a ball of radius less or greater than ϕ. Namely

$$-V = x\phi(t)^{-3}\chi_{(|x|<\phi(t))} + \frac{x}{|x|^3}\chi_{(|x|\geq\phi(t))}.$$

Moreover, observe that we can rewrite the function f in two main cases:

Steady state

$$f(x,v) = \left(\frac{4\pi^3}{3}\right)^{-1}(1-2E-F)_+^{-1/2}\chi_{(F<1)},$$

Camm-type steady states

$$f(x,v) := (E_0 - 2E - \gamma F)_+^* F^l.$$

Note that as Rein et al. showed in [RaRe18], it is possible to rescale the function f in the following way

$$f(t,x,v) = f_0\left(\frac{x}{\phi(t)}, v\phi(t) - \phi'(t)x\right).$$

This rescaling allows us to connect the *static* solution with the *periodic* solution (3.3.5).

Gravitational breathers for the vacuum Einstein equations

An iterative approach to obtain solutions of the vacuum Einstein equations when the metric tensor only depends on two variables, was developed by Belinski and Zhakarov, in an analogous way as the classical ISM [BeZa78]. It was established that, from the mathematical point of view, these solutions have the full status of solitons, although they possess a number of unusual features.

The nonstationary integrable metrics are

$$-ds^2 = f(t,z)(-dt^2 + dz^2) + g_{ab}(t,z)dx^a dx^b, \quad a,b = 1,2. \tag{3.3.6}$$

The function f is positive, the signature of g_{ab} is $(++)$, and the square of the norm of the time-like vector is negative. The Einstein equations for (3.3.6) in a vacuum give, for the matrix g (with components g_{ab} with $\alpha = \det(g)^{1/2}$) and coordinates $\zeta = \frac{1}{2}(z+t),\ \eta = \frac{1}{2}(z-t)$:

$$(\alpha g_\zeta g^{-1})_\eta + (\alpha g_\eta g^{-1})_\zeta = 0. \tag{3.3.7}$$

The trace of (3.3.7) gives the wave equation

$$\alpha_{\zeta\eta} = 0. \tag{3.3.8}$$

In addition to α, a second independent solution β of (3.3.8) is needed. These two solutions are

$$\alpha = a(\zeta) + b(\eta), \ \ \beta = a(\zeta) - b(\eta).$$

Now, to construct a one-soliton solution for (3.3.6), it is necessary to "dress" some background metric. Let us take this seed solution in diagonal form:

$$-ds^2 = f_0(-dt^2 + dz^2) + \alpha e^{u_0}(dx^1)^2 + \alpha e^{-u_0}(dx^2)^2, \tag{3.3.9}$$

where $\alpha(t, z)$ is a solution of (3.3.8) and $u_0(t, z)$, as follows from (3.3.7), should satisfy

$$(\alpha u_{0,\zeta})_\eta + (\alpha u_{o,\eta})_\zeta = 0.$$

The *one-solitonic metric* on this background or *gravisoliton* can be written in the form

$$g = \begin{pmatrix} g_{11} & g_{12} \\ g_{21} & g_{22} \end{pmatrix} = \frac{1}{\cosh(\rho)} \begin{pmatrix} (\mu^2 e^\rho + \alpha^2 e^{-\rho})e^{u_0} & \alpha^2 - \mu^2 \\ \alpha^2 - \mu^2 & (\mu^2 e^\rho + \alpha^2 e^{-\rho})e^{-u_0} \end{pmatrix}$$

$$f = f_0 \alpha^{1/2} \mu \cosh(\rho)(\alpha^2 - \mu^2)^{-1}.$$

From here the procedure can be repeated to obtain the *two-solitonic solution* for the metric (3.3.6) on the background (3.3.9) which indeed can be written as

$$\begin{aligned} g_{11} &= \left\{ 1 + \frac{1}{D} \left[\cosh(2\tau) + \frac{(\mu_1 + \mu_2)}{(\mu_1 - \mu_2)} \sinh(2\tau) \right.\right. \\ &\qquad \left.\left. + \cosh(2\sigma) - \frac{(\alpha^2 + \mu_1\mu_2)}{(\alpha^2 - \mu_1\mu_2)} \sinh(2\sigma) \right] \right\} \alpha e^{u_0}, \\ g_{22} &= \left\{ 1 + \frac{1}{D} \left[\cosh(2\tau) - \frac{(\mu_1 + \mu_2)}{(\mu_1 - \mu_2)} \sinh(2\tau) \right.\right. \\ &\qquad \left.\left. + \cosh(2\sigma) + \frac{(\alpha^2 + \mu_1\mu_2)}{(\alpha^2 - \mu_1\mu_2)} \sinh(2\sigma) \right] \right\} \alpha e^{-u_0}, \\ g_{12} &= \frac{2\alpha}{D} \left[\frac{(\mu_1 + \mu_2)}{(\mu_1 - \mu_2)} \sinh(\tau)\sinh(\sigma) \right. \\ &\qquad \left. + \frac{(\alpha^2 + \mu_1\mu_2)}{(\alpha^2 - \mu_1\mu_2)} \cosh(\tau)\cosh(\sigma) \right], \end{aligned} \tag{3.3.10}$$

$$f = \frac{f_0 \mu_1 \mu_2 D}{(\alpha^2 - \mu_1^2)(\alpha^2 - \mu_2^2)}, \tag{3.3.11}$$

where

$$D = 4\mu_1\mu_2 \left[\frac{\alpha^2 \cosh(\sigma)^2}{(\alpha^2 - \mu_1\mu_2)^2} + \frac{\sinh(\tau)^2}{(\mu_1 - \mu_2)^2} \right].$$

The pole trajectories μ_1, μ_2 follow from the quadratic equations

$$\begin{aligned} \mu_1^2 + 2(\beta - \omega_1)\mu_1 + \alpha^2 &= 0, \\ \mu_2^2 + 2(\beta - \omega_2)\mu_2 + \alpha^2 &= 0, \end{aligned} \tag{3.3.12}$$

where ω_1 and ω_2 are arbitrary constants. The functions σ, τ can be found by quadrature from the following differential equations:

$$\begin{aligned} \sigma_\zeta &= \frac{(\alpha^2 - \mu_1\mu_2)u_{0,\zeta}}{(\alpha - \mu_1)(\alpha - \mu_2)}, \\ \sigma_\eta &= \frac{(\alpha^2 - \mu_1\mu_2)u_{0,\eta}}{(\alpha + \mu_1)(\alpha + \mu_2)}, \\ \tau_\zeta &= \frac{\alpha(\mu_1 - \mu_2)u_{0,\zeta}}{(\alpha - \mu_1)(\alpha - \mu_2)}, \\ \tau_\eta &= -\frac{\alpha(\mu_1 - \mu_2)u_{0,\eta}}{(\alpha + \mu_1)(\alpha + \mu_2)}. \end{aligned}$$

In [BeZa78] it was shown that μ_1 and μ_2 can be either both real or complex conjugate to each other. Because β can be replaced by $\beta + const$ without any physical consequences, we can set $\omega_1 = \omega, \omega_2 = -\omega$ in (3.3.12), which now gives

$$\begin{aligned} \mu_1^2 + 2(\beta - \omega)\mu_1 + \alpha^2 &= 0, \\ \mu_2^2 + 2(\beta + \omega)\mu_2 + \alpha^2 &= 0, \end{aligned} \tag{3.3.13}$$

and for the real μ_1, μ_2 we should take real ω, while the complex-conjugate μ pair corresponds to the purely imaginary values of ω.

There also exists the third kind of two-soliton solution (3.3.10). It corresponds to the case when μ_1 and μ_2 form a complex-conjugate pair. For the spacelike u_0 this solution becomes the gravitational analogue of the breather, and currently named as *gravibreather*. In the case $\mu_2 = \bar{\mu}_1$ (complex conjugate), the variable τ and quantity $\mu_1 - \mu_2$, become purely imaginary; however, the metrics (3.3.10) and (3.3.11) remain real with correct physical signature. After the substitution $\tau = i\tau'$, we get the real variable τ' which is timelike, and two-soliton solutions (3.3.10) and (3.3.11) appear to be oscillating in time τ'. This solution corresponds to the purely imaginary values of the constant ω in (26). Thus, the gravibreather can be considered as the analytic continuation in ω of one of the two-soliton solutions with real μ_1 and μ_2. It is a simple task to prove that the gravibreather emerges in this way from the gravisoliton- antigravisoliton state only.

Following [BeZa78], the two-solitonic perturbation on the background (3.3.9) with the poles

$$\mu_1 = q(\sinh(z) - i)(1 - \cosh(t)), \quad \mu_2 = q(\sinh(z) + i)(1 - \cosh(t))$$

(they are solutions to Eqs. (3.3.13), in which $\omega = -iq$) gives the following exact *gravibreather* solution to the vacuum Einstein equations:

$$
\begin{aligned}
-ds^2 = & \frac{De^{k^2q^2\sinh(t)^2\cosh(z)^2}}{(q\sinh(t)\cosh(z))^{1/2}}(-dt^2+dz^2) \\
& + \frac{q}{D}\sinh(t)\cosh(z)\Big\{A_1 e^{2kq\cosh(t)\sinh(z)}(dx^1)^2 \\
& \qquad + A_2 e^{-2kq\cosh(t)\sinh(z)}(dx^2)^2 + A_3 dx^1 dx^2\Big\},
\end{aligned} \tag{3.3.14}
$$

where

$$
\begin{aligned}
A_1 &= (\cosh(t)\cosh(\sigma) - \sinh(\sigma))^2 + (\sinh(z)\sin(\tau) - \cos(\tau))^2, \\
A_2 &= (\cosh(t)\cosh(\sigma) + \sinh(\sigma))^2 + (\sinh(z)\sin(\tau) + \cos(\tau))^2, \\
A_3 &= 4[\cosh(t)\cosh(\sigma)\cos(\tau) - \sinh(\sigma)\sinh(z)\sin(\tau)], \\
D &= \sinh(t)^2\cosh(\sigma)^2 + \cosh(z)^2\sin(\tau)^2, \\
\tau &= 2kq(\cosh(t) - 1) + q\tau_0,
\end{aligned}
$$

and σ_0, τ_0 are arbitrary constants. Note that the substitution $t \to t/i\omega, z \to z/i\omega, q \to i\omega$ would replace (3.3.14) by the two-soliton solution.

The gravibreather (3.3.14) represents the inhomogeneous cosmological model which starts at the instant $t = 0$ by the big bang with anisotropic Kasner-like asymptotics and at $t \to +\infty$, approaches the background solution (3.3.9). In between, the model oscillates in time.

The logarithmic NLS equation and its breather solutions

Finally we show that the logarithmic NLS (log-NLS) equation

$$
iu_t + \Delta u = \varepsilon \log(|u|^2)u, \quad (t,x) \in \mathbb{R}\times\mathbb{R}^n, \quad \varepsilon = \pm 1, \tag{3.3.15}
$$

has breather solutions. This model was introduced in wave mechanics and nonlinear optics [BiMy76] (see also [HeRe81], [KEB00]).

For (3.3.15) mass and momentum are conserved and equal to the current NLS laws, namely $M[u](t) := \|u(t,\cdot)\|_{L^2}^2$ and $P[u](t) := \operatorname{Im}\int_{\mathbb{R}^n}\nabla_x u(t,\cdot)\bar{u}(t,x)dx$. Its energy is written as

$$
E[u](t) := \int_{\mathbb{R}}\left(|u_x|^2 + 2\varepsilon|u|^2\log(|u|^2)\right)dx = E[u](0).
$$

Note moreover that in (3.3.15) the nonlinear *log* term has a singularity at the origin, removing the local Lipschitz continuity. The Cauchy problem for (3.3.15) was fully described in [HaOy24, Th.1.1] (see also [CaGa18], [GLN10]): namely for initial data in a subspace

$$W^1(\mathbb{R}^n) := \{v \in H^1(\mathbb{R}^n), |v|^2 \ln |v|^2 \in L^1(\mathbb{R}^n)\} \subset H^1(\mathbb{R}^n),$$

there exists a unique solution $u \in (C \cap L^\infty)(\mathbb{R}, W^1)$ in the case $\varepsilon = -1$, and $u \in C(\mathbb{R}, W^1)$ in the case $\varepsilon = 1$.

With respect to explicit solutions of (3.3.15), standing wave solutions $u(t, x) = e^{i\omega t}G_\omega(x)$ are known, where

$$G_\omega(x) := e^{\frac{\omega+n}{2} - \frac{\varepsilon|x|^2}{2}}, \quad x \in \mathbb{R}^n,$$

is the so-called *Gausson*. Beside standing waves, breather solutions are also explicitly known (see [Fer20], [BiMy76]) in the case $\varepsilon = -1, n = 1$, namely

$$u^\alpha(t, x) := \sqrt{\frac{\alpha_r}{r_\alpha(t)}} \exp\left[\frac{1}{2} - i\Phi^\alpha(t) - \frac{x^2}{4r_\alpha(t)^2} + i\frac{\dot{r}_\alpha(t)}{r_\alpha(t)}\frac{x^2}{4}\right], \qquad t, x \in \mathbb{R},$$

with

$$\Phi^\alpha(t) := \frac{1}{2}\int_0^t \frac{1}{r_\alpha(s)^2} ds + \varepsilon \int_0^t \ln\left(\frac{r_\alpha(s)}{\alpha_0}\right) ds - \varepsilon t,$$

and where for any $\alpha \in \mathbb{C}^+$, the function r_α is defined by

$$\ddot{r}_\alpha = \frac{1}{r_\alpha^3} - \frac{2\varepsilon}{r_\alpha}, \qquad r_\alpha(0) = \operatorname{Re} \alpha := \alpha_r, \qquad \dot{r}_\alpha(0) = \operatorname{Im} \alpha := \alpha_i.$$

In fact the above ODE has a unique solution $r_\alpha \in C^\infty(\mathbb{R})$ with values in $(0, \infty)$, and it is periodic.

3.4 Exercises

(3.1) Show that B (3.2.1) solves the mKdV equation and that for $\alpha \gg \beta$,

$$B \approx 2\sqrt{2}\beta \cos(\alpha y_1)\operatorname{sech}(\beta y_2).$$

(3.2) Show that for $B_{\alpha,\beta,\mu}$ (3.2.2), if $\alpha \gg \beta \gg \mu$, then

$$B_{\alpha,\beta,\mu} \approx 2\sqrt{2}\beta \cos(\alpha y_1)\operatorname{sech}(\beta y_2).$$

(3.3) Graph (3.2.5) and describe how the SS breather changes with respect to the parameter η.

(3.4) (Open problem) Prove or disprove that 1D cubic NLS with nonzero background at infinity is globally well-posed.

Chapter 4

Breathers and their stability in the modified Korteweg-de Vries equation

In what follows we will describe the main properties of breather solutions to the modified KdV model already presented in Chapter 3. Indeed, from Subsection 3.2.1, the mKdV model is somehow the most simple model to exemplify, from the mathematical point of view, the inherent properties of breathers.

4.1 Introduction

The modified Korteweg-de Vries equation (mKdV) is the one-dimensional model described by the partial differential equation

$$\partial_t u + \partial_x(\partial_x^2 u + u^3) = 0. \tag{4.1.1}$$

In this case, $u = u(t, x)$ is a scalar, real-valued function, depending on a time $t \in \mathbb{R}$ and space $x \in \mathbb{R}$.

This equation is invariant under space and time shifts, that is, if $u(t, x)$ is a solution of (4.1.1), then

$$u(t + t_0, x + x_0) \tag{4.1.2}$$

is a new solution of (4.1.1), no matter what are $t_0, x_0 \in \mathbb{R}$. Additionally, but not less important, the equation remains invariant under the scaling

$$u_\lambda(t, x) := \lambda u(\lambda^3, \lambda x), \qquad \lambda > 0. \tag{4.1.3}$$

This scaling leaves the $H^{-1/2}(\mathbb{R})$ norm invariant, and it is a key information for the well-posedness theory, see also Subsection 2.7 for more details. Also, the L^1 integral

$$\int_{\mathbb{R}} u_\lambda(t, x)dx,$$

which will reveal to be important when we will study mKdV breathers. As we expressed above, the critical scaling regularity is $s = -\frac{1}{2}$, very far from the best local well-posedness result, which is only $s > \frac{1}{4}$, and it is sharp for (4.1.1), in the sense that the solution map is no longer uniformly continuous for $s < 1/4$. It is important to mention that mKdV has a conservation law at the L^2 level, *the mass*

$$\int_{\mathbb{R}} u^2(t,x)dx.$$

However, no local well-posedness result is available at this regularity, only the existence of a solution constructed by a compactness method.

4.1.1 *The Cauchy problem*

Let us consider the initial value problem associated to (4.1.1):

$$\partial_t u + \partial_x(\partial_x^2 u + u^3) = 0, \quad u(t=0) = u_0.$$

The initial datum is u_0, supposed in the Hilbert space H^s. If the initial datum is in the space H^s, with $s > \frac{1}{4}$, then the initial value problem for (4.1.1) is globally well-posed. These are results proved in a series of papers by Kenig-Ponce-Vega [KPV93], and Colliander, Keel, Staffilani, Takaoka and Tao [CKS03]. Below this regularity the problem becomes a very complicated one and only partial results are available. Indeed, the flow map is not uniformly continuous if $s < \frac{1}{4}$ [KPV01]. In order to prove this last result, Kenig, Ponce and Vega considered *breathers*. However, for our purposes we shall work at very good regularity, say H^1.

4.1.2 *Conservation laws*

Being an integrable model, mKdV also has infinitely many conservation laws. See Section 2.9 for further details on the meaning of these conserved quantities.

The first three conservation laws in the case of mKdV (4.1.1) at the H^1-level are the *mass*

$$M[u](t) := \frac{1}{2}\int_{\mathbb{R}} u^2(t,x)dx = M[u](0), \tag{4.1.4}$$

and *energy*

$$E[u](t) := \frac{1}{2}\int_{\mathbb{R}} u_x^2(t,x)dx - \frac{1}{4}\int_{\mathbb{R}} u^4(t,x)dx = E[u](0). \tag{4.1.5}$$

One also has the integral

$$I[u](t) := \int u(t,x)dx = I[u](0),$$

which is zero in the case of mKdV breathers. Additional to the previous ones, there is an important third conserved quantity, called F:

$$F[u](t) := \frac{1}{2}\int_{\mathbb{R}} u_{xx}^2(t,x)dx - \frac{5}{2}\int_{\mathbb{R}} u^2u_x^2(t,x)dx + \frac{1}{4}\int_{\mathbb{R}} u^6(t,x)dx \\ = F[u](0). \tag{4.1.6}$$

This conservation law is key for the stability problem of breathers. Indeed, it is the quantity that characterizes breathers, in a sense to be made precise below. Note that (4.1.6) is only well-defined for H^2 valued solutions of mKdV.

4.1.3 *The aperiodic mKdV breather*

Recall the mKdV breather already introduced in Subsection 3.2.1:

Definition 4.1.1. Let $\alpha,\beta > 0$ and $x_1, x_2 \in \mathbb{R}$. The breather solution of mKdV (4.1.1) is explicitly given by

$$B_{\alpha,\beta}(t,x) = B(t,x;\alpha,\beta,x_1,x_2) := 2\sqrt{2}\partial_x\left[\arctan\left(\frac{\beta}{\alpha}\frac{\sin(\alpha y_1)}{\cosh(\beta y_2)}\right)\right], \tag{4.1.7}$$

with

$$y_1 = x + \delta t + x_1, \quad y_2 = x + \gamma t + x_2, \tag{4.1.8}$$

and

$$\delta := \alpha^2 - 3\beta^2, \quad \gamma := 3\alpha^2 - \beta^2. \tag{4.1.9}$$

As we already mentioned before, this solution was discovered by Wadati in [Wad73]. Notice that $\alpha,\beta > 0$ is just a particular choice because $\alpha,\beta < 0$ are also possible choices. This is concordance with the fact that $u \longmapsto -u$ is an admissible map in mKdV. However, for the sake of simplicity, we reduce the space of parameters to the positive sector only.

The parameters α and β, usually called as *frequency and amplitude respectively,* are scaling parameters, x_1, x_2 are shifts, and $-\gamma$ represents the *velocity* of a breather (see Figs. 4.1 and 3.1).

4.1.4 *Elementary properties*

We first review some very simple but also essential properties satisfied by (4.1.7).

Note that breather solutions of the mKdV are periodic in time, but not in space. A simple but very important remark is that $\delta \neq \gamma$, for all values of α and β different from zero. This means that the variables $x+\delta t$ and $x+\gamma t$ are **always independent**,

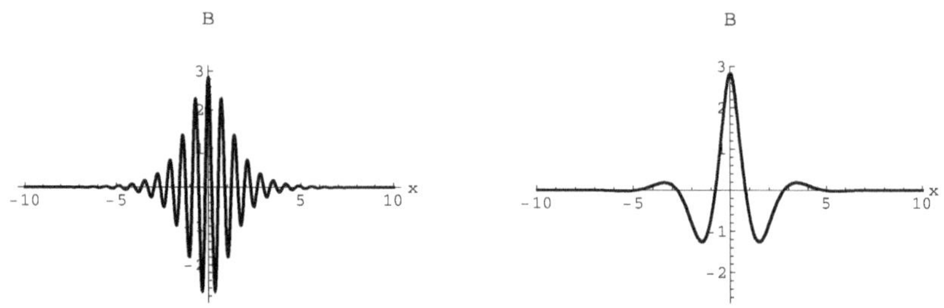

Fig. 4.1 Left: mKdV breather (4.1.7) with $\alpha = 9, \beta = 1$ at $t = 0$. Right: mKdV breather (4.1.7) with $\alpha = 1.5, \beta = 1$ at $t = 0$.

a property that characterizes breather solutions, which is not satisfied by standard solitons from Chapter 2.10.

Additionally, thanks to the scaling invariance (4.1.3), one can assume without loss of generality that $\beta = 1$. We have chosen not to perform this transformation since some important properties are better presented in terms of general β.

Using the velocity $-\gamma = \beta^2 - 3\alpha^2$, we can distinguish three different regimes for a breather, depending on the value of $\alpha > 0$:

(1) The case where $\alpha < \frac{\beta}{\sqrt{3}}$; in this case $-\gamma > 0$ and the breather moves to the right, as solitons do;
(2) The case where $\alpha = \frac{\beta}{\sqrt{3}}$; here $\gamma = 0$ and we are in presence of a stationary breather; and finally,
(3) The case where $\alpha > \frac{\beta}{\sqrt{3}}$; in this case $-\gamma < 0$ and the breather, unlike mKdV solitons, moves to the left.

Similarly, using the translation invariance in space and time (4.1.2), it is possible to choose $x_1 = x_2 = 0$ in (4.1.7). Consequently, in the rest of this chapter, we will work with the mKdV breather

$$B(t,x) := 2\sqrt{2}\partial_x \left[\arctan\left(\frac{\beta}{\alpha} \frac{\sin(\alpha(x + \delta t))}{\cosh(x + \gamma t)} \right) \right], \tag{4.1.10}$$

with $\gamma = 3\alpha^2 - \beta^2$, $\delta = \alpha^2 - 3\beta^2$.

4.1.5 *Breathers in mathematics and physics*

Breathers as in (4.1.7) are *oscillatory bound states*, because, as times evolves, they do not decouple into solitons.

For a detailed account of the physics of breathers see e.g. [AbCl91, Aub97, Ale12a, AlMu13] and references therein.

In Mathematics, breather solutions arise in different contexts. In a geometrical setting, mKdV breathers appear in the evolution of closed planar curves playing the role of smooth localized deformations traveling along the closed curve [Ale12a]. Moreover, as it was mentioned before, it is interesting to point out that mKdV breather solutions have also been considered by Kenig, Ponce and Vega in their proof of the non-uniform continuity of the mKdV flow in the Sobolev spaces H^s, $s < \frac{1}{4}$ [KPV01].

On the other hand, they should play an important role in the *soliton-resolution* conjecture for the mKdV equation, according to the analysis developed by Schuur in [Sch86] and as it was recently shown in [ChLi21] for initial conditions in some weighted Sobolev spaces. Indeed, according to the IST theory, any suitable[1] solution of mKdV should decompose, as time goes to infinity, into solitons, breathers and radiation.

4.1.6 *Mass of breather solutions*

Let us describe breathers in a more qualitative way. In order to do this, we need to understand the mass and energy of a breather. We start with a simple result.

Lemma 4.1.2. *mKdV breathers as in* (4.1.10) *have zero mean, namely*

$$\int_{\mathbb{R}} B = 0.$$

Proof. Direct from Definition 4.1.1. □

Lemma 4.1.2 has deep implications in the dynamics of breathers. Among the most important ones, one has the possibility to work with a localized integral of B, called $\widetilde{B}$ (or kink profile) along these pages, which will have an important role in many results. This function cannot be properly defined (i.e., as a localized function) in other more general contexts.

Lemma 4.1.3 (Mass of mKdV breathers). *mKdV breathers satisfy*

$$M[B] := \frac{1}{2}\int_{\mathbb{R}} B^2 = 4\beta. \tag{4.1.11}$$

Remark 4.1.4. It is interesting to notice that the mass of a breather does not depend on the scaling α (frequency), which reveals a sort of degeneracy of the L^2 norm, probably related to the lack of well-posedness for L^2 solutions for mKdV.

[1]Here suitable means "with sufficient decay at infinity" such that IST works properly.

The proof of Lemma 4.1.3 can be done using a scientific computing program, or by hand [AlMu13]. Computing by hand is always, in the case of breathers, quite difficult and cumbersome. However, in the case of mKdV breather mass, it can be done in some simple steps.

4.2 Nonlinear stationary equations satisfied by breather profiles

The purpose of this section is to put in evidence that breathers may satisfy intricate systems of elliptic equations, a fact that reveals a deep variational structure underlying the existence and stability of these particular solutions.

In the following, without loss of generality, we will assume that the mKdV breather $B_{\alpha,\beta}$ has scalings $\alpha,\beta > 0$.

The question that naturally arises is the following: is there any analogous elliptic equation satisfied by breather solutions, as solitons in (2.10.2)? For the mKdV breather and for almost every breather treated now one has that the answer is positive. Indeed, any mKdV breather (4.1.7) $B = B_{\alpha,\beta}$ satisfies a nonlinear stationary equation:

Theorem 4.2.1 ([AlMu13]). *Let* $B = B_{\alpha,\beta}$ *be any mKdV breather* (4.1.7) *with scalings* $\alpha,\beta > 0$. *Then, for any fixed time* $t \in \mathbb{R}$, *the breather* B *satisfies the nonlinear stationary equation*

$$B_{4x} - 2(\beta^2 - \alpha^2)(B_{xx} + B^3) + (\alpha^2 + \beta^2)^2 B + 5BB_x^2 + 5B^2 B_{xx} + \frac{3}{2}B^5 = 0. \quad (4.2.1)$$

This identity can be seen as the *nonlinear elliptic equation* satisfied by any mKdV breather profile, and therefore it is independent in time and translation parameters $x_1, x_2 \in \mathbb{R}$. One can compare with the soliton profile $Q_c(x - ct - x_0)$, which satisfies the standard elliptic equation (2.10.2), obtained as the first variation of the H^1 functional (2.10.4).

Remark 4.2.2. The fact that multisolitons satisfy intricate elliptic equations is not new at all. For similar results in the case of KdV soliton solutions, see e.g. Lax [Lax68] and Maddocks-Sachs [MaSa93].

From Theorem 4.2.1, some deep questions remain unsolved. For instance: are breathers, modulo invariances, the unique solutions to (4.2.1)? Notice that (4.2.1) is a fourth order model. Note also that for any x_1, x_2 shifts, the breather *profile* $B(0, x; \alpha, \beta, x_1, x_2)$ is solution to (4.2.1). In other words, there is a sort of two dimensional set of invariances for (4.2.1). Some progress in that question has been made in [FFM24].

Another question is related to the mKdV hierarchy: the set of integrable equations associated to mKdV, see [Ale24, Ale23, Ale24] for results in this direction. It turns out that any breather of the mKdV hierarchy satisfy the same equation, namely (4.2.1). This is an impressive result, so far. See Chapter 12 for further details.

4.3 Breather profiles

Proving (4.2.1) analytically is quite involved, but very easy with a computer if one has the right hand side in (4.2.1). Let us follow the first method, that will be useful to other problems appearing in the stability question. We will first prove an auxiliary result, which is in some sense the equivalent to (2.10.2). But first we introduce new definitions.

We introduce now the notion of breather profile. Given parameters $x_1, x_2 \in \mathbb{R}$ and $\alpha, \beta > 0$, we consider y_1 and y_2 defined by

$$y_1 := x + x_1, \quad y_2 := x + x_2. \tag{4.3.1}$$

Let $\widetilde{B}$ be the kink profile

$$\widetilde{B} = \widetilde{B}(x; \alpha, \beta, x_1, x_2) := 2\sqrt{2} \arctan\left(\frac{\beta}{\alpha}\frac{\sin(\alpha y_1)}{\cosh(\beta y_2)}\right), \tag{4.3.2}$$

and with a slight abuse of notation, we redefine

$$B = B(x; \alpha, \beta, x_1, x_2) := \widetilde{B}_x(x; \alpha, \beta, x_1, x_2). \tag{4.3.3}$$

Now we introduce the directions associated to the shifts x_1 and x_2. Given a breather profile of parameters α, β, x_1 and x_2, we define

$$B_1 = B_1(x; \alpha, \beta, x_1, x_2) := \partial_{x_1} B,$$

$$B_2 = B_2(x; \alpha, \beta, x_1, x_2) := \partial_{x_2} B,$$

and for δ and γ defined in (4.1.9),

$$\widetilde{B}_t := \delta B_1 + \gamma B_2. \tag{4.3.4}$$

We also have from (4.4.2),

$$\widetilde{B}_t + B_{xx} + B^3 = 0. \tag{4.3.5}$$

4.4 Partial mass of the breather

Since the mass of the breather in Lemma 4.1.3 resulted a very clean result, it is clear that a deeper structure is present. Let us define the partial mass of the breather

$$\mathcal{M} = \mathcal{M}(x;\alpha,\beta,x_1,x_2) := \frac{1}{2}\int_{-\infty}^{x} B^2(s;\alpha,\beta,x_1,x_2)ds.$$

Then one has

$$\lim_{x\to-\infty} \mathcal{M}(x;\alpha,\beta,x_1,x_2) = 0, \qquad \lim_{x\to+\infty} \mathcal{M}(x;\alpha,\beta,x_1,x_2) = 4\beta.$$

Lemma 4.4.1 (Partial mass of the mKdV breather). *One has*

$$\begin{aligned}&\mathcal{M}(x;\alpha,\beta,x_1,x_2)\\ &= \frac{2[\alpha^2+1+\alpha\sin(2\alpha y_1)-\cos(2\alpha y_1)+\alpha^2(\sinh(2y_2)+\cosh(2y_2))]}{\alpha^2+1+\alpha^2\cosh(2y_2)-\cos(2\alpha y_1)},\end{aligned} \tag{4.4.1}$$

with y_1 and y_2 defined in (4.3.1).

The proof of this result is left as an exercise to the reader, and it can be obtained directly from the proof of Lemma 4.1.3.

Corollary 4.4.2.

$$B_x^2 + \frac{1}{2}B^4 + 2B\widetilde{B}_t - 2\mathcal{M}_t = 0. \tag{4.4.2}$$

This last identity is obtained by integration by parts in (4.3.5), after multiplication by B_x.

From the above Corollary, it is clear that we will need an expression for $\mathcal{M}_t$.

Lemma 4.4.3.

$$\frac{1}{3}\mathcal{M}_t = \frac{4\alpha^2\tilde{h}_\alpha}{3(\alpha^2+1+\alpha^2\cosh(2y_2)-\cos(2\alpha y_1))^2}, \tag{4.4.3}$$

with

$$\begin{aligned}\tilde{h}_\alpha &:= \gamma\alpha^2-\delta+(\alpha^2+1)(\delta\cos(2\alpha y_1)+\gamma\cosh(2y_2))\\ &\quad +(\delta\alpha^2-\gamma)\cos(2\alpha y_1)\cosh(2y_2)-\alpha(\delta+\gamma)\sin(2\alpha y_1)\sinh(2y_2).\end{aligned} \tag{4.4.4}$$

Proof. From (4.4.1) we have

$$\mathcal{M}_t = 2\frac{(f_\alpha)_t g_\alpha - f_\alpha (g_\alpha)_t}{g_\alpha^2},$$

where, with a slight abuse of notation, f_α and g_α are given now by

$$f_\alpha = \alpha^2 + 1 + \alpha\sin(2\alpha y_1) - \cos(2\alpha y_1) + \alpha^2(\sinh(2y_2) + \cosh(2y_2)),$$

and

$$g_\alpha = \alpha^2 + 1 + \alpha^2\cosh(2y_2) - \cos(2\alpha y_1). \tag{4.4.5}$$

It is clear that

$$(f_\alpha)_t = 2\alpha[\alpha\delta\cos(2\alpha y_1) + \delta\sin(2\alpha y_1) + \alpha\gamma\cosh(2y_2) + \alpha\gamma\sinh(2y_2)],$$

and

$$(g_\alpha)_t = 2\alpha[\delta\sin(2\alpha y_1) + \alpha\gamma\sinh(2y_2)].$$

Therefore

$$(f_\alpha)_t g_\alpha - f_\alpha(g_\alpha)_t = 2\alpha^2\tilde{h}_\alpha,$$

and

$$\begin{aligned}\tilde{h}_\alpha := \gamma\alpha^2 - \delta + (\alpha^2+1)(\delta\cos(2\alpha y_1) + \gamma\cosh(2y_2))\\ + (\delta\alpha^2 - \gamma)\cos(2\alpha y_1)\cosh(2y_2) - \alpha(\delta+\gamma)\sin(2\alpha y_1)\sinh(2y_2).\end{aligned}$$

In conclusion, we get (4.4.3). □

4.5 The equivalent to the soliton equation

We present now what is probably the closest second order ODE satisfied by a breather profile. This equation is nonlocal in principle, but it can be understood as the equivalent to the soliton equation $Q_c'' + Q_c^3 = cQ_c$. In order to get a full local ODE, one needs a fourth order model.

Lemma 4.5.1. *Let $B = B(x;\alpha,\beta,x_1,x_2)$ be any mKdV breather profile. Then*

$$B_{xt} + 2\mathcal{M}_t B = 2(\beta^2 - \alpha^2)\widetilde{B}_t + (\alpha^2+\beta^2)^2 B. \tag{4.5.1}$$

Proof. We make use of the explicit expression of the breather.

First of all, from (4.3.2) and (4.3.4),

$$\widetilde{B}_t = 4\sqrt{2}\alpha\beta\left(\frac{\alpha\delta\cos(\alpha y_1)\cosh(\beta y_2) - \beta\gamma\sin(\alpha y_1)\sinh(\beta y_2)}{\alpha^2+\beta^2+\alpha^2\cosh(2\beta y_2) - \beta^2\cos(2\alpha y_1)}\right). \tag{4.5.2}$$

From (4.3.3) we have

$$B = 2\sqrt{2}\alpha\beta\left(\frac{\alpha\cos(\alpha y_1)\cosh(\beta y_2) - \beta\sin(\alpha y_1)\sinh(\beta y_2)}{\alpha^2\cosh^2(\beta y_2) + \beta^2\sin^2(\alpha y_1)}\right) \tag{4.5.3}$$

$$= 4\sqrt{2}\alpha\beta\left(\frac{\alpha\cos(\alpha y_1)\cosh(\beta y_2) - \beta\sin(\alpha y_1)\sinh(\beta y_2)}{\alpha^2+\beta^2+\alpha^2\cosh(2\beta y_2) - \beta^2\cos(2\alpha y_1)}\right). \tag{4.5.4}$$

Therefore, from (4.5.3) and (4.4.3),

$$\mathcal{M}_t B = \frac{16\sqrt{2}\alpha^3\beta^3\tilde{\theta}(t,x)}{(\alpha^2+\beta^2+\alpha^2\cosh(2\beta y_2)-\beta^2\cos(2\alpha y_1))^3}$$

where, with the definition of $\tilde{h}_\alpha$ in (4.4.4),

$$\tilde{\theta}(t,x) := \tilde{h}_\alpha(\alpha\cos(\alpha y_1)\cosh(\beta y_2)-\beta\sin(\alpha y_1)\sinh(\beta y_2)).$$

Let us compute B_{xt}. First we have from (4.5.3),

$$B_x = \frac{4\sqrt{2}\alpha\beta h_1(t,x)}{(\alpha^2+\beta^2+\alpha^2\cosh(2\beta y_2)-\beta^2\cos(2\alpha y_1))^2},$$

where

$$\begin{aligned}h_1(t,x) &:= -(\alpha^2+\beta^2)\cosh(\beta y_2)\sin(\alpha y_1)[\alpha^2+\beta^2+\alpha^2\cosh(2\beta y_2)-\beta^2\cos(2\alpha y_1)]\\ &\quad -2\alpha\beta[\alpha\cos(\alpha y_1)\cosh(\beta y_2)-\beta\sin(\alpha y_1)\sinh(\beta y_2)][\beta\sin(2\alpha y_1)+\alpha\sinh(2\beta y_2)].\end{aligned}$$

Then,

$$B_{xt} = \frac{4\sqrt{2}\alpha\beta h_2}{(\alpha^2+\beta^2+\alpha^2\cosh(2\beta y_2)-\beta^2\cos(2\alpha y_1))^3},$$

where, using g_α previously defined in (4.4.5),

$$h_2(t,x) := -4\alpha\beta\Big(\beta\delta\sin(2\alpha y_1)+\alpha\gamma\sinh(2\beta y_2)\Big)h_{21}+g_\alpha h_{22},$$

and

$$\begin{aligned}&h_{21}(t,x)\\ &:= -(\alpha^2+\beta^2)\cosh(\beta y_2)\sin(\alpha y_1)g_\alpha\\ &\quad -2\alpha\beta(\alpha\cos(\alpha y_1)\cosh(\beta y_2)-\beta\sin(\alpha y_1)\sinh(\beta y_2))(\beta\sin(2\alpha y_1)+\alpha\sinh(2\beta y_2)),\end{aligned}$$

$$\begin{aligned}&h_{22}(t,x)\\ &:= g_\alpha\\ &\Big(-g_\alpha\Big(\alpha\delta(\alpha^2+\beta^2)\cos(\alpha y_1)\cosh(\beta y_2)+\beta\gamma(\alpha^2+\beta^2)\sin(\alpha y_1)\sinh(\beta y_2)\Big)\\ &\quad -2\alpha\beta(\alpha^2+\beta^2)\sin(\alpha y_1)\cosh(\beta y_2)\Big(\alpha\gamma\sinh(2\beta y_2)+\delta\beta\sin(2\alpha y_1)\Big)\\ &\quad -2\alpha\beta\Big(-\alpha^2\delta\sin(\alpha y_1)\cosh(\beta y_2)+\alpha\beta\gamma\cos(\alpha y_1)\sinh(\beta y_2)\\ &\quad -\beta\alpha\delta\cos(\alpha y_1)\sinh(\beta y_2)-\beta^2\gamma\sin(\alpha y_1)\cosh(\beta y_2)\Big)\Big(\alpha\sinh(2\beta y_2)+\beta\sin(2\alpha y_1)\Big)\\ &\quad -4\alpha^2\beta^2\Big(\alpha\cos(\alpha y_1)\cosh(\beta y_2)-\beta\sin(\alpha y_1)\sinh(\beta y_2)\Big)\Big(\gamma\cosh(2\beta y_2)+\delta\cos(2\alpha y_1)\Big)\Big).\end{aligned}$$

Then,

$$B_{xt} + 2(\mathcal{M})_t B = \frac{4\sqrt{2}\alpha\beta\left(h_2 + 8\alpha^2\beta^2\tilde{\theta}\right)}{(\alpha^2+\beta^2+\alpha^2\cosh(2\beta y_2)-\beta^2\cos(2\alpha y_1))^3}, \tag{4.5.5}$$

and recalling that

$$\begin{aligned} g_\alpha^2 &= (\alpha^2+\beta^2+\alpha^2\cosh(2\beta y_2)-\beta^2\cos(2\alpha y_1))^2 \\ &= (\alpha^2+\beta^2)^2 + (\alpha^4\cosh(2\beta y_2)^2 + \beta^4\cos(2\alpha y_1)^2 - 2\alpha^2\beta^2\cosh(2\beta y_2)\cos(2\alpha y_1)) \\ &\quad -2(\alpha^2+\beta^2)\left(\alpha^2\cosh(2\beta y_2) - \beta^2\cos(2\alpha y_1)\right), \end{aligned}$$

collecting terms in (4.5.5) and taking into account (4.5.4) and (4.5.2), after some calculations we have

$$\begin{aligned} &\text{r.h.s. of (4.5.5)} \\ &= \frac{4\sqrt{2}\alpha\beta}{(\alpha^2+\beta^2+\alpha^2\cosh(2\beta y_2)-\beta^2\cos(2\alpha y_1))^3}\times \\ &\quad \Big\{g_\alpha^2 \times \Big(2(\beta^2-\alpha^2)\left(\delta\alpha\cos(\alpha y_1)\cosh(\beta y_2) - \gamma\beta\sin(\alpha y_1)\sinh(\beta y_2)\right) \\ &\quad + (\alpha^2+\beta^2)^2\left(\alpha\cos(\alpha y_1)\cosh(\beta y_2) - \beta\sin(\alpha y_1)\sinh(\beta y_2)\right)\Big)\Big\} \\ &= 2(\beta^2-\alpha^2)(\widetilde{B}_{\alpha,\beta})_t + (\alpha^2+\beta^2)^2 B_{\alpha,\beta}. \end{aligned}$$

□

It turns out that with (4.5.1) one can directly prove Theorem 4.2.1.

4.6 Breathers and elliptic equations

Now we prove Theorem 4.2.1. Let B be any breather profile. From (4.3.5) and (4.4.2), one has

$$\begin{aligned} &\text{l.h.s. of (4.2.1)} \\ &= -(B_t + B^3)_{xx} + 2(\beta^2-\alpha^2)\widetilde{B}_t + (\alpha^2+\beta^2)^2 B + 5BB_x^2 + 5B^2B_{xx} + \frac{3}{2}B^5 \\ &= -B_{tx} - BB_x^2 + 2B^2B_{xx} + 2(\beta^2-\alpha^2)\widetilde{B}_t + (\alpha^2+\beta^2)^2 B + \frac{3}{2}B^5 \\ &= -B_{tx} + B\left(\frac{1}{2}B^4 + 2B\widetilde{B}_t - 2\mathcal{M}_t\right) - 2B^2(\widetilde{B}_t + B^3) + \frac{3}{2}B^5 \\ &\quad + 2(\beta^2-\alpha^2)\widetilde{B}_t + (\alpha^2+\beta^2)^2 B \\ &= -[B_{tx} + 2\mathcal{M}_t B] + 2(\beta^2-\alpha^2)\widetilde{B}_t + (\alpha^2+\beta^2)^2 B = 0. \end{aligned}$$

In the last line we have used (4.5.1). This ends the proof.

A beautiful corollary of the previous result is the following useful one:

Corollary 4.6.1. *Let t belong to a given open interval. For any $x_1(t), x_2(t) \in \mathbb{R}$ smooth, one has that the modified profile $B(x;\alpha,\beta,x_1(t),x_2(t))$ solves the fourth order ODE* (4.2.1).

4.7 Energy of breathers

Now we turn to the energy of breathers, where the energy functional E is given in (4.1.5). This quantity is essential to understand the movement of breathers.

Lemma 4.7.1 (Energy of breathers). *Any mKdV breather B satisfies*

$$E[B] = \frac{4}{3}\gamma\beta. \tag{4.7.1}$$

Remark 4.7.2. Note that the energy (4.7.1) takes into account the velocity of the breather itself, in the sense that negative energy breathers are characterized by positive velocities, zero energy breathers are static, and finally, positive energy breathers move to the left. Let us also remark that the sign of the energy is dictated by the sign of the velocity γ.

Proof of Lemma 10.3.5. First of all, let us prove the following reduction

$$E[B](t) = \frac{1}{3}\int_{\mathbb{R}} (\mathcal{M})_t(t,x)dx. \tag{4.7.2}$$

Indeed, we multiply (4.3.5) by B and integrate in space: we get

$$\int_{\mathbb{R}} B_x^2 = \int_{\mathbb{R}} B\widetilde{B}_t + \int_{\mathbb{R}} B^4.$$

On the other hand, integrating (4.4.2),

$$\int_{\mathbb{R}} B_x^2 + \frac{1}{2}\int_{\mathbb{R}} B^4 + 2\int_{\mathbb{R}} B\widetilde{B}_t - 2\int_{\mathbb{R}} \mathcal{M}_t = 0.$$

From these two identities, we get

$$\int_{\mathbb{R}} B^4 = \frac{4}{3}\int_{\mathbb{R}} \mathcal{M}_t - 2\int_{\mathbb{R}} B\widetilde{B}_t,$$

and therefore

$$\int_{\mathbb{R}} B_x^2 = \frac{4}{3}\int_{\mathbb{R}} \mathcal{M}_t - \int_{\mathbb{R}} B\widetilde{B}_t.$$

Finally, replacing the last two identities in (4.1.5), we get (4.7.2), as desired.

Secondly, we split the aforementioned $\tilde{h}_\alpha$ (4.4.4) into two pieces, according to the parameter γ. From the definition of γ and δ, we have

$$\tilde{h}_\alpha = \gamma h_\alpha + 2(\alpha^2 + 1)\hat{h}_\alpha,$$

where

$$\begin{aligned} h_\alpha := \alpha^2 - 1 &+ (\alpha^2 + 1)(\cos(2\alpha y_1) + \cosh(2y_2)) \\ &+ (\alpha^2 - 1)\cos(2\alpha y_1)\cosh(2y_2) - 2\alpha\sin(2\alpha y_1)\sinh(2y_2), \end{aligned}$$

and

$$\hat{h}_\alpha := 1 - (\alpha^2 + 1)\cos(2\alpha y_1) - \alpha^2\cos(2\alpha y_1)\cosh(2y_2) + \alpha\sin(2\alpha y_1)\sinh(2y_2).$$

Note that from (4.4.1),

$$\int_{\mathbb{R}} \frac{4\alpha^2\gamma h_\alpha}{3(\alpha^2 + 1 + \alpha^2\cosh(2y_2) - \cos(2\alpha y_1))^2} = \frac{4}{3}\gamma = 2\gamma|E[Q]|,$$

then, in order to conclude, from (4.7.2) we are reduced to prove that

$$\int_{\mathbb{R}} \frac{\hat{h}_\alpha}{(\alpha^2 + 1 + \alpha^2\cosh(2y_2) - \cos(2\alpha y_1))^2} = 0. \tag{4.7.3}$$

We prove this identity, noting that $(\frac{1}{2\alpha}\sin(2\alpha y_1))_x = \cos(2\alpha y_1)$, and from (4.4.5),

$$\begin{aligned} &g_\alpha\left(\frac{1}{2\alpha}\sin(2\alpha y_1)\right)_x - \frac{1}{2\alpha}\sin(2\alpha y_1)(g_\alpha)_x = \\ &= \cos(2\alpha y_1)[\alpha^2 + 1 + \alpha^2\cosh(2y_2) - \cos(2\alpha y_1)] \\ &\qquad - \sin(2\alpha y_1)[\sin(2\alpha y_1) + \alpha\sinh(2y_2)] \\ &= -(\cos^2(2\alpha y_1) + \sin^2(2\alpha y_1)) + (\alpha^2 + 1)\cos(2\alpha y_1) + \alpha^2\cos(2\alpha y_1)\cosh(2y_2) \\ &\qquad - \alpha\sin(2\alpha y_1)\sinh(2y_2) \\ &= -\hat{h}_\alpha. \end{aligned}$$

Therefore,

$$\text{l.h.s. of (4.7.3)} = \lim_{x\to+\infty} \frac{-\sin(2\alpha y_1)}{2\alpha(\alpha^2 + 1 + \alpha^2\cosh(2y_2) - \cos(2\alpha y_1))} = 0.$$

□

4.8 Stability of breathers

We arrive to what we believe is the most important mathematical and physical question about breathers: are they stable?

4.8.1 *Main ideas behind the stability of breathers*

In the previous section we proved that mKdV breathers satisfy intricate elliptic models. These elliptic equations imply that breathers can be understood as critical points of a suitable Lyapunov functional.

However, it will be important to note that this functional should be defined in a *proper* subspace of the so-called *energy space*, namely the space where the standard variational structure of solitons is placed. Instead, we must look for functionals described at the leading order by an additional conservation law.

Thanks to the integrability of the equation, such a quantity is always available. However, the addition of a new conservation law carries together a rigidity structure, in the sense that breathers should not exist in nonintegrable models (a fact that actually does not happen, as we show in Section 3.3).

Since these particular solutions satisfy a well-defined variational characterization, their dynamical stability reduces to the study of the spectrum of a suitable linear operator defined in a subspace of the energy space. Except for some simple cases as the already mentioned, this task is far from being simple. Sometimes, numerical analysis is needed to guess the good outcome.

However, the extension of this stability result to a bigger class of initial perturbations seems not possible via the aforementioned variational structure. Indeed, one needs a sort of rigidity or dynamical structure that gives a new kind of stability result for more general Sobolev spaces.

4.8.2 *Orbital stability*

Before proving that mKdV breathers are indeed stable, we need a suitable notion of stability for these structures. As in the case of solitons, we will use a modified version of the orbital stability property.

Definition 4.8.1. We will say that the mKdV breather is orbitally stable in H^s, $s \geq 1$, if the following is satisfied. For any $\alpha, \beta > 0$, $x_1^0, x_2^0 \in \mathbb{R}$, there are $C_0 > 0$ and $\eta_0 > 0$ such that for all $\eta \in (0, \eta_0)$, if $u_0 \in H^2(\mathbb{R})$ satisfies

$$\left\| u_0 - B(0, \cdot; \alpha, \beta, x_1^0, x_2^0) \right\|_{H^s(\mathbb{R})} < \eta,$$

then there are $x_1(t)$, $x_2(t) \in \mathbb{R}$ such that

$$\sup_{t \in \mathbb{R}} \| u(t) - B(t, \cdot; \alpha, \beta, x_1(t), x_2(t)) \|_{H^s(\mathbb{R})} < C_0 \eta.$$

Otherwise, we will say that the breather is unstable.

The term $C_0\eta$ in the last inequality can be weakened to any $\varepsilon > 0$, provided $\eta(\varepsilon) > 0$ is chosen small enough. Also, the sup in time can be taken either to the positive side, or the negative one, or both. This is because the dynamics is reversible in time. Sometimes (this is not the case), one must consider only one direction of time.

Theorem 4.8.2. *mKdV breathers are orbitally stable in H^2, in the sense of Definition 4.8.1.*

Numerical computations (see Gorria-Alejo-Vega [AGV13]) show that mKdV breathers are *numerically* stable. Schuur [Sch86], using IST methods, pointed out that breathers may appear from general but rapidly decaying initial data. However, a rigorous proof of orbital stability for breathers for initial data in Sobolev spaces was missing [AlMu13].

We should also remark that the original seed of these ideas is certainty not new and it was introduced in a seminal paper by Lax [Lax68], in the particular case of the two-soliton solution of the Korteweg-de Vries equation. This method has been generalized to several equations with soliton solutions [MaSa93, HPZ11, Kap07, NeLo06]. However, no previous result was available in the case of breathers, apparently because of their dynamics, which do not resemble any type of simple, decoupled 2-soliton solution. Compared with the previous results, proofs in [AlMu13] are more involved, since there is no mass splitting as $t \to +\infty$. Further results where this technique has been successfully applied to the understanding of stability of different soliton solutions are the works [PeSh14, GaPe15, JoPe15], see also references therein.

4.8.3 *Sketch of proof of orbital stability*

Since mKdV is an integrable equation, it has infinitely many conserved quantities. For the proof of Theorem 4.8.2 we will need the additional H^2-conserved quantity (4.1.6)

$$F[u](t) := \frac{1}{2}\int_{\mathbb{R}} u_{xx}^2(t) - \frac{5}{2}\int_{\mathbb{R}} u^2 u_x^2(t) + \frac{1}{4}\int_{\mathbb{R}} u^6(t) = F[u](0), \tag{4.8.1}$$

well-defined for H^2 initial data. See Fig. 4.2 for a schematic view of the stability via F, in terms of formally defined Lagrange multipliers.

We will show that mKdV breathers are critical points of the Lyapunov functional

$$\mathcal{H}[u](t) := F[u](t) + 2(\beta^2 - \alpha^2)E[u](t) + (\alpha^2 + \beta^2)^2 M[u](t). \tag{4.8.2}$$

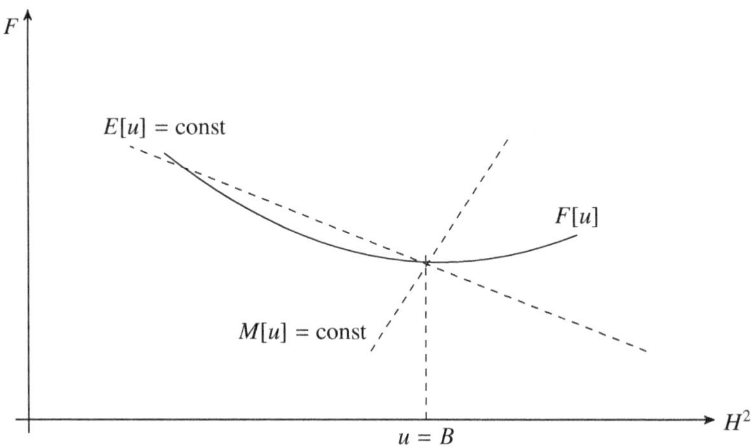

Fig. 4.2 Schematic representation of Theorem 4.8.2. Each breather has a variational characterization; it is a (local) minimizer of the functional F defined in (4.8.1) under suitable constraints on its mass and energy.

Clearly $\mathcal{H}$ is conserved for initial data $u_0 \in H^2(\mathbb{R})$.

Now the proof uses the fact that $\mathcal{H}$ defined in (4.8.2) controls the dynamics of perturbations of breathers. Indeed, any perturbation of a breather solution of the form

$$u(t,x) = B(t,x;\alpha,\beta,x_1(t),x_2(t)) + z(t,x),$$

with z *small* and $x_1(t)$, $x_2(t)$ to be chosen later, must satisfy the expansion

$$\mathcal{H}[u](t) = \mathcal{H}[B] + \int_{\mathbb{R}} G[B](t)z + \frac{1}{2}\int_{\mathbb{R}} z(t)\mathcal{L}z(t) + O(\|z(t)\|^3_{H^2(\mathbb{R})}). \tag{4.8.3}$$

Moreover,

$$G[B] := B_{4x} - 2(\beta^2-\alpha^2)(B_{xx}+B^3) + (\alpha^2+\beta^2)^2 B + 5BB_x^2 + 5B^2B_{xx} + \frac{3}{2}B^5,$$

exactly the left hand side of (4.2.1). Thanks to (4.2.1), no matter what are x_1 and x_2, one has

$$\int_{\mathbb{R}} G[B](t)z \equiv 0.$$

This last fact establishes that breathers are critical points of the functional $\mathcal{H}$.

Additionally, $\mathcal{L}$ in (4.8.3) denotes the self-adjoint operator with domain $H^4(\mathbb{R})$:

$$\begin{aligned}\mathcal{L}z := z_{4x} - 2(\beta^2 - \alpha^2)z_{xx} + (\alpha^2 + \beta^2)^2 z + 5B^2 z_{xx} + 10BB_x z_x \\ + (5B_x^2 + 10BB_{xx} + \frac{15}{2}B^4 - 6(\beta^2 - \alpha^2)B^2)z.\end{aligned} \tag{4.8.4}$$

Therefore, the stability question is reduced to the study of spectral properties of the linearized operator $\mathcal{L}$. This operator has a kernel, and negative directions. The big surprise in the proof of stability in [AlMu13] is the fact that, despite having two free scaling parameters α and β, mKdV breathers have **only one negative eigenvalue**.

4.8.4 *Spectral analysis*

A beautiful outcome of (4.2.1) is the following important property

$$\mathcal{L}\partial_{x_1} B = \mathcal{L}\partial_{x_2} B = 0.$$

Namely, the kernel of $\mathcal{L}$ is at least of dimension 2. This is naturally related to the fact that mKdV breathers have two shifts. But the natural question is: are these the only linearly independent members of the kernel?

In the case of second order elliptic ODEs or PDEs, the question is solved using standard techniques. In the case of fourth order ODEs or PDEs, this is far from obvious, since several important techniques, such as maximum principles, are lost in this case.

In the mKdV case, the fact that these two directions span the entire L^2 kernel of $\mathcal{L}$ is fortunately not difficult to check. Indeed, the equation $\mathcal{L}z = 0$ is a fourth order ODE whose solutions are spanned by four linearly independent solutions. Their asymptotic at positive infinity are given by the forms

$$e^{\pm\beta x}\sin(\alpha x), \quad e^{\pm\beta x}\cos(\alpha x),$$

so unless $\beta = 0$, which is impossible by hypothesis, we only have two linearly independent localized solutions, which coincide with $\partial_{x_1} B$ and $\partial_{x_2} B$ above.

On the other hand, using the Weyl theorem we have that the continuum spectrum is given by the intervals $[(\alpha^2 + \beta^2)^2, \infty)$ if $\beta \geq \alpha$, and $[4\alpha^2\beta^2, \infty)$ if $\beta \leq \alpha$.

We finally consider the problem of counting the number of negative eigenvalues. This is not an easy task, mainly because we deal with a fourth order ODE. The idea is to use the work by L. Greenberg [Gre91], who shows that the number of negative eigenvalues,

$$\#\text{ negative eigenvalues} = \sum_{x\in\mathbb{R}} \dim\ker W[\partial_{x_1} B, \partial_{x_2} B](t, x),$$

where $W[\partial_{x_1} B, \partial_{x_2} B]$ is the Wronskian matrix associated to $\partial_{x_1} B$ and $\partial_{x_2} B$.

The best way for understanding this identity is by considering the same problem for the case of a soliton solution. We have in this case

$$\#\text{ negative eigenvalues} = \sum_{x\in\mathbb{R}} \dim\ker Q'(x),$$

where we understand Q' as a linear operator in one dimension. Since $Q'(x) = 0$ only for $x = 0$, we have that $\sum_{x\in\mathbb{R}} \dim Q'(x)$ is finite and equals one (i.e., we have just one negative eigenvalue).

After some tedious computations we will obtain

$$\det W[\partial_{x_1} B, \partial_{x_2} B](t,x) = \frac{16\alpha^3\beta^3(\alpha^2+\beta^2)(\alpha\sinh(2\beta y_2) - \beta\sin(2\alpha y_1))}{(\alpha^2+\beta^2+\alpha^2\cosh(2\beta y_2) - \beta^2\cos(2\alpha y_1))^2},$$

so we have nontrivial kernel if and only if

$$\alpha\sinh(2\beta y_2) = \beta\sin(2\alpha y_1). \tag{4.8.5}$$

Given x_1, x_2, t, α and β fixed, there is only one point $x = x_0 \in \mathbb{R}$ (depending on the previous parameters) for which this last identity holds. Indeed, fix x_1, x_2, t, α and β. We will look for $\tilde{y}_2$ solution of

$$\sinh(\tilde{y}_2) = \frac{\beta}{\alpha}\sin\left(\frac{\alpha}{\beta}\tilde{y}_2 + \tilde{x}_{12}\right), \tag{4.8.6}$$

where

$$\tilde{x}_{12} := 2\alpha[(\delta-\gamma)t + x_1 - x_2], \quad \tilde{y}_2 := 2\beta y_2.$$

(Recall that $y_1 = x + \delta t + x_1$, $y_2 = x + \gamma t + x_2$.) If $|\tilde{y}_2|$ is large enough, say $> M$, there is no solution for this equation. If now $|\tilde{y}_2| \le M$, note that the function

$$[-M,M] \ni \tilde{y}_2 \mapsto \sinh(\tilde{y}_2) - \frac{\beta}{\alpha}\sin\left(\frac{\alpha}{\beta}\tilde{y}_2 + \tilde{x}_{12}\right) \in \mathbb{R} \tag{4.8.7}$$

changes its sign on $[-M,M]$, M large, so it has a root. Moreover, if $\tilde{x}_{12} \neq 2k\pi$, $k \in \mathbb{Z}$, such a root is unique ((4.8.7) has positive derivative). Now, if $\tilde{x}_{12} = 2k\pi$ for some $k \in \mathbb{Z}$, we will have

$$\sinh(\tilde{y}_2) = \frac{\beta}{\alpha}\sin\left(\frac{\alpha}{\beta}\tilde{y}_2\right),$$

for which $\tilde{y}_2 = 0$ is a root. If there is another one, it must be unique, by the same reason as before. We conclude that there a unique root $\tilde{y}_2$ of (4.8.6). In particular, there is unique x_0 satisfying (4.8.5).

At the point x_0 we will have

$$1 \le \dim\ker W[\partial_{x_1} B, \partial_{x_2} B](t, x_0) \le 2,$$

but it is easy to see that the dimension cannot be 2 since the Wronskian matrix at that point is never identically zero.

We conclude that $\mathcal{L}$ has a unique negative eigenvalue. It is possible to prove that such an eigenvalue is always far from zero, uniformly in time. Even better, there is $c_0 > 0$ only depending on α and β such that for all $z \in H^2(\mathbb{R})$, if

$$\int_{\mathbb{R}} z\partial_{x_1} B = \int_{\mathbb{R}} z\partial_{x_2} B = 0,$$

then

$$\int_{\mathbb{R}} z\mathcal{L}z \geq c_0 \|z\|_{H^2(\mathbb{R})}^2 - \frac{1}{c_0}\left|\int_{\mathbb{R}} Bz\right|^2. \tag{4.8.8}$$

Using the conservation of mass we can estimate the last term above:

$$\begin{aligned}\left|\int_{\mathbb{R}} Bz(t)\right| &\leq \left|\int_{\mathbb{R}} Bz(0)\right| + \sup_{t\geq 0} \|z(t)\|_{L^2(\mathbb{R})}^2 \\ &\leq \|z(0)\|_{L^2(\mathbb{R})} + \sup_{t\geq 0} \|z(t)\|_{L^2(\mathbb{R})}^2.\end{aligned}$$

Replacing in (4.8.8), and using (4.8.3), we conclude.

4.9 Extended proof

Now we explain in detail the previous arguments, in the mKdV case. Since the same argument works for other breathers, we will only sketch the proof of stability in the forthcoming sections.

4.9.1 *Weinstein inequalities*

Let us start with the following generalized Weinstein conditions:

Corollary 4.9.1. *Let $B = B_{\alpha,\beta}$ be any mKdV breather. Given $t \in \mathbb{R}$ fixed, let*

$$\Lambda_\alpha B := \partial_\alpha B, \quad \text{and} \quad \Lambda_\beta B := \partial_\beta B. \tag{4.9.1}$$

Then both functions $\Lambda_\alpha B$ and $\Lambda_\beta B$ are in the Schwartz class for the spatial variable, and satisfy the identities

$$\partial_\alpha M[B] = \int_{\mathbb{R}} B\Lambda_\alpha B = 0, \tag{4.9.2}$$

and

$$\partial_\beta M[B] = \int_{\mathbb{R}} B\Lambda_\beta B = 4 > 0, \tag{4.9.3}$$

independently of time.

Proof. By simple inspection, one can see that, given t fixed, $\Lambda_\alpha B$ and $\Lambda_\beta B$ are well-defined Schwartz functions. The proof of (4.9.2) and (4.9.3) is a consequence of (4.1.11), and the definition of mass (4.1.4). □

4.9.2 Detailed spectral analysis

Let $z = z(x)$ be a function to be specified in the following lines. Let $B = B_{\alpha,\beta}$ be any breather solution, with shift parameters x_1, x_2. Recall the linear operator introduced in (4.8.4):

$$\begin{aligned}\mathcal{L}z(x;t) := {}& z_{4x}(x) - 2(\beta^2 - \alpha^2)z_{xx}(x) \\ & + (\alpha^2 + \beta^2)^2 z(x) + 5B^2 z_{xx}(x) + 10BB_x z_x(x) \\ & + \left(5B_x^2 + 10BB_{xx} + \frac{15}{2}B^4 - 6(\beta^2 - \alpha^2)B^2\right)z(x).\end{aligned}$$

Let us describe the spectrum of this operator. More precisely, our main purpose is to find a suitable coercivity property, independently of the nature of scaling parameters. The main result of this section is contained in Proposition 4.9.15. Part of the analysis carried out in this section has been previously introduced by Lax [Lax68], and Maddocks and Sachs [MaSa93], so we follow their arguments adapted to the breather case, sketching several proofs.

Lemma 4.9.2. *$\mathcal{L}$ is a linear, unbounded operator in $L^2(\mathbb{R})$, with dense domain $H^4(\mathbb{R})$. Moreover, $\mathcal{L}$ is self-adjoint.*

It is a surprising fact that $\mathcal{L}$ is actually self-adjoint, due to the non-constant terms appearing in the definition of $\mathcal{L}$. From standard spectral theory of unbounded operators with rapidly decaying coefficients, it is enough to prove that $\mathcal{L}^* = \mathcal{L}$ in $H^4(\mathbb{R})^2$.

Proof. Let $z, w \in H^4(\mathbb{R})$. Integrating by parts, one has

$$\begin{aligned}&\int_{\mathbb{R}} w\mathcal{L}z \\ &= \int_{\mathbb{R}} w\left(z_{4x} - 2(\beta^2 - \alpha^2)z_{xx} + (\alpha^2 + \beta^2)^2 z + 5B^2 z_{xx} + 10BB_x z_x\right) \\ &\quad + \int_{\mathbb{R}} \left(5B_x^2 + 10BB_{xx} + \frac{15}{2}B^4 - 6(\beta^2 - \alpha^2)B^2\right)zw \\ &= \int_{\mathbb{R}} \left(w_{4x} - 2(\beta^2 - \alpha^2)w_{xx} + (\alpha^2 + \beta^2)^2 w + (5B^2 w)_{xx} - (10BB_x w)_x\right)z \\ &\quad + \int_{\mathbb{R}} \left(5B_x^2 + 10BB_{xx} + \frac{15}{2}B^4 - 6(\beta^2 - \alpha^2)B^2\right)zw = \int_{\mathbb{R}} \mathcal{L}wz.\end{aligned}$$

Finally, it is clear that $D(\mathcal{L}^*)$ can be identified with $D(\mathcal{L}) = H^4(\mathbb{R})$. □

A consequence of the above result is the fact that the spectrum of $\mathcal{L}$ is real-valued. Furthermore, the following result describes the continuous spectrum of $\mathcal{L}$.

Lemma 4.9.3. *Let $\alpha,\beta > 0$. The operator $\mathcal{L}$ is a compact perturbation of the constant coefficients operator*

$$\mathcal{L}_0 z := z_{4x} - 2(\beta^2 - \alpha^2)z_{xx} + (\alpha^2 + \beta^2)^2 z.$$

In particular, the continuous spectrum of $\mathcal{L}$ is the closed interval $[(\alpha^2+\beta^2)^2, +\infty)$ in the case $\beta \geq \alpha$, and $[4\alpha^2\beta^2, +\infty)$ in the case $\beta < \alpha$. No embedded eigenvalues are contained in this region.

Proof. This result is a consequence of the Weyl Theorem on continuous spectrum. Let us note that the nonexistence of embedded eigenvalues (or resonances) is a consequence of the rapidly decreasing character of the potentials involved in the definition of $\mathcal{L}$. □

Remark 4.9.4. Note that the condition $\alpha = \beta$ is equivalent to the identity $\partial_\beta E[B] = 0$. Solitons do not satisfy this last identity.

We introduce now two directions associated to spatial translations. We define

$$\begin{aligned} B_1(t; x_1, x_2) &:= \partial_{x_1} B_{\alpha,\beta}(t; x_1, x_2), \\ B_2(t; x_1, x_2) &:= \partial_{x_2} B_{\alpha,\beta}(t; x_1, x_2). \end{aligned} \tag{4.9.4}$$

It is clear that, for all $t \in \mathbb{R}$ $\alpha,\beta > 0$ and $x_1, x_2 \in \mathbb{R}$, both B_1 and B_2 are real-valued functions in the Schwartz class, exponentially decreasing in space. Moreover, it is not difficult to see that they are *linearly independent* as functions of the x-variable, for all time t fixed.

Lemma 4.9.5. *For each $t \in \mathbb{R}$, one has*

$$\ker \mathcal{L} = \operatorname{span}\{B_1(t; x_1, x_2), B_2(t; x_1, x_2)\}.$$

Proof. From the elliptic equation satisfied by the breather, and taking derivative with respect to x_1 and x_2, we obtain

$$\mathcal{L}B_1(t; x_1, x_2) = \mathcal{L}B_2(t; x_1, x_2) = 0, \tag{4.9.5}$$

with $\mathcal{L}$ the linearized operator defined in (4.8.4) and B_1, B_2 defined in (4.9.4). A direct analysis involving ordinary differential equations shows that the null space of $\mathcal{L}_0$ is spawned by functions of the type

$$e^{\pm\beta x}\cos(\alpha x), \quad e^{\pm\beta x}\sin(\alpha x), \quad \alpha,\beta > 0,$$

(note that this set is linearly independent). Among these four functions, there are only two L^2-integrable ones in the semi-infinite line $[0, +\infty)$. Therefore, the null space of $\mathcal{L}|_{H^4(\mathbb{R})}$ is spanned by at most two L^2-functions. Finally, comparing with (4.9.5), we have the desired conclusion. □

A direct consequence of the above identities and Corollary 4.9.1 is the following result:

Corollary 4.9.6. *With the notation of Lemma 8.11.4, let*

$$B_0 := \frac{\alpha\Lambda_\beta B + \beta\Lambda_\alpha B}{8\alpha\beta(\alpha^2+\beta^2)}. \tag{4.9.6}$$

Then B_0 is Schwartz and satisfies $\mathcal{L}B_0 = -B$,

$$\begin{aligned}\int_{\mathbb{R}} B_0 B &= \frac{1}{2\beta(\alpha^2+\beta^2)} > 0,\\ \frac{1}{2}\int_{\mathbb{R}} B_0\mathcal{L}B_0 &= -\frac{1}{4\beta(\alpha^2+\beta^2)} < 0.\end{aligned} \tag{4.9.7}$$

Remark 4.9.7. In other words, B_0 is also a negative direction. Moreover, it is not orthogonal to the breather itself. Note additionally that the constant involved in (4.9.7) is independent of time.

It turns out that the most important consequence of (8.11.2) is the fact that $\mathcal{L}$ possesses, for all time, *only one negative eigenvalue*. Indeed, in order to prove that result, we follow the Greenberg and Maddocks-Sachs strategy [Gre91, MaSa93], applied this time to the linear, *oscillatory* operator $\mathcal{L}$. More specifically, we will use the following

Lemma 4.9.8 (Uniqueness criterion, see also [Gre91, MaSa93]). *Let $B = B_{\alpha,\beta}$ be any mKdV breather, and B_1, B_2 the corresponding kernel of the operator $\mathcal{L}$. Then $\mathcal{L}$ has*

$$\sum_{x\in\mathbb{R}} \dim\ker W[B_1, B_2](t;x)$$

negative eigenvalues, counting multiplicity. Here, W is the Wronskian matrix of the functions B_1 and B_2,

$$W[B_1, B_2](t;x) := \begin{bmatrix} B_1 & B_2 \\ (B_1)_x & (B_2)_x \end{bmatrix}(t,x). \tag{4.9.8}$$

Proof. This result is essentially contained in [Gre91, Theorem 2.2], where the finite interval case was considered. As shown in several articles (see e.g. [MaSa93, HPZ11]), the extension to the real line is direct and does not require additional efforts. We skip the details. □

In what follows, we compute the Wronskian (4.9.8). Contrary to the 2-soliton case, where the decoupling of both solitons at infinity simplifies the proof, here we have carried the computations by hand, because of the coupled character of the breather. The surprising fact is the following greatly simplified expression for (4.9.8):

Lemma 4.9.9. *Let $B = B_{\alpha,\beta}$ be any mKdV breather, and B_1, B_2 the corresponding kernel elements defined in* (4.9.4). *Then*

$$\det W[B_1, B_2](t;x) = \frac{16\alpha^3\beta^3(\alpha^2+\beta^2)[\alpha\sinh(2\beta y_2) - \beta\sin(2\alpha y_1)]}{(\alpha^2+\beta^2+\alpha^2\cosh(2\beta y_2) - \beta^2\cos(2\alpha y_1))^2}. \tag{4.9.9}$$

Remark 4.9.10. Since the computation of (4.9.9) involves only partial derivatives on the x-variable, the result above is still valid for the case of breathers with parameters x_1, x_2 depending on time. We skip the details.

Proposition 4.9.11. *The operator $\mathcal{L}$ defined in* (4.8.4) *has a unique negative eigenvalue $-\lambda_0^2 < 0$, of multiplicity one, and $\lambda_0 = \lambda_0(\alpha,\beta,x_1,x_2,t)$.*

Proof. We compute the determinant (4.9.8) required by Lemma 4.9.8. From Lemma 4.9.9, after a standard translation argument, we just need to consider the behavior of the function

$$f(y_2) = f_{t,\alpha,\beta,\tilde{x}_2}(y_2) := \alpha\sinh(2\beta y_2) - \beta\sin(2\alpha(y_2 + (\delta-\gamma)t + \tilde{x}_2)), \tag{4.9.10}$$

for $\tilde{x}_2 := x_1 - x_2 \in \mathbb{R}$, and $\delta - \gamma = -2(\alpha^2+\beta^2)$.

A simple argument shows that for $y_2 \in \mathbb{R}$ such that $|\sinh(2\beta y_2)| > \frac{\beta}{\alpha}$, f has no root. Moreover, there exists $R_0 = R_0(\alpha,\beta) > 0$ such that, for all $y_2 > R_0$ one has $f(y_2) > 0$ and for all $y_2 < -R_0$, $f(y_2) < 0$. Therefore, since f is continuous, there is a root $y_0 = y_0(t,\alpha,\beta,\tilde{x}_2) \in [-R_0, R_0]$ for f. Additionally, if $y_2 \neq 0$,

$$f'(y_2) = 2\alpha\beta[\cosh(2\beta y_2) - \cos(2\alpha(y_2 - 2(\alpha^2+\beta^2)t + \tilde{x}_2))] > 0,$$

by simple inspection. Therefore, if $y_0 \neq 0$ then it is unique and then

$$\sum_{x\in\mathbb{R}} \dim\ker W[B_1,B_2](t;x) = \dim\ker W[B_1,B_2](t; y_0 - \gamma t - x_2) = 1,$$

since B_1 or $(B_1)_x$ are not zero at that time. Indeed, it is enough to show that $W[B_1,B_2](t,x)$ is not identically zero, then $\dim\ker W[B_1,B_2] < 2$. In order to prove this fact, note that B_1 solves, for $t, x_1, x_2 \in \mathbb{R}$ fixed, a second order linear ODE with source term $-(\widetilde{B})_t$. Therefore, by standard well-posedness results, both B_1 and $(B_1)_x$ cannot be identically zero at the same point.

Now, let us assume that $y_2 = 0$ is a zero of f. We give a different proof of the same result proved above. From (4.9.10), $t = t_k$ must satisfy the condition

$$-2(\alpha^2+\beta^2)t + \tilde{x}_2 = \frac{k\pi}{2\alpha}, \quad k \in \mathbb{Z}.$$

In terms of the variables y_1 and y_2, one has $y_1 = \frac{k\pi}{2\alpha}$, $k \in \mathbb{Z}$, and $y_2 = 0$. Recall that

$$\begin{aligned} B_1(t,x) = -2\sqrt{2}\alpha^2\beta\Bigg(& \frac{\alpha\sin(\alpha y_1)\cosh(\beta y_2) + \beta\cos(\alpha y_1)\sinh(\beta y_2)}{\alpha^2\cosh^2(\beta y_2) + \beta^2\sin^2(\alpha y_1)} \\ & + 2\beta^2\sin(\alpha y_1)\cos(\alpha y_1) \\ & \times \frac{[\alpha\cos(\alpha y_1)\cosh(\beta y_2) - \beta\sin(\alpha y_1)\sinh(\beta y_2)]}{(\alpha^2\cosh^2(\beta y_2) + \beta^2\sin^2(\alpha y_1))^2}\Bigg). \end{aligned} \tag{4.9.11}$$

Replacing in (4.9.11), we get

$$B_1(t_k,x) = \frac{-2\sqrt{2}\alpha^3\beta\sin(\frac{k}{2}\pi)}{\alpha^2 + \beta^2\sin^2(\frac{k}{2}\pi)} = \begin{cases} \neq 0, & k \text{ odd}, \\ 0, & k \text{ even}. \end{cases}$$

For the first case above we conclude as in the previous one. Finally, if t_k satisfies

$$-2(\alpha^2 + \beta^2)t + \tilde{x}_2 = \frac{k\pi}{\alpha}, \quad k \in \mathbb{Z},$$

one has $y_1 = k\pi$, $y_2 = 0$ and $B_1(t,x) = 0$, but from (4.9.11), after a direct computation, $(B_1)_x$ is given now by the quantity

$$(B_1)_x(t_k,x) = -2\sqrt{2}\alpha^2\beta\left(1 + \frac{\beta^2}{\alpha^2}\cos(k\pi) + \frac{2\beta^2\alpha^2}{\alpha^4}\cos^2(k\pi)\right) \neq 0$$

for all $k \in \mathbb{Z}$. Therefore $\sum_{x\in\mathbb{R}} \dim\ker W[B_1,B_2](t_k;x) = 1$. In conclusion, for all $t \in \mathbb{R}$, $\mathcal{L}$ has just one negative eigenvalue, of multiplicity one. □

We consider now some standard remarks. We can reduce the spectral problem to another independent of time. Indeed, after translation and redefinition of the shift parameters x_1 and x_2, we can assume that

$$B = B_{\alpha,\beta}(0,x;x_1,0), \quad x_1 \in \left[0, \frac{2\pi}{\alpha}\right].$$

In what follows we assume that B is given by the previous formula.

Corollary 4.9.12. *There exists a continuous function $f_0 = f_0(\alpha,\beta)$, well-defined for all $\alpha,\beta > 0$, and such that*

$$-\lambda_0^2 < -f_0(\alpha,\beta) < 0,$$

for all $\alpha,\beta > 0$, and all $t, x_1, x_2 \in \mathbb{R}$.

Proof. From the variational characterization of the first eigenvalue via a minimization problem we have

$$-\lambda_0^2 \leq \frac{\int_{\mathbb{R}} \mathcal{L}\Lambda_\beta B \Lambda_\beta B}{\int_{\mathbb{R}} (\Lambda_\beta B)^2} = -\frac{32\alpha^2\beta}{\int_{\mathbb{R}} (\Lambda_\beta B)^2}.$$

Since $\int_{\mathbb{R}} \Lambda_\beta B^2$ is continuous on x_1, $x_1 \in \left[0, \frac{2\pi}{\alpha}\right]$, we conclude. □

Let $z \in H^2(\mathbb{R})$, and $B = B_{\alpha,\beta}$ be any mKdV breather. Let us consider the quadratic form associated to $\mathcal{L}$:

$$\begin{aligned}
Q[z] := \int_{\mathbb{R}} z\mathcal{L}z := \int_{\mathbb{R}} z_{xx}^2 + 2(\beta^2 - \alpha^2) \int_{\mathbb{R}} z_x^2 \\
+ (\alpha^2 + \beta^2)^2 \int_{\mathbb{R}} z^2 - 5 \int_{\mathbb{R}} B^2 z_x^2 \\
+ 5 \int_{\mathbb{R}} B_x^2 z^2 + 10 \int_{\mathbb{R}} BB_{xx} z^2 \\
+ \frac{15}{2} \int_{\mathbb{R}} B^4 z^2 - 6(\beta^2 - \alpha^2) \int_{\mathbb{R}} B^2 z^2.
\end{aligned} \tag{4.9.12}$$

Remark 4.9.13. From the definition of Q and Lemma 4.9.5, it is clear that $Q[B_1] = Q[B_2] = 0$. Moreover, inequality (8.11.2) means that $\Lambda_\alpha B$ is actually a positive direction for Q, a completely unexpected result. Additionally, from (4.9.12) Q is bounded below, namely

$$Q[z] \geq -c_{\alpha,\beta} \|z\|_{H^2(\mathbb{R})}^2,$$

for some positive constant $c_{\alpha,\beta}$ depending on α and β only.

Let $B_{-1} \in \mathcal{S}\backslash\{0\}$ be an eigenfunction associated to the unique negative eigenvalue of the operator $\mathcal{L}$, as stated in Proposition 4.9.11. We assume that B_{-1} has unit L^2-norm, so B_{-1} is now unique. In particular, one has $\mathcal{L}B_{-1} = -\lambda_0^2 B_{-1}$. It is clear from Proposition 4.9.11 and Lemma 4.9.5 that the following result holds.

Lemma 4.9.14. *The eigenvalue zero is isolated. Moreover, there exists a continuous function $\nu_0 = \nu_0(\alpha,\beta)$, well-defined and positive for all $\alpha, \beta > 0$ and such that, for all $z_0 \in H^2(\mathbb{R})$ satisfying*

$$\int_{\mathbb{R}} z_0 B_{-1} = \int_{\mathbb{R}} z_0 B_1 = \int_{\mathbb{R}} z_0 B_2 = 0, \tag{4.9.13}$$

then

$$Q[z_0] \geq \nu_0 \|z_0\|_{H^2(\mathbb{R})}^2. \tag{4.9.14}$$

Proof. The isolatedness of the zero eigenvalue is a direct consequence of standard elliptic estimates for the eigenvalue problem associated to $\mathcal{L}$, corresponding uniform convergence on compact subsets of $\mathbb{R}$, and the nondegeneracy of the kernel associated to $\mathcal{L}$.

On the other hand, the existence of a *positive* constant $\nu_0 = \nu_0(\alpha, \beta, x_1)$ such that (4.9.14) is satisfied is now clear. Moreover, thanks to the periodic character of the variable x_1, and the nondegeneracy of the kernel, we obtain a uniform, positive bound independent of x_1, x_2 and t, still denoted ν_0. The proof is complete. □

It turns out that B_{-1} is hard to manipulate; we need a more tractable version of the previous result. (4.9.13) will be changed to (4.9.15) as better orthogonality conditions.

Proposition 4.9.15. *Let $B = B_{\alpha,\beta}$ be any mKdV breather, and B_1, B_2 the corresponding kernel of the associated operator $\mathcal{L}$. There exists $\mu_0 > 0$, depending on α, β only, such that, for any $z \in H^2(\mathbb{R})$ satisfying*

$$\int_{\mathbb{R}} B_1 z = \int_{\mathbb{R}} B_2 z = 0, \tag{4.9.15}$$

one has

$$Q[z] \geq \mu_0 \|z\|^2_{H^2(\mathbb{R})} - \frac{1}{\mu_0}\left(\int_{\mathbb{R}} zB\right)^2. \tag{4.9.16}$$

Proof. This is a standard result, but we include it for the sake of completeness. Indeed, it is enough to prove that, under the conditions (4.9.15) and the additional orthogonality condition $\int_{\mathbb{R}} zB = 0$, one has

$$Q[z] \geq \mu_0 \|z\|^2_{H^2(\mathbb{R})}.$$

In what follows we prove that we can replace B_{-1} by the breather B in Lemma 4.9.14 and the result essentially does not change. Indeed, note that from (4.9.6), the function B_0 satisfies $\mathcal{L}B_0 = -B$, and from (4.9.7),

$$\int_{\mathbb{R}} B_0 B = -\int_{\mathbb{R}} B_0 \mathcal{L} B_0 = -Q[B_0] > 0. \tag{4.9.17}$$

The next step is to decompose z and B_0 in $\text{span}(B_{-1}, B_1, B_2)$ and the corresponding orthogonal subspace. One has

$$z = \tilde{z} + mB_{-1}, \quad B_0 = b_0 + nB_{-1} + p_1 B_1 + p_2 B_2, \quad m, n, p_1, p_2 \in \mathbb{R},$$

where

$$\int_{\mathbb{R}} \tilde{z} B_{-1} = \int_{\mathbb{R}} \tilde{z} B_1 = \int_{\mathbb{R}} \tilde{z} B_2 = \int_{\mathbb{R}} b_0 B_{-1} = \int_{\mathbb{R}} b_0 B_1 = \int_{\mathbb{R}} b_0 B_2 = 0.$$

Note in addition that

$$\int_{\mathbb{R}} B_{-1} B_1 = \int_{\mathbb{R}} B_{-1} B_2 = 0.$$

From here and the previous identities we have

$$Q[z] = \int_{\mathbb{R}} (\mathcal{L}\tilde{z} - m\lambda_0^2 B_{-1})(\tilde{z} + mB_{-1}) = Q[\tilde{z}] - m^2 \lambda_0^2. \tag{4.9.18}$$

Now, since $\mathcal{L}B_0 = -B$, one has

$$\begin{aligned}
0 &= \int_{\mathbb{R}} zB = -\int_{\mathbb{R}} z\mathcal{L}B_0 = \int_{\mathbb{R}} \mathcal{L}(\tilde{z} + mB_{-1})B_0 \\
&= \int_{\mathbb{R}} (\mathcal{L}\tilde{z} - m\lambda_0^2 B_{-1})(b_0 + nB_{-1} + p_1 B_1 + p_2 B_2) \\
&= \int_{\mathbb{R}} \mathcal{L}\tilde{z}b_0 - mn\lambda_0^2. \qquad (4.9.19)
\end{aligned}$$

On the other hand, from Corollary 4.9.6,

$$\begin{aligned}
\int_{\mathbb{R}} B_0 B &= -\int_{\mathbb{R}} B_0 \mathcal{L}B_0 \\
&= -\int_{\mathbb{R}} (b_0 + nB_{-1})(\mathcal{L}b_0 - n\lambda_0^2 B_{-1}) = -Q[b_0] + n^2\lambda_0^2.
\end{aligned} \qquad (4.9.20)$$

Replacing (4.9.19) and (4.9.20) into (4.9.18), we get

$$Q[z] = Q[\tilde{z}] - \frac{\left(\int_{\mathbb{R}} b_0 \mathcal{L}\tilde{z}\right)^2}{\int_{\mathbb{R}} B_0 B + Q[b_0]}. \qquad (4.9.21)$$

Note that from (4.9.17) and (4.9.14) both quantities in the denominator are positive. Additionally, note that if $\tilde{z} = \lambda b_0$, with $\lambda \neq 0$, then

$$\left(\int_{\mathbb{R}} b_0 \mathcal{L}\tilde{z}\right)^2 = Q[\tilde{z}]Q[b_0].$$

In particular, if $\tilde{z} = \lambda b_0$,

$$\frac{\left(\int_{\mathbb{R}} b_0 \mathcal{L}\tilde{z}\right)^2}{\int_{\mathbb{R}} B_0 B + Q[b_0]} \leq a\, Q[\tilde{z}], \quad 0 < a < 1. \qquad (4.9.22)$$

In the general case, using the orthogonal decomposition induced by the scalar product $(\mathcal{L}\cdot, \cdot)_{L^2}$ on span(B_{-1}, B_1, B_2), we get the same conclusion as before. Therefore, we have proved (4.9.22) for all possible $\tilde{z}$.

Finally, replacing in (4.9.21) and (4.9.18), $Q[z] \geq (1-a)Q[\tilde{z}] \geq 0$, and $Q[\tilde{z}] \geq m^2\lambda_0^2$. We have, for some $C > 0$,

$$\begin{aligned}
Q[z] &\geq (1-a)Q[\tilde{z}] \geq \frac{1}{2}(1-a)Q[\tilde{z}] + (1-a)m^2\lambda_0^2 \\
&\geq \frac{1}{C}\left(2\|\tilde{z}\|_{H^2(\mathbb{R})}^2 + 2m^2\|B_{-1}\|_{H^2(\mathbb{R})}^2\right) \geq \frac{1}{C}\|z\|_{H^2(\mathbb{R})}^2.
\end{aligned}$$

□

4.9.3 *Lyapunov functional*

In this section we introduce a new Lyapunov functional for equation (4.1.1), which will be well-defined at the natural H^2 level. Indeed, let $u_0 \in H^2(\mathbb{R})$ and let $u = u(t) \in H^2(\mathbb{R})$ be the corresponding local-in-time solution of the Cauchy problem associated to (4.1.1), with initial condition $u(0) = u_0$. Let us remember the H^2-functional (4.1.6)

$$F[u](t) := \frac{1}{2}\int_{\mathbb{R}} u_{xx}^2(t,x)dx - \frac{5}{2}\int_{\mathbb{R}} u^2(t,x)u_x^2(t,x)dx + \frac{1}{4}\int_{\mathbb{R}} u^6(t,x)dx. \quad (4.9.23)$$

Then one has

Lemma 4.9.16. *Given u local H^2-solution of* (4.1.1) *with initial data u_0, the functional $F[u](t)$ is a conserved quantity. In particular, u is a global-in-time H^2-solution.*

The existence of this last conserved quantity is a deep consequence of the *integrability property*. In particular, it is not present in a general, nonintegrable gKdV equation (2.8.1). The verification of Lemma 4.9.16 is a direct computation.

Using the functional $F[u]$ (4.9.23), we remember the new Lyapunov functional (4.8.2), specifically related to the breather solution. Let $B = B_{\alpha,\beta}$ be a mKdV breather, and $t \in \mathbb{R}$, and $M[u]$ and $E[u]$ given in (4.1.4), (4.1.5). Then

$$\begin{aligned} &\mathcal{H}[u](t) := F[u](t) + 2(\beta^2 - \alpha^2)E[u](t) + (\alpha^2 + \beta^2)^2 M[u](t), \\ &\alpha, \beta \text{ scaling parameters.} \end{aligned} \quad (4.9.24)$$

It is clear that $\mathcal{H}[u]$ represents a real-valued conserved quantity, well-defined for H^2-solutions of (4.1.1). Moreover, one has the following

Lemma 4.9.17. *Let $z \in H^2(\mathbb{R})$ be any function with sufficiently small H^2-norm, and $B = B_{\alpha,\beta}$ be any breather solution. Then, for all $t \in \mathbb{R}$, one has*

$$\mathcal{H}[B+z] - \mathcal{H}[B] = \frac{1}{2}Q[z] + \mathcal{N}[z], \quad (4.9.25)$$

with Q being the quadratic form defined in (4.9.12), *and $\mathcal{N}[z]$ satisfying $|\mathcal{N}[z]| \leq K\|z\|^3_{H^2(\mathbb{R})}$.*

Proof. We compute:

$$
\begin{aligned}
&\mathcal{H}[B+z]\\
&= \frac{1}{2}\int_{\mathbb{R}}(B+z)_{xx}^2 - \frac{5}{2}\int_{\mathbb{R}}(B+z)^2(B+z)_x^2 + \frac{1}{4}\int_{\mathbb{R}}(B+z)^6\\
&+ (\beta^2-\alpha^2)\int_{\mathbb{R}}(B+z)_x^2 - \frac{1}{2}(\beta^2-\alpha^2)\int_{\mathbb{R}}(B+z)^4 + \frac{1}{2}(\alpha^2+\beta^2)^2\int_{\mathbb{R}}(B+z)^2\\
&= \frac{1}{2}\int_{\mathbb{R}}B_{xx}^2 - \frac{5}{2}\int_{\mathbb{R}}B^2B_x^2 + \frac{1}{4}\int_{\mathbb{R}}B^6\\
&+ (\beta^2-\alpha^2)\int_{\mathbb{R}}B_x^2 - \frac{1}{2}(\beta^2-\alpha^2)\int_{\mathbb{R}}B^4 + \frac{1}{2}(\alpha^2+\beta^2)^2\int_{\mathbb{R}}B^2\\
&+ \int_{\mathbb{R}}\left(B_{4x} - 2(\beta^2-\alpha^2)(B_{xx}+B^3) + (\alpha^2+\beta^2)^2B + 5B_x^2B + 5B^2B_{xx} + \frac{3}{2}B^5\right)z\\
&\quad + \frac{1}{2}\left(\int_{\mathbb{R}}z_{xx}^2 + 2(\beta^2-\alpha^2)\int_{\mathbb{R}}z_x^2 + (\alpha^2+\beta^2)^2\int_{\mathbb{R}}z^2 - 5\int_{\mathbb{R}}B^2z_x^2\right.\\
&\quad \left. + 5\int_{\mathbb{R}}B_x^2z^2 + 10\int_{\mathbb{R}}BB_{xx}z^2 + \frac{15}{2}\int_{\mathbb{R}}B^4z^2 - 6(\beta^2-\alpha^2)\int_{\mathbb{R}}B^2z^2\right)\\
&\quad + 5\int_{\mathbb{R}}B^3z^3 - 2(\beta^2-\alpha^2)\int_{\mathbb{R}}Bz^3 + \frac{5}{3}\int_{\mathbb{R}}B_{xx}z^3 - 5\int_{\mathbb{R}}Bz_x^2z\\
&\quad + \frac{15}{4}\int_{\mathbb{R}}B^2z^4 - \frac{1}{2}(\beta^2-\alpha^2)\int_{\mathbb{R}}z^4 - \frac{5}{2}\int_{\mathbb{R}}z^2z_x^2 + \frac{3}{2}\int_{\mathbb{R}}Bz^5 + \frac{1}{4}\int_{\mathbb{R}}z^6.
\end{aligned}
$$

Therefore, we have the decomposition

$$
\mathcal{H}[B+z] = \mathcal{H}[B] + \int_{\mathbb{R}}G[B]z(t) + \frac{1}{2}Q[z] + \mathcal{N}[z],
$$

where Q is defined in (4.9.12), and

$$
G[B] = B_{4x} - 2(\beta^2-\alpha^2)(B_{xx}+B^3) + (\alpha^2+\beta^2)^2B + 5B_x^2B + 5B^2B_{xx} + \frac{3}{2}B^5.
$$

Recall that $G[B]\equiv 0$ from (4.2.1). Finally, the term $N[z]$ is given by

$$
\begin{aligned}
\mathcal{N}[z] :=&5\int_{\mathbb{R}}B^3z^3 - 2(\beta^2-\alpha^2)\int_{\mathbb{R}}Bz^3 + \frac{5}{3}\int_{\mathbb{R}}B_{xx}z^3 - 5\int_{\mathbb{R}}Bz_x^2z\\
&+ \frac{15}{4}\int_{\mathbb{R}}B^2z^4 - \frac{1}{2}(\beta^2-\alpha^2)\int_{\mathbb{R}}z^4 - \frac{5}{2}\int_{\mathbb{R}}z^2z_x^2 + \frac{3}{2}\int_{\mathbb{R}}Bz^5 + \frac{1}{4}\int_{\mathbb{R}}z^6.
\end{aligned}
$$

Therefore, from direct estimates one has $\mathcal{N}[z] = O(\|z\|_{H^2(\mathbb{R})}^3)$, as desired. □

The previous Lemma is the key step to the proof of the main result of this paper, that we develop in the next section.

4.9.4 *Proof of stability*

In this section we prove a detailed version of Theorem 9.9.4.

Theorem 4.9.18 (H^2-stability of mKdV breathers). *Let $\alpha, \beta > 0$. There exist parameters η_0, A_0, depending on α and β, such that the following holds. Consider $u_0 \in H^2(\mathbb{R})$, and assume that there exists $\eta \in (0, \eta_0)$ such that*

$$\|u_0 - B_{\alpha,\beta}(0; 0, 0)\|_{H^2(\mathbb{R})} \leq \eta. \tag{4.9.26}$$

Then there exist $x_1(t), x_2(t) \in \mathbb{R}$ such that the solution $u(t)$ of the Cauchy problem for the mKdV equation (4.1.1), *with initial data u_0, satisfies*

$$\sup_{t\in\mathbb{R}} \left\|u(t) - B_{\alpha,\beta}(t; x_1(t), x_2(t))\right\|_{H^2(\mathbb{R})} \leq A_0\eta, \tag{4.9.27}$$

with

$$\sup_{t\in\mathbb{R}} |x_1'(t)| + |x_2'(t)| \leq KA_0\eta, \tag{4.9.28}$$

for some constant $K > 0$.

Remark 4.9.19. The initial condition (4.9.26) can be replaced by any initial breather profile of the form $\widetilde{B} := B_{\alpha,\beta}(t_0; x_1^0, x_2^0)$, with $t_0, x_1^0, x_2^0 \in \mathbb{R}$, thanks to the invariance of the equation under translations in time and space. In addition, a similar result is available for the *negative* breather $-B_{\alpha,\beta}$, which is also a solution of (4.1.1).

Proof of Theorem 4.9.18. Let $u_0 \in H^2(\mathbb{R})$ satisfying (4.9.26), and let $u \in C(\mathbb{R}; H^2(\mathbb{R}))$ be the associated solution of the Cauchy problem (4.1.1), with initial data $u(0) = u_0$. In what follows, we denote

$$B = B_{\alpha,\beta},$$

if no confusion arises.

We prove the theorem only for positive times, since the negative time case is completely analogous. From the continuity of the mKdV flow for $H^2(\mathbb{R})$ data, there exists a time $T_0 > 0$ and continuous parameters $x_1(t), x_2(t) \in \mathbb{R}$, defined for all $t \in [0, T_0]$, and such that the solution $u(t)$ of the Cauchy problem for the mKdV equation (4.1.1), with initial data u_0, satisfies

$$\sup_{t\in[0,T_0]} \left\|u(t) - B(t; x_1(t), x_2(t))\right\|_{H^2(\mathbb{R})} \leq 2\eta. \tag{4.9.29}$$

The idea is to prove that $T_0 = +\infty$. In order to do this, let $K^* > 2$ be a constant, to be fixed later. Let us suppose, by contradiction, that the *maximal time of stability* T^*, namely

$$T^* := \sup\Big\{T > 0 \,\Big|\, \text{ for all } t \in [0, T], \text{ there exist } \tilde{x}_1(t), \tilde{x}_2(t) \in \mathbb{R} \text{ such that}$$
$$\sup_{t\in[0,T]} \left\|u(t) - B(t; \tilde{x}_1(t), \tilde{x}_2(t))\right\|_{H^2(\mathbb{R})} \leq K^*\eta\Big\}, \tag{4.9.30}$$

is finite. It is clear from (4.9.29) that T^* is a well-defined quantity. Our idea is to find a suitable contradiction to the assumption $T^* < +\infty$. By taking η_0 smaller, if necessary, we can apply a well-known theory of modulation for the solution $u(t)$.

Lemma 4.9.20 (Modulation). *There exists $\eta_0 > 0$ such that, for all $\eta \in (0, \eta_0)$, the following holds. There exist C^1 functions $x_1(t)$, $x_2(t) \in \mathbb{R}$, defined for all $t \in [0, T^*]$, and such that*

$$z(t) := u(t) - B(t), \quad B(t, x) := B_{\alpha,\beta}(t, x; x_1(t), x_2(t)) \tag{4.9.31}$$

satisfies, for $t \in [0, T^*]$,

$$\int_{\mathbb{R}} B_1(t; x_1(t), x_2(t))z(t) = \int_{\mathbb{R}} B_2(t; x_1(t), x_2(t))z(t) = 0. \tag{4.9.32}$$

Moreover, one has

$$\|z(t)\|_{H^2(\mathbb{R})} + |x_1'(t)| + |x_2'(t)| \leq KK^*\eta, \quad \|z(0)\|_{H^2(\mathbb{R})} \leq K\eta, \tag{4.9.33}$$

for some constant $K > 0$, independent of K^*.

Lemma 4.9.20 is key to choose the good directions (4.9.32) where perturbation of breathers can be controlled.

Proof. The proof of this result is a classical application of the Implicit Function Theorem. Let

$$J_j(u(t), x_1, x_2) := \int_{\mathbb{R}} (u(t, x) - B(t, x; x_1, x_2))B_j(t, x; x_1, x_2)dx, \quad j = 1, 2.$$

It is clear that $J_j(B(t; x_1, x_2), x_1, x_2) \equiv 0$, for all $x_1, x_2 \in \mathbb{R}$. On the other hand, one has for $j, k = 1, 2$,

$$\partial_{x_k} J_j(u(t), x_1, x_2)\Big|_{(B(t),0,0)} = -\int_{\mathbb{R}} B_k(t, x; 0, 0)B_j(t, x; 0, 0)dx.$$

Let J be the 2×2 matrix with components $J_{j,k} := (\partial_{x_k} J_j)_{j,k=1,2}$. From the identity above, one has

$$\det J = -\left(\int_{\mathbb{R}} B_1^2 \int_{\mathbb{R}} B_2^2 - \left(\int_{\mathbb{R}} B_1 B_2\right)^2\right)(t; 0, 0),$$

which is different from zero from the Cauchy-Schwarz inequality and the fact that B_1 and B_2 are not parallel for all time. Therefore, in a small H^2 neighborhood of $B(t; 0, 0)$, $t \in [0, T^*]$ (given by the definition of (4.9.30)), it is possible to write the decomposition (4.9.31)-(4.9.32).

Now we look at the bounds (4.9.33). The first bounds are consequence of the decomposition itself and the equations satisfied by the derivatives of the scaling parameters, after taking time derivative in (4.9.32) and using that $\det J \neq 0$. The last bound in (4.9.33) is consequence of (4.9.26). □

Now, we apply Lemma 4.9.17 to the function $u(t)$. Since $z(t)$ defined by (4.9.31) is small, we get from (4.9.25) and Corollary 4.6.1:

$$\mathcal{H}[u](t) = \mathcal{H}[B](t) + \frac{1}{2}Q[z](t) + N[z](t). \tag{4.9.34}$$

Note that $|N[z](t)| \leq K\|z(t)\|^3_{H^1(\mathbb{R})}$. On the other hand, by the translation invariance in space,

$$\mathcal{H}[B](t) = \mathcal{H}[B](0) = \text{constant}.$$

Indeed, first of all we have

$$B(t, x; x_1(t), x_2(t)) = B(t - t_0(t), x - x_0(t)),$$

for some specific t_0, x_0. Since $\mathcal{H}$ involves integration in space of polynomial functions on B, B_x and B_{xx}, we have

$$\mathcal{H}[B(t, \cdot; x_1(t), x_2(t))] = \mathcal{H}[B(t - t_0(t), \cdot - x_0(t); 0, 0)] = \mathcal{H}[B(t - t_0(t), \cdot; 0, 0)].$$

Finally, $\mathcal{H}[B(t - t_0(t), \cdot; 0, 0)] = \mathcal{H}[B(\cdot, \cdot; 0, 0)](t - t_0(t))$. Taking time derivative,

$$\partial_t\mathcal{H}[B(t, \cdot; x_1(t), x_2(t))] = \mathcal{H}'[B(\cdot, \cdot; 0, 0)](t - t_0(t)) \times (1 - t_0'(t)) \equiv 0,$$

hence $\mathcal{H}[B]$ is constant in time. Now we compare (4.9.34) at times $t = 0$ and $t \leq T^*$. From Lemma 4.9.16 and (4.9.24) we have

$$Q[z](t) \leq Q[z](0) + K\|z(t)\|^3_{H^2(\mathbb{R})} + K\|z(0)\|^3_{H^2(\mathbb{R})} \leq K\|z(0)\|^2_{H^2(\mathbb{R})} + K\|z(t)\|^3_{H^2(\mathbb{R})}.$$

Additionally, from (4.9.15)-(4.9.16) applied this time to the time-dependent function $z(t)$, which satisfies (4.9.32), we get

$$\begin{aligned}\|z(t)\|^2_{H^2(\mathbb{R})} &\leq K\|z(0)\|^2_{H^2(\mathbb{R})} + K\|z(t)\|^3_{H^2(\mathbb{R})} + K\left|\int_{\mathbb{R}} B(t)z(t)\right|^2 \\ &\leq K\eta^2 + K(K^*)^3\eta^3 + K\left|\int_{\mathbb{R}} B(t)z(t)\right|^2.\end{aligned} \tag{4.9.35}$$

Conclusion of the proof. Using the conservation of mass (4.1.4), we have, after expanding $u = B + z$,

$$\begin{aligned}\left|\int_{\mathbb{R}} B(t)z(t)\right| &\leq K\left|\int_{\mathbb{R}} B(0)z(0)\right| + K\|z(0)\|^2_{H^2(\mathbb{R})} + K\|z(t)\|^2_{H^2(\mathbb{R})} \\ &\leq K(\eta + (K^*)^2\eta^2), \qquad \text{for each } t \in [0, T^*].\end{aligned}$$

Replacing this last identity in (4.9.35), we get

$$\|z(t)\|^2_{H^2(\mathbb{R})} \leq K\eta^2(1 + (K^*)^2\eta^3) \leq \frac{1}{2}(K^*)^2\eta^2,$$

by taking K^* large enough. This last fact contradicts the definition of T^* and therefore the stability property (4.9.27) holds true. Finally, (4.9.28) is a consequence of (4.9.33). □

4.10 Exercises

(4.1) Prove Lemma 4.1.2, i.e. $\int_{\mathbb{R}} B = 0$.
(4.2) Prove by hand Lemma 4.1.3.
(4.3) Prove by hand Lemma 4.4.1. (See [AlMu13] for some help.)
(4.4) Prove Lemma 4.9.16.

Chapter 5

Breathers and rigorous integrability theory

5.1 Introduction

It is well-known that mKdV (1.4.1)-(4.1.1) is an integrable model. Many aspects and properties of breathers and other mKdV nonlinear objects are strong consequences of this fundamental property.

Usually, the most powerful technique associated to an integrable system is the Inverse Scattering Method (ISM). It is not the purpose of this monograph to explain this technique, since many books have been devoted to explain in detail many applications, see e.g. [Yan10] for a detailed account. This method very often needs data in a stronger space than the energy space, usually weighted spaces. Also, the solution obtained via the Cauchy problem may not be directly related to the solution obtained via the ISM, since in general both problems are not directly related.

Precisely, one of the most important elements in inverse scattering theory is the *scattering data* produced by the initial data. This is essentially a sort of nonlinear Fourier transform of the initial data, which under the evolution of the flow evolves in a simple fashion, exactly as the Fourier transform evolves in classical linear equations. Usually scattering data needs to satisfy certain hypotheses in order to run the full inverse scattering theory argument, and if these are satisfied, the outcome is very detailed and complete. For several years, rigorous analysis on the validity of the hypotheses needed for such arguments to be valid have appeared, and here we mention the ones related to the mKdV case.

However, an important result on breathers was recently proved by Chen and Liu [ChLi21]. They presented a description of the long time behavior of focusing mKdV solutions using the precise integrability of the equation. Previous results were essentially focused on the defocusing case, where no solitons nor breathers seem to be present. Also, Chen and Liu consider the problem of asymptotic stability for solitons and breathers.

5.2 Classification of regions

In order to explain better this result, some description of the real line must be taken into account. First, let us consider the following regions.

- Oscillatory region: $x < 0, |x/t| = O(1)$ as $t \to \infty$. In this region, there exist breathers traveling to the left direction.
- Self-similar region: $|x/t^{1/3}| \leq c$ as $t \to \infty$. This region, does not have breathers and solitons as $t \to \infty$.
- Soliton region: $x > 0, |x/t| = O(1)$ as $t \to \infty$. In this region, there exist breathers and solitons traveling to the right direction.

Remark 5.2.1. The long time asymptotics for overlaps of the regions have been studied in a previous paper [ChLi22, Theorem 1.6]. There are no solitons and breathers in those overlap regions. They will cover the left portion of the real line ($\{x < 0\}$), by using the technique known as the nonlinear steepest descent approach to oscillatory Riemann-Hilbert problems.

5.3 Soliton resolution for mKdV

Chen-Liu proved the following detailed long-time asymptotics of the solution to the focusing mKdV. In fact, they proved the *soliton resolution* for the focusing mKdV case, giving detailed asymptotic analysis for the Cauchy problem of the focusing mKdV with generic data. By this, we mean the following: long time data can be written as a superposition of solitons, breathers and radiation in the following explicit way:

$$u(t,x) = \sum_{l=1}^{N_1} u_l^{\text{sol}} + \sum_{l=1}^{N_2} u_l^{\text{bre}} + R(t,x).$$

Here, u_l^{sol} means solitons, u_l^{bre} means breathers, and R is a remainder. They understood the interaction among solitons, breathers and radiation in different regions in the following sense. Recall the weighted Sobolev space $H^{s,r}(\mathbb{R})$, defined as the completion of $C_0^\infty(\mathbb{R})$ with the norm

$$\|u\|_{H^{s,r}} := \left(\|\langle x\rangle u\|_{L^2}^2 + \|u\|_{\dot{H}^s}^2\right)^{\frac{1}{2}}. \tag{5.3.1}$$

For the purposes of this book, we shall work (5.3.1) with the values $s = 2$ and $r = 1$. This means that, roughly speaking, we shall work with data u in H^2 such that $\langle x \rangle u \in L^2$. Notice that this is clearly a subspace of the energy space H^1. Since the results by Chen and Liu are highly nontrivial, we shall separate the outcomes in different results. First of all, we state some hypotheses.

We shall assume that the initial data $u_0 \in H^{2,1}(\mathbb{R})$ satisfies:

(1) u_0 is generic in the sense of [ChLi21, Definition 1.5].
(2) Suppose that the initial data produce the *scattering data*

$$S = \left\{ r(z),\ \{z_k, c_k\}_{k=1}^{N_1},\ \ \{z_j, c_j\}_{j=1}^{N_2} \right\} \in H^{2,1}(\mathbb{R}) \oplus \mathbb{C}^{2N_1} \oplus \mathbb{C}^{2N_2}.$$

Here r is the so-called reflection coefficient, while z_k, c_k are scaling parameters for solitons and breathers.

Let us write $z_j = \xi_j + i\eta_j,\ \ \xi_j, \eta_j > 0$, and suppose for some $1 \leq l_0 \leq N_2$ one has the following ordering in speeds

$$4\eta_1^2 - 12\xi_1^2 < \dots 4\eta_{l_0}^2 - 12\xi_{l_0}^2 < 0 < 4\eta_{l_0+1}^2 - 12\xi_{l_0+1}^2 < \cdots < 4\eta_{N_2}^2 - 12\xi_{N_2}^2. \quad (5.3.2)$$

Notice that the zero speed is not allowed. These parameters will represent breathers. On the other hand, we decompose $z_k = i\zeta_k,\ \zeta_k > 0$ as

$$0 < \zeta_1 < \cdots < \zeta_{N_1}.$$

Therefore, ζ_k will be purely imaginary complex numbers representing solitons. Finally, denote

$$\tau = z_0^3 t, \quad z_0 = \sqrt{\frac{-x}{12t}}. \quad (5.3.3)$$

Theorem 5.3.1. *Let u be the solution of the focusing mKdV equation*

$$\partial_t u + \partial_x^3 u + 6u^2 \partial_x u = 0, \quad (t, x) \in \mathbb{R}^+ \times \mathbb{R}, \quad (5.3.4)$$

with initial data u_0 satisfying the hypotheses above described. Then the solution u can be written as the superposition of solitons, breathers and radiation as following:

$$u(t, x) = \sum_{l=1}^{N_1} u_l^{sol} + \sum_{l=1}^{N_2} u_l^{br} + R(t, x). \quad (5.3.5)$$

Now we describe in more detail the members in the decomposition (5.3.5). It will take some time to describe them, and we will need some additional definitions.

5.3.1 Detailed description of the breather part

Recall the parameter l_0 in (5.3.2) that separates positive from negative speeds. The breather region is decomposed in two parts, depending on the positive or negative speed.

Theorem 5.3.2. *In Theorem 5.3.1,*

$$u_l^{br}(t,x) = -4\frac{\eta_l}{\xi_l}\frac{\xi_l\cosh(v_2+\tilde{\omega}_2)\sin(v_1+\tilde{\omega}_1)+\eta_l\sinh(v_2+\tilde{\omega}_2)\cos(v_1+\tilde{\omega}_1)}{\cosh^2(v_2+\tilde{\omega}_2)+(\eta_l/\xi_l)^2\cos^2(v_1+\tilde{\omega}_1)}, \tag{5.3.6}$$

where

- *if $l \leq l_0$, the profile variables are*

$$v_1 = 2\xi_l\left(x+4(\xi_l^2-3\eta_l^2)t\right), \quad v_2 = 2\eta_l\left(x-4(\eta_l^2-3\xi_l^2)t\right), \tag{5.3.7}$$

  and the shifts are given by

$$\tan(\tilde{\omega}_1) = \frac{\tilde{B}\xi_l-\tilde{A}\eta_l}{\tilde{A}\xi_l+\tilde{B}\eta_l}, \quad e^{\tilde{\omega}_2} = \left|\frac{\xi_l}{2\eta_l}\right|\sqrt{\frac{\tilde{B}^2+\tilde{A}^2}{\xi_l^2+\eta_l^2}}. \tag{5.3.8}$$

  Here $\tilde{A}, \tilde{B}$ are given as

$$\tilde{c}_l = \frac{c_l}{\delta(z_l)^2} =: \tilde{A}+i\tilde{B}, \tag{5.3.9}$$

  where the scalar function $\delta(z)$ is given by

$$\delta(z) = \left(\prod_{k=1}^{N_1}\frac{z-\bar{z}_k}{z-z_k}\right)\left(\prod_{z_j\in\mathcal{B}_l}\frac{z-\bar{z}_j}{z-z_j}\right)\left(\prod_{z_j\in\mathcal{B}_l}\frac{z-z_j}{z+\bar{z}_j}\right)\left(\frac{z-z_0}{z+z_0}\right)^{i\kappa}e^{\chi(z)},$$

  with z_0 defined in (5.3.3),

$$\chi(z) = \frac{1}{2i\pi}\int_{-z_0}^{z_0}\log\left(\frac{1+|r(\zeta)|^2}{1+|r(\zeta_0)|^2}\right)\frac{d\zeta}{\zeta-z},$$
$$\kappa = -\frac{1}{2\pi}\log(1+|r(\zeta_0)|^2),$$

  and

$$\mathcal{B}_l = \left\{z_j = \xi_j+i\eta_j:\ 4\eta_j^2-12\xi_j^2 > 4\eta_l^2-12\xi_l^2\right\}. \tag{5.3.10}$$

- *If $l_0+1 \leq l$, we have the same expression as* (5.3.6), (5.3.8), (5.3.7), (5.3.9) *but with $\tilde{A}$ and $\tilde{B}$ are given as*

$$\tilde{c}_l = \frac{c_l}{\psi(z_l)^2} = \tilde{A}+i\tilde{B},$$

where the function ψ is defined by

$$\psi(z) = \left(\prod_{z_k\in\mathcal{B}_{l,s}} \frac{z-\bar{z}_k}{z-z_k}\right)\left(\prod_{z_j\in\mathcal{B}_{l,b}} \frac{z-\bar{z}_j}{z-z_j}\right)\left(\prod_{z_j\in\mathcal{B}_{l,b}} \frac{z+z_j}{z+\bar{z}_j}\right),$$

and

$$\begin{aligned}\mathcal{B}_l = &\left\{z_j = \xi_j + i\eta_j :\ \ 4\eta_j^2 - 12\xi_j^2 > 4\eta_l^2 - 12\xi_l^2\right\}\\ &\cup \left\{z_k = i\zeta_k :\ \ 4\zeta_k^2 > 4\eta_l^2 - 12\xi_l^2\right\} := \mathcal{B}_{l,b} \cup \mathcal{B}_{l,s}.\end{aligned}$$

Notice that the solution (5.3.6) is nothing but the classical mKdV breather solution described in previous chapters.

5.3.2 *Description of the soliton region*

We somehow have a clean description of the soliton region.

Theorem 5.3.3. *In Theorem 5.3.1, for the soliton part, we have*

$$u_l^{(so)}(t,x) = 2\zeta_l\varepsilon_{\pm,l}\,\mathrm{sech}\left(-2\zeta_l(x-4\zeta_l^2 t) + \omega_l\right),$$

with shifts

$$\begin{aligned}\omega_l = \log\left(\frac{|c_l|}{2\zeta_l}\right) &+ 2\sum_{z_k\in\mathcal{S}_{l,s}} \log\left|\frac{z_l - z_k}{z_l - \bar{z}_k}\right|\\ &+ 2\sum_{z_k\in\mathcal{S}_{l,b}} \log\left|\frac{z_l - z_j}{z_l - \bar{z}_j}\right| + 2\sum_{z_k\in\mathcal{S}_{l,b}} \log\left|\frac{z_l + \bar{z}_j}{z_l + z_j}\right|,\end{aligned}$$

and where

$$\begin{aligned}\mathcal{S}_l = &\left\{z_j = \xi_j + i\eta_j :\ \ 4\eta_j^2 - 12\xi_j^2 > 4\zeta_l^2\right\} \cup \left\{z_k = i\zeta_k :\ \ 4\zeta_k^2 > 4\zeta_l^2\right\}\\ &=: \mathcal{S}_{l,b} \cup \mathcal{S}_{l,s}.\end{aligned}$$

In this regime, one easily recovers the solitons from the mKdV solution. Notice that the number of solitons is finite. This is not true in general, when one works in the energy space.

5.3.3 *Description of the radiation term R*

Now we describe the radiation term R. It turns out that this description depends on the considered spatial region.

Theorem 5.3.4. *In Theorem 5.3.1, the radiation term R obeys the following asymptotics:*

(1) *In the soliton region, (i.e. Region III), we have*

$$|R(t,x)| \lesssim \frac{1}{t}.$$

(2) *In the self similar region (i.e. Region II), one has for* $4 < p < \infty$

$$R(t,x) = \frac{1}{(3t)^{\frac{1}{3}}} P\left(\frac{x}{(3t)^{\frac{1}{3}}}\right) + O\left(t^{\frac{2}{3p}-\frac{1}{2}}\right),$$

where P is a solution of the **Painlevé II equation**

$$P''(s) - sP'(s) + 2P^3(s) = 0,$$

determined by $r(0)$.

(3) *Finally, in the oscillatory region, (i.e. Region I), there are two separate cases.*

(a) *If we choose the frame* $x = vt$ *with* $v = 4\eta_1^2 - 12\xi_1^2 < 0$, *then one can strongly see the influence of the breather on the radiation. One has*

$$R(t,x) = u_{as}(t,x) + O\left((z_0 t)^{-\frac{3}{4}}\right),$$

where z_0 *is given in* (5.3.3),

$$u_{as}(t,x) = \frac{1}{\sqrt{48tz_0}}\left(\left(m_{11}^{(br)}(-z_0)^2(i\delta_A^0)^2\bar{\beta}_{12}\right) + m_{12}^{(br)}(-z_0)^2\left((i\delta_A^0)^2\bar{\beta}_{21}\right)\right)$$
$$+ \frac{1}{\sqrt{48tz_0}}\left(\left(m_{11}^{(br)}(z_0)^2(i\delta_B^0)^2\beta_{12}\right) - m_{12}^{(br)}(z_0)^2\left((i\delta_B^0)^2\beta_{21}\right)\right),$$

with some explicit constants $\left(m_{11}^{(br)}(-z_0), m_{11}^{(br)}(z_0), m_{12}^{(br)}(-z_0), m_{12}^{(br)}(z_0)\right)$ *from the breather matrix,*

$$\beta_{12} = \frac{\sqrt{2\pi}e^{i\pi/4}e^{-\pi\kappa}}{r(z_0)\Gamma(-i\kappa)}, \quad \beta_{21} = -\frac{\sqrt{2\pi}e^{-i\pi/4}e^{-\pi\kappa}}{\overline{r(z_0)}\Gamma(i\kappa)},$$

$$\delta_A^0 = (192\tau)^{i\kappa/2}e^{-8i\tau}e^{\aleph(-z_0)}, \quad \delta_B^0 = (192\tau)^{-i\kappa/2}e^{8i\tau}e^{\aleph(z_0)},$$

and

$$\eta_0(\pm z_0) = \left(\prod_{z_k=1}^{N_1} \frac{\pm z_0 - \bar{z}_k}{\pm z_0 - z_k}\right)\left(\prod_{z_j\in\mathcal{B}_l} \frac{\pm z_0 - \bar{z}_j}{\pm z_0 - z_j}\right)\left(\prod_{z_j\in\mathcal{B}_l} \frac{\pm z_0 + z_j}{\pm z_0 + \bar{z}_j}\right),$$

where $\mathcal{B}_l$ *was defined in* (5.3.10).

(ii) If $v \neq 4\eta_j^2 - 12\xi_j^2$ *for* $1 \leq j \leq N_2$ *then we have*

$$R(t,x) = u_{as}(t,x) + O\left((z_0 t)^{-\frac{3}{4}}\right),$$

where

$$u_{as}(t,x) = \left(\frac{\kappa}{3tz_0}\right)^{\frac{1}{2}} \cos\left(16tz_0^3 - \kappa\log(192tz_0^3) + \phi(z_0)\right),$$

with

$$\phi(z_0) = \arg\Gamma(i\kappa) - \frac{\pi}{4} - arg\ r(z_0) + \frac{1}{\pi}\int_{-z_0}^{z_0} \log\left(\frac{1+|r(\zeta)|^2}{1+|r(z_0)|^2}\right)\frac{d\zeta}{\zeta - z_0}$$
$$- 4\left(\sum_{k=1}^{N_1} arg(z_0 - z_k) + \sum_{z_j\in\mathcal{B}_l} arg(z_0 - z_j) + \sum_{z_j\in\mathcal{B}_l}' arg(z_0 + \bar{z}_j)\right).$$

Theorems 5.3.2, 5.3.3 and 5.3.4 conclude the description of the terms appearing in Theorem 5.3.1, and in particular the breather description for data in $H^{2,1}$. For the proof of this result, the reader can consult [ChLi21].

5.4 Stability of breathers in weighted spaces

The asymptotic the stability of the breather traveling to the left can be stated as a corollary of the previous result. Notice that here $\hat{z}_0$ is a free parameter, and not the term defined in (5.3.3).

Corollary 5.4.1. *Let* $u^{(br)}(t,x;z_0)$ *be a breather with discrete scattering data* (z_0, c_0) *such that* $\hat{z}_0 = \xi_0 + i\eta_0$, $\xi_0, \eta_0 > 0$ *with* $-12\xi_0^2 + 4\eta_0^2 < 0$ *(negative speed). Suppose that*

$$||R(0)||_{H^{\frac{1}{4},1}(\mathbb{R})} < \varepsilon \tag{5.4.1}$$

for $\varepsilon > 0$ *small enough. Let us consider the solution* u *to the focusing mKdV equation* (5.3.4) *with the initial data*

$$u_0 = u^{(br)}(x,0;\hat{z}_0,c_0) + R(0).$$

Let $r(z)$ *be the reflection coefficient computed from* u_0. *Then the following are satisfied:*

(1) There exist $z_1 = \xi_1 + i\eta_1$ *and* c_1 *such that*

$$|z_1 - \hat{z}_0| + |c_1 - c_0| \lesssim \varepsilon.$$

(2) As $t \to \infty$

$$u = u^{(br)}(t, x; z_1, c_1) + R(t, x)$$

where the radiation term $R(t, x)$ has a particular decaying asymptotics as time tends to infinity.

Let us discuss in some detail the asymptotics of R, which is similar to the one already described in Theorem 5.3.4.

(1) In the soliton region, i.e. Region III, we have

$$|R(t, x)| \lesssim \frac{1}{t}.$$

(2) In the self similar region, i.e. Region II, for $4 < p < \infty$, one has

$$R(t, x) = \frac{1}{(3t)^{1/3}} P\left(\frac{x}{(3t)^{1/3}}\right) + O\left(t^{\frac{2}{3p} - \frac{1}{2}}\right),$$

where P is a solution of the Painlevé II equation

$$P''(s) - sP'(s) + 2P^3(s) = 0,$$

determined by $r(0)$.

(3) In the oscillatory region, i.e. Region I, the asymptotics for $R(t, x)$ are more involved.

(i) If we choose the frame $x = vt$ with $v = 4\eta_1^2 - 12\xi_1^2 < 0$, one has

$$R(t, x) = u_{as}(t, x) + O\left((z_0 t)^{-\frac{3}{4}}\right)$$

where

$$u_{as}(t, x) = \frac{1}{\sqrt{48 t z_0}} \left(\left(m_{11}^{(br)}(-z_0)^2 (i\delta_A^0)^2 \bar{\beta}_{12} \right) + m_{12}^{(br)}(-z_0)^2 \left((i\delta_A^0)^2 \bar{\beta}_{21} \right) \right)$$
$$+ \frac{1}{\sqrt{48 t z_0}} \left(\left(m_{11}^{(br)}(z_0)^2 (i\delta_B^0)^2 \beta_{12} \right) - m_{12}^{(br)}(z_0)^2 \left((i\delta_B^0)^2 \beta_{21} \right) \right),$$

with some explicit constants $\left(m_{11}^{(br)}(-z_0), m_{11}^{(br)}(z_0), m_{12}^{(br)}(-z_0), m_{12}^{(br)}(z_0)\right)$ from the breather matrix,

$$\beta_{12} = \frac{\sqrt{2\pi} e^{i\pi/4} e^{-\pi\kappa}}{r(z_0)\Gamma(-i\kappa)}, \quad \beta_{21} = -\frac{\sqrt{2\pi} e^{-i\pi/4} e^{-\pi\kappa}}{r(\bar{z}_0)\Gamma(i\kappa)},$$

$$\delta_A^0 = (192\tau)^{i\kappa/2} e^{-8i\tau} e^{\aleph(-z_0)}, \quad \delta_B^0 = (192\tau)^{-i\kappa/2} e^{8i\tau} e^{\aleph(z_0)}.$$

(ii) If $v \neq 4\eta_1^2 - 12\xi_1^2$ then we have

$$R(t,x) = u_{as}(t,x) + O\left((z_0 t)^{-\frac{3}{4}}\right),$$

where

$$u_{as}(t,x) = \left(\frac{\kappa}{3tz_0}\right)^{\frac{1}{2}} \cos\left(16tz_0^3 - \kappa\log(192tz_0^3) + \phi(z_0)\right),$$

with

$$\phi(z_0) = \arg\Gamma(i\kappa) - \frac{\pi}{4} - \arg r(z_0) + \frac{1}{\pi}\int_{-z_0}^{z_0} \log\left(\frac{1+|r(\zeta)|^2}{1+|r(z_0)|^2}\right)\frac{d\zeta}{\zeta - z_0}.$$

Corollary 5.4.1 gives the asymptotic stability of the breather solution in the weighted space defined by (5.4.1). We finish this section by remarking that condition (5.4.1) ensures the nonexistence of more solitons and breathers as time evolves. In the case of data in the energy space, this is not satisfied.

5.5 Exercises

(5.1) Transform the mKdV equation (5.3.4) into (4.1.1).

Chapter 6

Recent developments for mKdV and related breathers

In this chapter we present three recent developments related to the existence and stability of breathers in several models. For a more detailed account of new findings, the reader can consult a forthcoming monograph strictly devoted to dealing with several new advances in this area.

The three recent developments are the following: the existence and uniqueness of multi-breathers in mKdV models, the stability of the positive negative mKdV breather on the left half-line, and finally, the possible existence of breathers in the modified Whitham model. Other topics such as nonexistence of breathers in mKdV and Zakharov-Kuznetsov models, asymptotic stability of breathers, breathers in other models, and other results, will be treated elsewhere.

6.1 Existence and uniqueness of mKdV multi-breathers

Semenov and Côte [Sem22, Sem23, CoSe23] have analysed in detail the existence and uniqueness of mKdV multi-breathers, specially in the energy spaces H^1 and H^2. Recall that solitons and breathers play a key role in the long time dynamics of mKdV. In fact, as it was shown in the previous chapter, the soliton-breather resolution states that any generic solution can be approached by a sum of solitons and breathers when $t \to \infty$.

In [OhWa82], there exists an explicit formula for a multi-breather, obtained by ISM, that in some sense already gives the existence of a mKdV multi-breather. However, it seems unclear how to derive from this formula the convergence to a sum of solitons and breathers uniformly in space and how to show that this convergence is exponential (local convergence around the center of mass of each object is already rather involved).

In this section, a different approach to prove the existence of a multi-breather is presented. In fact, it is possible to show that the constructed multi-breather converges to the corresponding sum of solitons and breathers in H^s and that this convergence is exponentially fast in time, and that the constructed multi-breather is smooth.

These results follow by adapting the arguments given by Martel and Merle [MaMe06], by Martel [Mar05] and by Côte and Friederich [CoFr20] to the context of breathers. To do so, one needs to understand the variational structure of breathers, in the same fashion as Weinstein did in [Wei86] for (NLS) solitons. Such results were obtained by Alejo and Muñoz in [AlMu13]: a breather is a critical point of a Lyapunov functional at the 2 level, whose Hessian is coercive up to several (but finitely many) orthogonal conditions. As in [AlMu13], the 2 regularity level is the most natural setting to study breathers, and the 1 regularity level is natural for the study of solitons, as in [Mar05, MaMe06]. One important issue to face is therefore to understand soliton variational structure at 2 level, and to adapt the Lyapunov functional in [AlMu13] to accommodate for a sum of breathers (and solitons). Notice that arguments based on monotonicity may be adapted only assuming that all the considered velocities are positive. Because [MaMe06, CoFr20] are not based on monotonicity (these are results for (NLS) which is not well suited for monotonicity), it is possible to adapt their arguments to obtain existence and uniqueness results for this case without any condition on the sign of velocities. The uniqueness result obtained in this setting is however weaker than the one that is obtained with monotonicity arguments.

In parallel with the definition of *multisoliton*, i.e. a solution $r(t)$ of mKdV such that there exists $0 < c_1 < c_2 < \cdots < c_N, \kappa_1, \ldots, \kappa_N \in \{-1, 1\}$ and $x_1, \ldots, x_N \in \mathbb{R}$ such that

$$\lim_{t\to+\infty} \left\| r(t) - \sum_{j=1}^{N} R_{c_j,\kappa_j}(t, \cdot; x_j) \right\|_{H^1(\mathbb{R})} = 0,$$

it is possible to define *multi-breathers*: any sum of solitons and breathers with distinct velocities, for which there exists a solution of mKdV whose difference with this sum tends to zero when time goes to infinity. More precisely,

Let $J \in \mathbb{N}$ and $K, L \in \mathbb{N}$ such that $J = K + L$. We will consider a set of L solitons and K breathers:

- the breather parameters are $\alpha_k > 0, \beta_k > 0, x^0_{1,k} \in \mathbb{R}$ and $x^0_{2,k} \in \mathbb{R}$ for $1 \leq k \leq K$.
- the solitons parameters are $c_l > 0, \kappa_l \in \{-1, 1\}$ and $x^0_{0,l} \in \mathbb{R}$ for $1 \leq l \leq L$.

We define for $1 \leq k \leq K$, the k-th breather as follows:

$$B_k(t, x) := B_{\alpha_k,\beta_k}(t, x; x^0_{1,k}, x^0_{2,k}), \tag{6.1.1}$$

and for $1 \leq l \leq L$, the l-th soliton:

$$R_l(t,x) := R_{c_l,\kappa_l}(t,x;x_{0,k}^0). \tag{6.1.2}$$

We now define the velocity of our objects. Recall that for $1 \leq k \leq K$, the velocity of B_k is

$$v_k^b := -\gamma_k = \beta_k^2 - 3\alpha_k^2, \tag{6.1.3}$$

and for $1 \leq l \leq L$, the velocity of R_l is

$$v_l^s := c_l. \tag{6.1.4}$$

The most important assumption we make is that all these velocities are distinct:

$$\forall k \neq k', \ v_k^b \neq v_{k'}^b, \quad \forall l \neq l', \ v_l^s \neq v_{l'}^s, \quad \forall k,l, \ v_k^b \neq v_l^s. \tag{6.1.5}$$

These imply for any two of these objects to be far from each other when time is large, and this assumption is essential in our analysis.

It will be useful to order breathers and solitons by increasing velocities. As these are distinct, we can define an increasing function

$$v : \{1,\ldots,J\} \to \{v_k^b, 1 \leq k \leq K\} \cup \{v_l^s, 1 \leq l \leq L\}.$$

The set $\{v_1,\ldots,v_J\}$ is thus the (ordered) set of all possible velocities of the objects. We define P_j, for $1 \leq j \leq J$, as the object (either a soliton R_l or a breather B_k) that corresponds to the velocity v_j. Hence, $P_1,\ldots,P_J$ are the considered objects ordered by increasing velocity. The following notations will be needed: the indexation by k and l, and the indexation by j, and we will keep these notations to avoid ambiguity.

Denoting by x_j the center of mass of P_j, that is

- if $P_j = B_k$ is a breather, we set $x_j(t) := -x_{2,k}^0 + v_j t$;
- if $P_j = R_l$ is a soliton, we set $x_j(t) := -x_{0,k}^0 + v_j t$;

We denote:

$$R = \sum_{l=1}^{L} R_l, \quad B = \sum_{k=1}^{K} B_k, \quad P = R + B = \sum_{j=1}^{J} P_j. \tag{6.1.6}$$

We can now define a multi-breather: as solitons are objects which can be studied naturally in $H^1(\mathbb{R})$, it turns out that breathers are best studied in $H^2(\mathbb{R})$; therefore, it is in this latter space that we develop our analysis.

Definition 6.1.1. A multi-breather associated to the sum P given in (6.1.6) of solitons and breathers is a solution $p \in C([T^*,+\infty), H^2(\mathbb{R}))$, for a constant $T^* > 0$, of mKdV such that

$$\lim_{t\to+\infty} \|p(t) - P(t)\|_{H^2(\mathbb{R})} = 0.$$

It is possible to establish the following two results. The first one is the existence and the regularity of a multi-breather, the second one is the uniqueness of a multi-breather in the case when all velocities are positive.

Theorem 6.1.2 (Theorem 1). *Given solitons and breathers* (6.1.2), (6.1.1) *whose velocities* (6.1.4) *and* (6.1.3) *satisfy* (6.1.5), *there exists a multi-breather p associated to P given in* (6.1.6). *Moreover,* $p \in C^\infty(\mathbb{R} \times \mathbb{R}) \cap C^\infty(\mathbb{R}, H^s(\mathbb{R}))$ *for any* $s > 0$ *and there exists* $\theta > 0$ *such that for any* $s > 0$, *there exists* $A_s > 1$ *and* $T^* > 0$ *such that,*

$$\|p(t) - P(t)\|_{H^s(\mathbb{R})} \leq A_s e^{-\theta t}, \quad \forall t \geq T^*.$$

Theorem 6.1.3 (Theorem 2). *Given the same set of solitons and breathers as in Theorem 6.1.2 whose velocities satisfy* (6.1.5) *and* $v_1 > 0$ *(so that all velocities are positive), the multi-breather p associated to P by Theorem 6.1.2 in the sense of Definition 6.1.1 is unique.*

It is also possible to prove a version of Theorem 6.1.2 that does not have any assumption on the sign of the considered velocities. However, the uniqueness is obtained in a narrower class.

Proposition 6.1.4. *Given the same set of solitons and breathers as in Theorem 6.1.3 whose velocities satisfy* (6.1.5), *there exists* $N > 0$ *large enough such that the multi-breather p associated to P by Theorem 6.1.3 is the unique solution* $u \in C([T_0, +\infty), H^2(\mathbb{R}))$ *of mKdV such that*

$$\|u(t) - P(t)\|_{H^2(\mathbb{R})} = O\left(\frac{1}{t^N}\right), \qquad as \quad t \to +\infty.$$

6.2 Stability of mKdV breathers on the half-line

An interesting problem is to describe the dynamics of breathers in mKdV posed in the half line [ACC22], since now static breathers may not be present. In particular, in this setting the mKdV breather is not an exact solution, except in a very particular case of boundary data that matches the value of the breather itself.

Many physical problems naturally arise as initial boundary value problems (IBVP), because of the local character of the corresponding phenomenon [ZaGa71]. However, the IBVP for the mKdV equation has been considerably less studied than the corresponding IVP in the full line. For example, there are at least two interesting IBVP for mKdV still in unbounded domains: the one posed on the right half-line, and a second one posed on the left portion of the line.

Alejo, Cavalcante and Corcho [ACC22] considered the nonlinear stability of mKdV breathers of the focusing modified Korteweg-de Vries (mKdV) equation posed on the *left half-line* $\mathbb{R}^- := (-\infty, 0)$:

$$\begin{cases} \partial_t u + \partial_x(\partial_x^2 u + u^3) = 0, & (t,x) \in \mathbb{R}^- \times (0,T), \\ u(x,0) = u_0(x), & x \in \mathbb{R}^-, \\ u(0,t) = f(t), & t \in (0,T), \\ \partial_x u(0,t) = f_1(t), & t \in (0,T). \end{cases} \tag{6.2.1}$$

In the half-line case, results by Holmer [Hol06] for the quadratic case (KdV) are foundational. In the cubic case, consider the solution u posed on the space

$$\begin{aligned} &u \in C\big(\mathbb{R}^+; H^2(\mathbb{R}^-)\big) \quad \text{and} \\ &\partial_x^j u \in C\big(\mathbb{R}_x^+; H^{(3-j)/3}(0,T)\big) \ \text{ for } \ j = 0,1,2,3. \end{aligned} \tag{6.2.2}$$

Under (6.2.2), one can construct a local and global theory well-adapted to the space $H^2(\mathbb{R}^-)$, see [ACC22] for precise details.

Let $\overleftarrow{B}_{\alpha,\beta}(t,x;x_1,x_2)$ be the mKdV breather (4.1.7), *moving leftwards* in space, and therefore when its velocity $-\gamma < 0$ or equivalently, from (4.1.9), when $\beta < \sqrt{3}\alpha$. We consider a breather solution on the left half-line as the restriction on $\mathbb{R}^-$ of classical breathers posed on the whole line, namely

$$B_{\alpha,\beta} = \overleftarrow{B}_{\alpha,\beta}\Big|_{\mathbb{R}^-}.$$

We highlight that the above breather on the left half-line is not an exact solution for the IBVP (6.2.1), except for very particular boundary conditions $f(t)$ and $f_1(t)$. More precisely, breathers induce the natural traces given by

$$f(t) = \overleftarrow{B}_{\alpha,\beta}(x=0,t;x_1,x_2) \quad \text{and} \quad f_1(t) = \partial_x \overleftarrow{B}_{\alpha,\beta}(x=0,t;x_1,x_2).$$

In [ACC22] it was proved that any classical mKdV breather solution, restricted to the left half-line $\mathbb{R}^-$, and placed far enough from the origin $x = 0$, is stable in $H^2(\mathbb{R}^-)$ under perturbations that preserve the zero boundary conditions. More precisely,

Theorem 6.2.1. *Let $\alpha, \beta > 0$, and $B_{\alpha,\beta}$ a restricted breather (such that it satisfies zero boundary conditions). Assuming that $\beta \le \alpha$ (breathers moving leftwards), there exist parameters η_0, A_0 and L_0, depending on α and β, such that for all $L > L_0$ and $\eta \in (0, \eta_0)$ the following holds: consider $u_0 \in H^2(\mathbb{R}^-)$ such that*

$$\big\|u_0 - B_{\alpha,\beta}(\cdot,0;0,L)\big\|_{H^2(\mathbb{R}^-)} \le \eta.$$

Then there exist continuous functions $\rho_1(t), \rho_2(t) \in \mathbb{R}$ *such that the solution* $u(\cdot, t)$ *of the IBVP* (6.2.1) *with initial data* u_0 *and homogeneous boundary conditions* $f(t) = f_1(t) \equiv 0$*, satisfies*

$$\sup_{t\in\mathbb{R}} \left\| u(t) - B_{\alpha,\beta}\Big(\cdot, t; \rho_1(t), \rho_2(t) + L\Big) \right\|_{H^2(\mathbb{R}^-)} \leq A_0\eta + Ke^{-\beta L}$$

for some constant $K > 0$.

The proof involves the construction and use of almost conserved Lyapunov functionals a la Martel-Merle-Tsai [MMT02, CaMu19], for which it is necessary to control some boundary terms. In addition, one has some error contributions that appear because the restricted breather is not an exact solution for the initial boundary value problem.

The case of the IBVP on the right half-line remains as a challenging open problem. This problem imposes several new conditions with respect to the left half-line case, as for instance, that the breather speed $-\gamma > 0$ or that a second boundary condition to the corresponding IBVP can not be imposed.

6.3 The cubic Whitham equation

The Korteweg-de Vries (KdV) equation (2.7.1)

$$u_t + (u_{xx} + u^2)_x = 0, \quad u = u(t,x) \in \mathbb{R}, \quad (t,x) \in \mathbb{R}^2,$$

is a simplified model equation for waves at the surface of an inviscid incompressible fluid. The equation includes terms that describe the effects of nonlinearity and dispersion, and the balance between these two is the theoretical mechanism which allows the existence of coherent structures and patterns like solitary-wave solutions and periodic traveling waves.

The KdV equation can also be used as a model for waves at the interface in a two-fluid system although it has been noted that the so-called extended KdV (eKdV) equation, also known as the Gardner equation (1.4.2), is more suitable for internal waves as it more closely describes the typically broader solitary waves [StOs98]. In the two-layer case, the unknown function $\eta(x,t)$ corresponds to the deflection of the interface from its rest position.

Following an idea of G.B. Whitham [Whi65], one may use the full representation of the dispersion relation for a two-fluid system. In particular for a lower layer of depth h_1 and density ρ_1 and an upper layer of depth h_2 and density ρ_2, the equation will have the form

$$\eta_t + \alpha_1\eta\eta_x + \alpha_2\eta^2\eta_x + K_I * \eta_x = 0$$

where the coefficients are given in terms of $c_0^2 = gh_1h_2\frac{\rho_1-\rho_2}{\rho_1h_2+\rho_2h_1}$ by

$$\alpha_1 = \frac{3}{2}c_0\frac{1}{h_1h_2}\frac{\rho_1h_2^2-\rho_2h_1^2}{\rho_1h_2+\rho_2h_1}$$

and

$$\alpha_2 = \frac{3c_0}{(h_1h_2)^2}\left(\frac{7}{8}\left(\frac{\rho_1h_2^2-\rho_2h_1^2}{\rho_1h_2+\rho_2h_1}\right)^2 - \frac{\rho_1h_2^3+\rho_2h_1^3}{\rho_1h_2+\rho_2h_1}\right).$$

Here K_I is the convolution operator and it is explicitly given by the kernel

$$K_I = \mathcal{F}^{-1}\left(\sqrt{\frac{g(\rho_1-\rho_2)\tanh(kh_1)\tanh(kh_2)}{\rho_1\tanh(kh_2)+\rho_2\tanh(kh_1)}}\right), \tag{6.3.1}$$

where $\mathcal{F}^{-1}$ is the inverse Fourier transform, defined by (see (2.3.1))

$$\mathcal{F}^{-1}U(x) = \frac{1}{2\pi}\int_{-\infty}^{\infty} U(k)e^{ikx}\,dk.$$

(see [KoBu81, StOs98, CGK05, KaNg10]). In the case when $\rho_1h_2^2 \sim \rho_2h_1^2$, the quadratic coefficient vanishes, and the pure cubic equation (without the quadratic term) appears as the most appropriate model. The equation can be normalized by taking the depth of the lower fluid h_1 as the unit of distance, and h_1/c_0 as a unit of time.

In this section, as in [ACK22], the so-called cubic Whitham equation is considered, which is a simplified version of (6.3.1) given by

$$\eta_t + \eta^2\eta_x + K * \eta_x = 0 \tag{6.3.2}$$

with

$$K = \mathcal{F}^{-1}\sqrt{\frac{\tanh k}{k}}.$$

Equation (6.3.2) has the advantage of being closer to the original (quadratic) Whitham equation (see [MKD15, KLP18, Car18, KhAb18] and references therein),

$$\eta_t + \eta\eta_x + K * \eta_x = 0, \tag{6.3.3}$$

and it can also be easily compared with the modified KdV equation

$$\eta_t + \eta^2\eta_x + \eta_{xxx} = 0. \tag{6.3.4}$$

As we have already shown in Chapter 4, the mKdV equation is an integrable model equation, and solutions may be found with the help of the ISM, or using the Miura transform which maps[1] solutions of the mKdV equation into solutions of the KdV

[1] See Exercises (6.2)-(6.3) for further details.

equation [Miu68]. On the other hand, the cubic Whitham equation (6.3.2) is not integrable as can be seen explicitly by examining the interaction of two solitary waves (see [ACK22, Fig.1]). The existence of traveling-wave solutions and the existence and stability of solitary wave solutions to the cubic Whitham equation (6.3.2) was studied in [EhKa09, EGW12, EhKa13].

More recently the formation of shocks has been proved in [KSW22]. Approximate solitary wave solutions can be found numerically using a manual cleaning process such as explained in [BoKa00]. As shown in [ACK22, Fig.1], the interaction between two solitary waves in the cubic Whitham equation creates a small dispersive tail after the interaction, while the same code applied to two solitary waves of the mKdV equation shows a clean interaction. The appearance of small dispersive ripples after the interaction is usually a strong indicator of non-integrability.

The cubic Whitham equation (6.3.2) admits the following conserved quantities:

$$I_1 = \int \eta dx, \quad I_2 = \int \eta^2 dx, \quad I_3 = \int \left(\eta K_1 * \eta + \frac{1}{4}\eta^4\right) dx,$$

where the integration is over the real line, or over one wavelength depending on the situation. Since the equation is nonintegrable, it is unlikely that further conserved integrals can be found.

One of the most prominent examples of a breather solution is the Peregrine breather (see Section 3.2.4) which is an exact solution of the NLS equation [Per83]. The Peregrine breather can be obtained by perturbing the exponential Stokes-wave solution of the NLS equation with specially chosen functions of polynomial decay. In most of the models mentioned above, in particular the mKdV and Gardner equations, breather solutions share common features such as having two free parameters which can be interpreted as the amplitude and the rate of the inner oscillation of the wave packet. In particular, the mKdV equation (6.3.4) has a closed-form breather solution given in terms of the *phase velocity* $\gamma = 3\alpha^2 - \beta^2$ and the *crest velocity* $\delta = \alpha^2 - 3\beta^2$.

However, in [AlMu13], breathers have been characterized as solutions of a fourth-order ordinary differential equation (see (4.2.1)) which implies that breather solutions are critical points of a second-order Lyapunov functional (see (4.8.2)), composed of a linear combination of low order conserved quantities [AlMu13, AlMu15, AMP17b]. Although some progress has been recently done in the last years (see Section 3.3), it remains a question whether breathers exist more generally as stable structures in nonintegrable evolution equations.

For the cubic Whitham equation (6.3.2) (as for the quadratic Whitham equation) no closed-form solutions are known. In particular, it is not known whether (6.3.2) features breather-type solutions. In [ACK22], some numerical simulations are presented, which strongly suggest that (6.3.2) does admit breather solutions. In order to find a breather, one may consider the evolution of initial data in the form of a Gaussian, or possibly a modulated Gaussian. As the solution develops, in many cases, solitary waves and breathers appear naturally.

For the purpose of approximating solutions of (6.3.2), a Fourier-collocation method is used. This method has the advantage that the symbol of the linear operator appearing in (6.3.2) can be represented exactly. The discretization and numerical implementation are similar to methods used in [EhKa13]. The code used in [ACK22] is versatile enough so that it can be used for KdV (2.7.1), Whitham and mKdV (4.1.1) and cubic Whitham (6.3.2) equations by simply changing coefficients.

In order to find coherent structures for the Whitham equation, an mKdV breather solution is inserted as initial data into the fully discrete scheme. As a test, firstly solve the mKdV equation itself and observe that the breather propagates as it should (see [ACK22, Fig.2]). Since this solution has an exact form, this procedure was also used to test the implementation of the numerical code. After that, it is possible to insert a perturbed breather into the quadratic and cubic Whitham equation (6.3.2). In the quadratic Whitham equation (6.3.3), a solitary wave and a dispersive shock wave are formed (see [ACK22, Fig.3]). In the cubic Whitham equation (6.3.2), the initial data evolve into a positive solitary wave, a negative solitary wave, a breather-like structure and a dispersive tail (see [ACK22, Fig.4]).

In order to extract numerical approximations to a breather, a cleaning approach needs to be used such as explained in [BoKa00]. In other words, at some intervals, the dispersive ripples are simply zeroed out by hand, and the code is restarted. If the coherent structure (solitary wave of breather) is stable, and the small oscillations are propagating at a different speed than the main structure, then this process eventually leads to a close numerical approximation of the target structure, with small ripples essentially absent almost to machine precision. The time evolution of three different breathers in the cubic Whitham equation (6.3.2) is depicted in [ACK22, Fig.5]. [ACK22, Fig.6] shows the first breather from [ACK22, Fig.5] after evolving for 10 periods, and no difference can be discerned visually between the starting waveform and the evolved wave.

The three breathers shown in [ACK22, Fig.5] suggest that a rich variety of different breathers exist in the cubic Whitham equation (6.3.2). Indeed, comparing the breather shown in the upper and middle panel, the breather shown in the middle panel appears slightly more oscillatory, but has a much larger total wave height (defined as the elevation of the highest crest minus the elevation of lowest trough in the wave packet) than the breather in the upper panel. On the other hand, the breather in the lowest panel has a similar wave height to the one in the middle panel, but features many more oscillations. This suggests that there is probably a two-parameter family of breather solutions, in analogy with the mKdV equation.

The question of the existence of breather solutions for nonintegrable equations was raised in [KoKo74, DHB75]. These authors provided an asymptotic development suggesting the existence of a breather solution in the so-called Klein-Gordon φ^4 equation. However, it was ultimately shown that this expansion did not converge, and the purported breather solution actually features a tiny amount of radiation, though the rate of radiation is so small that it can only be detected using asymptotics beyond all orders [Seg87, SoWe99].

Since no radiation was detected with the numerical code used in [ACK22], these numerical results suggest the mathematical existence of true breather solutions. The solutions constructed in [ACK22] are also different from the SG breather in the sense that they are propagating, and have a well-defined phase and crest velocity, similar to the two-parameter family of mKdV breathers. In contrast, the SG breather is a one-parameter family of stationary breathers. Nevertheless, it is possible that the breathers found in the present work also feature very small radiation which may be below machine precision, and in this case the breathers found here would also be approximate in the mathematical sense, i.e. decaying after a very long time. On the other hand, the question whether the breather is an exact solution or a slowly radiating meta-stable state may not be of much importance in practice. Indeed while it has been shown for example that the Peregrine breather is unstable to virtually all perturbations [KlHa17, Mu17], breathers are nevertheless observable in the laboratory [CNH12]. Moreover asymptotic model equations such as the KdV and Whitham equations are generally valid physical models only on a time scale inversely proportional to the amplitude [Lan13, KLP18, SaWa21, Eme21].

It's worth to mention that there are some works discussing mathematical existence of structures more similar to the SG breathers (see Section 3.3 for further reading). In [BCB11], [MaSc21] and [Hir19] nonexplicit real-valued, time-periodic and spatially localized solutions are constructed for nonlinear nonintegrable wave equations with periodic potentials using variational methods and spatial dynamics and center manifold reductions. These solutions are of breather type, but are stationary, and not propagating, so they are more similar to the breather of the

SG equation. It will be interesting to see whether breather solutions are possible in other fully dispersive equations such as the fully dispersive Gardner equation [Car22] and the fully dispersive NLS equation [TKDV00].

6.4 Exercises

(6.1) Transform the mKdV equation appearing in this chapter (6.3.4), into the corresponding one in (1.4.1)-(4.1.1).

(6.2) (*Miura Transform* [GKM68]) Check the (*complex*) Miura transform

$$v = i\frac{3}{\sqrt{2}}u_x + \frac{3}{2}u^2.$$

Namely, prove that if we assume that u is a solution of the *focusing* mKdV equation (4.1.1), then v is a solution of the KdV equation (2.7.1).

(6.3) Check the (*real*) Miura transform

$$v = \frac{3}{\sqrt{2}}u_x - \frac{3}{2}u^2.$$

Namely, prove that if u is a solution of the *defocusing* mKdV equation

$$u_t + (u_{xx} - u^3)_x = 0, \tag{6.4.1}$$

then v is a solution of the KdV equation (2.7.1).

(6.4) Build, if possible, the equivalent Miura transform mapping solutions of the cubic Whitham equation (6.3.2) to solutions of the Whitham equation (6.3.3)

$$\eta_t + \eta\eta_x + K * \eta_x = 0.$$

Chapter 7

Numerical study of the mKdV breather linear spectrum

7.1 First (and incomplete) ideas

The purpose of this section is to explain how one can recover the discrete spectrum of a mKdV breather-like self-adjoint operator by using some standard numerical schemes. Since mKdV is invariant under suitable scaling symmetries (see (2.10.1)), we can always assume that $\beta = 1$. Similarly, thanks to the invariance by shifts in space, we can also assume $x_2 = 0$. Therefore, for spectral considerations, we are finally reduced to the study of the breather profile

$$B_0(x) := 2\sqrt{2}\partial_x\left(\arctan\left(\frac{1}{\alpha}\frac{\sin(\alpha(x+x_1))}{\cosh x}\right)\right).$$

Thanks to the periodicity of the sine function, we can also assume $x_1 \in [0, \frac{2\pi}{\alpha}]$. This breather profile is a solution to (4.2.1) with $\beta = 1$, and from standard linear arguments [AlMu13], the associated selfadjoint linear operator

$$\begin{aligned}\mathcal{L}z &:= z_{4x} - 2(1-\alpha^2)z_{xx} + (1+\alpha^2)^2 z \\ &\quad + 5B_0^2 z_{xx} + 10B_0(B_0)_x z_x \\ &\quad + \left(5(B_0)_x^2 + 10B_0(B_0)_{xx} + \frac{15}{2}B_0^4 - 6(1-\alpha^2)B_0^2\right)z,\end{aligned} \tag{7.1.1}$$

has an expected two dimensional kernel.

It is more difficult to prove that $\mathcal{L}$ has, regardless the values of $\alpha \in (0,\infty)$ and $x_1 \in [0, \frac{2\pi}{\alpha}]$, an always negative, unique eigenvalue $\lambda_1(\alpha, x_1) < 0$, which is uniformly in x_1 away from zero [AlMu13, Proposition 4.8 and Corollary 4.9]. Additionally, the spectrum of $\mathcal{L}$ is isolated and the continuum spectrum is always positive and away from zero, depending on α and β. We would like to recover numerically all the properties mentioned before.

7.2 Numerical scheme for mKdV

In order to obtain a numerical approximation of the eigenvalues of $\mathcal{L}$, we follow the approach based on the approximate Galerkin method for the Hilbert space $L^2(\mathbb{R})$. First, we consider the subspace $\mathbb{V}_N$ generated by a finite dimensional subset of an orthonormal basis of $L^2(\mathbb{R})$:

$$\mathbb{V}_N := \text{span}\{f_0, f_1, f_2, \dots, f_N\} \subset L^2(\mathbb{R}).$$

For practical and computational purposes, the best candidate for an orthonormal basis is the one generated by the Hermite functions:

$$f_n(x) := p_n(x)e^{-x^2/2}, \quad x \in \mathbb{R}, \quad n \geq 0, \qquad \int_{\mathbb{R}} f_n^2(x)dx = 1.$$

Here $p_n(x)$ are the Hermite polynomials suitably normalized such that the f_n have unit norm.

We will approximate the spectrum of $\mathcal{L}$ by the spectrum of the finite dimensional operator

$$\mathcal{L}_N := \mathbb{P}_N \mathcal{L} \mathbb{P}_N,$$

where $\mathbb{P}_N$ is the projection operator into the subspace $\mathbb{V}_N$. It is expected that, as long as N is large enough, $\mathcal{L}_N$ will approach the discrete spectrum, in particular the eigenvalues of $\mathcal{L}$, any spectral gap present in the exact operator $\mathcal{L}$, and even the multiplicity of the eigenvalues (seen as two very close approximate eigenvalues).[1]

The above scheme reduces matters to understanding the eigenvalues of the matrix

$$(\mathcal{M}_N)_{i,j} := \int_{\mathbb{R}} f_i \mathcal{L} f_j, \quad i, j = 0, \dots, N, \quad f_i, f_j \in \mathbb{V}_N.$$

Note that this $(N+1) \times (N+1)$ matrix is symmetric, so its eigenvalues are real-valued. Using some symbolic computations done by *Mathematica* for some simple cases (periodic mKdV), or using *Matlab* for more demanding and time-consuming cases (aperiodic mKdV), one can compute $\mathcal{M}_N$ and its eigenvalues.

For the operator (7.1.1), it was important to understand the first four eigenvalues, ordered by size. We recover the two-dimensional kernel, along with the unique negative eigenvalue. Additionally, at each computation performed, we recovered the spectral gap between the kernel and the bottom of the positive spectrum. We compared our results with some standard linear operators that have continuous

---

[1] Additionally, as far as N is large, the continuous spectrum is also discretely approximated on compact intervals.

spectrum $[0, +\infty)$ and resonances at the origin, recovering the absence of a spectral gap by having plenty of numerical eigenvalues near zero as N was larger and larger. As it was revealed in the numerical computations, for α large the code becomes unstable, and a larger N is needed to recover the expected results.

7.3 First results

First of all, Table 7.1 represents our findings for the case of a breather with parameters $\beta = 1$, $\alpha = 0.5$, and different values for the shift x_1.

Table 7.1 The first four eigenvalues of $\mathcal{L}$ for $\beta = 1$, $\alpha = 0.5$, and x_1 varying. All computations were made with $N = 160$ test functions. The third and fourth columns are the approximate kernel of $\mathcal{L}$, and the fourth eigenvalue is immediately followed by a big amount of closer and larger eigenvalues, which represent a discrete approximation of the continuous spectrum.

x_1	1st eig.	2nd	3rd	4th
0.09	-4.226	0.028	0.028	1.776
0.81	-4.191	-0.039	0.081	1.862
1.53	-3.491	-0.005	0.006	1.886
2.15	-2.557	-0.000	0.000	1.901
3.14	-1.507	0.007	0.004	1.915

In Fig. 7.1, we plot the behavior of the only negative eigenvalue with respect to the shift x_1. In order to ensure the best possible approximations, we have chosen $N = 160$ Hermite test functions to compute the corresponding eigenvalues.

A second test seeks for the description of the eigenvalues for a higher oscillatory breather, with $\alpha = 1.5$. For this case, we obtain the results described in Table 7.2 and Fig. 7.2.

For $\alpha = 2.5$, computations become more difficult to perform, due to the oscillatory behavior of the breather solution, and $N \geq 160$ is probably needed; an example is the case where $x_1 = 0.05$ and $x_1 = 2$, where we performed computations with $N = 166$. Our results for the first four eigenvalues are, respectively,

$$\{-71.66, \ -0.043, \ 0.003, \ 28.090\}, \qquad \{-72.335, \ -0.050, \ 0.055, \ 27.926\}.$$

We conclude from numerical simulations that $x_1 = 0$ represents a critical point for the negative eigenvalue $\lambda_1(\alpha, x_1)$, either a maximum or a minimum, depending on the size of the oscillatory scaling α. Additionally, when computing λ_1 numerically, some additional, unexpected positive error appears at $x_1 = 0$ (see e.g. Fig. 7.2), that we have not yet fully understood.

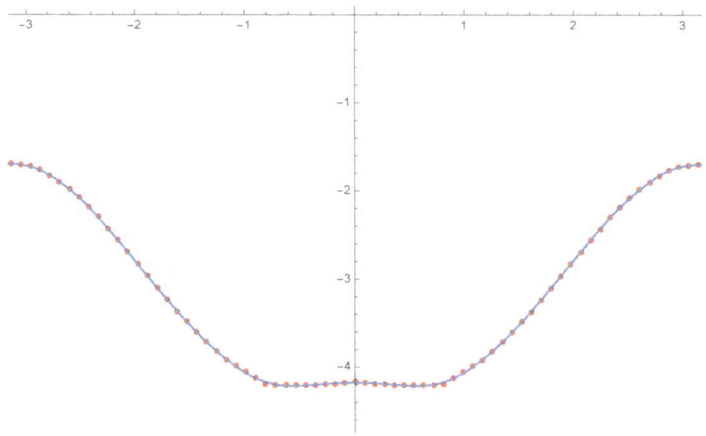

Fig. 7.1 Red dots: Variation of the negative eigenvalue of $\mathcal{L}$ with $\beta = 1$, $\alpha = 0.5$ and x_1 varying from -3 to 3. All negative eigenvalues are between -1 and -5. The blue line corresponds to a best polynomial fit of the data. The small upward bump appearing near the center is believed to be error of numerical origin.

Table 7.2 The first four eigenvalues of $\mathcal{L}$ for $\beta = 1$, $\alpha = 1.5$, and x_1 varying. All computations were made with $N = 164$ test functions. The third and fourth columns represent the approximate kernel of $\mathcal{L}$.

x_1	1st eig.	2nd	3rd	4th
0.00	-22.067	-0.031	-0.000	10.043
0.99	-22.571	-0.000	0.000	10.086
1.57	-22.423	-0.007	0.002	10.247
2.51	-22.354	-0.003	0.007	10.222
3.14	-22.571	-0.000	0.000	10.084

Finally, we look for the behavior of the minimal eigenvalue with respect to the oscillatory parameter α. We fix $x_1 = 0$ and $\beta = 1$, and we plot the resulting minimum eigenvalue and the second eigenvalue as a function of α. We get the results shown in Figure 7.3. Note that the larger α is, the less accurate is the second eigenvalue (believed to be zero), but up to $\alpha = 2$, the relative error stays bounded below 10^{-1}.

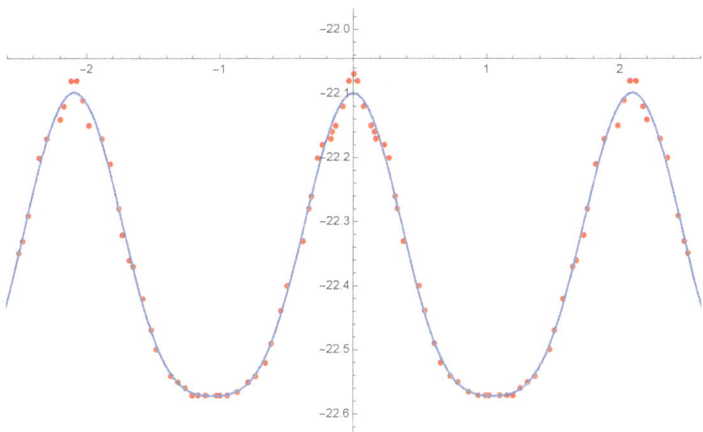

Fig. 7.2 Red dots: Variation of the negative eigenvalue of $\mathcal{L}$ with $\beta = 1$, $\alpha = 1.5$ and x_1 varying from -2.6 to 2.6. All data is below -22.05 and above -22.6. The blue line corresponds to a suitable best polynomial fit of the data, and distance of numerical negative eigenvalues (red points) to this curve are of relative error $\sim 10^{-2}$.

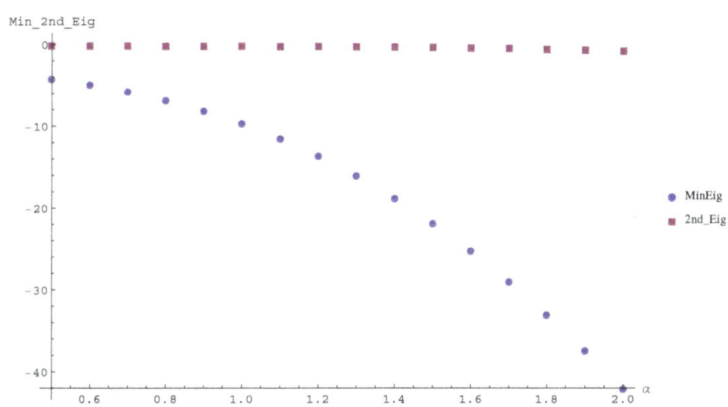

Fig. 7.3 Two first minimal eigenvalues of the linearized mKdV operator around the mKdV breather (4.1.1) with $x_1 = 0$ and $\beta = 1$ while α is varying from 0.5 to 2.0. A simple log plot reveals that the absolute value of the minimal eigenvalue depends approximately on the second power of α.

7.4 Conclusions

We can conclude, from the previous numerical analysis, that the rigorous description obtained in [AlMu13] is certainly recovered in a satisfactory way.

Chapter 8

The Gardner model

8.1 Introduction

There is a second KdV-like model with explicit breather solutions, and in some sense more general than the mKdV breather (4.1.7) presented in Chapter 3. It was already introduced in Subsection 3.2.2, but in this chapter we will further describe its breathers and their main properties.

This model is a mix between the KdV and mKdV nonlinearities, and is usually referred to as the Gardner equation

$$\begin{aligned} &\partial_t w + \partial_x(\partial_x^2 w + \mu w^2 + w^3) = 0, \\ &\mu > 0, \quad w(t,x) \in \mathbb{R}, \quad (t,x) \in \mathbb{R}^2. \end{aligned} \tag{8.1.1}$$

The Gardner equation (8.1.1) is the prototype of integrable model counterexample to several desired properties in nonintegrable models. Indeed, the key characteristic of the Gardner equation (8.1.1) is that it contains a nonlinear part composed of a Korteweg-de Vries (KdV) quadratic term $(w^2)_x$ and a positive modified KdV (mKdV) cubic term $(w^3)_x$. The competition between this nonlinear part and the linear dispersive term w_{xxx} allows the existence of intricate soliton, multisolitons as well as exact real-valued *breather* solutions (see (8.4.1)).

The Gardner equation (8.1.1) is a well-known *completely integrable* model [GKM68, AbCl91, Lam80], with infinitely many conservation laws. As a physical model, (8.1.1) describes large-amplitude internal solitary waves, showing a dynamic which can look rather different from the KdV form. Solutions of (8.1.1) are invariant under space and time translations. Indeed, for any $t_0, x_0 \in \mathbb{R}$, $w(t - t_0, x - x_0)$ is also a solution. Note that (8.1.1) is not scaling invariant. Moreover, (8.1.1) is closely related to the modified Korteweg-de Vries (mKdV) equation already studied in (4.1.1), through the search of L^∞-solutions. In fact, it is easy to see by substitution, that the following property holds:

Lemma 8.1.1. *Let u be a solution of the mKdV equation* (4.1.1) *with a nonvanishing boundary value or condition (NVBC)* $\mu \in \mathbb{R}\backslash\{0\}$ *at* $\pm\infty$. *Then*

$$w(t,x) := u\left(t, x + \frac{\mu^2}{3}t\right) - \frac{\mu}{3}$$

is a solution of the Gardner equation (8.1.1).

Lemma 8.1.1 provides a useful tool, since this relationship between these two models will appear several times. For this reason, we will shortly explain each result on the Gardner model, leaving the reader the main details in the bibliography.

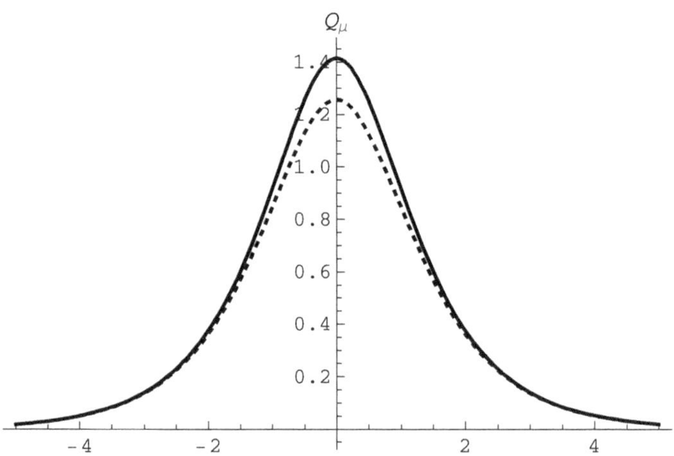

Fig. 8.1 Comparison between the mKdV soliton (full line, $\mu = 0, c = 1$) and the Gardner soliton (dashed) (8.2.1) with $\mu = 0.25, c = 1$.

8.2 Solitons

As well as mKdV, Gardner has solitons and breathers, but their description is more complicated. The profile of the soliton solution (see also Section 2.10) is explicitly given by:

$$\begin{aligned} w(t,x) &:= Q_{c,\mu}(x - ct), \quad c, \mu > 0, \\ Q_{c,\mu}(s) &:= \frac{c}{\frac{\mu}{3} + \sqrt{\frac{\mu^2}{9} + \frac{c}{2}} \cosh\left(\sqrt{c}s\right)}. \end{aligned} \tag{8.2.1}$$

The reader can also check that $Q_{c,\mu} > 0$ satisfies the nonlinear elliptic equation

$$Q''_{c,\mu} - c\,Q_{c,\mu} + \mu Q^2_{c,\mu} + Q^3_{c,\mu} = 0, \quad Q_{c,\mu} > 0, \quad Q_{c,\mu} \in H^1(\mathbb{R}).$$

In fact, in the limit $\mu \to 0$, one recovers the classical mKdV soliton.

8.3 Conserved quantities

In the case of the Gardner equation (8.1.1), the mass has the same definition as in mKdV and the energy is given by

$$E_\mu[w](t) := \frac{1}{2}\int_{\mathbb{R}} w_x^2 - \frac{\mu}{3}\int_{\mathbb{R}} w^3 - \frac{1}{4}\int_{\mathbb{R}} w^4 = E_\mu[w](0). \tag{8.3.1}$$

As well as mKdV, one has a third conserved quantity, given by

$$\begin{aligned} F_\mu[w](t) :=& \frac{1}{2}\int_{\mathbb{R}} w_{xx}^2 - \frac{5}{3}\mu\int_{\mathbb{R}} ww_x^2 + \frac{5}{18}\mu^2\int_{\mathbb{R}} w^4 \\ &- \frac{5}{2}\int_{\mathbb{R}} w^2w_x^2 + \frac{\mu}{2}\int_{\mathbb{R}} w^5 + \frac{1}{4}\int_{\mathbb{R}} w^6 = F_\mu[w](0). \end{aligned} \tag{8.3.2}$$

This conserved quantity is key for the stability proof, as well as in mKdV.

8.4 Gardner breathers

For the Gardner equation (8.1.1), the breather solution can be obtained by using different methods (e.g. Inverse Scattering (ISM), Hirota method, etc), and it is characterized by the introduction of the parameter μ previously shown.

Definition 8.4.1. Let $\alpha, \beta, \mu \in \mathbb{R}\backslash\{0\}$ such that $\Delta = \alpha^2 + \beta^2 - \frac{2\mu^2}{9} > 0$, and $x_1, x_2 \in \mathbb{R}$. The real breather solution of the Gardner equation (8.1.1) is given explicitly by the formula

$$B_{\alpha,\beta,\mu}(t,x) := \partial_x \widetilde{B} := 2\sqrt{2}\partial_x\left(\arctan\left(\frac{G_{\alpha,\beta,\mu}(t,x)}{F_{\alpha,\beta,\mu}(t,x)}\right)\right), \tag{8.4.1}$$

with

$$G_{\alpha,\beta,\mu}(t,x) := \frac{\beta(\alpha\sin(\alpha y_1) - \beta\cos(\alpha y_1))}{\alpha\sqrt{\Delta}} - \frac{\sqrt{2}\mu\beta(\cosh(\beta y_2) + \sinh(\beta y_2))}{3\Delta},$$

$$F_{\alpha,\beta,\mu}(t,x) := \cosh(\beta y_2) - \frac{\sqrt{2}\mu\beta}{3\alpha\sqrt{\Delta}}\cos(\alpha y_1),$$

with y_1 and y_2 defined in (4.1.8).

Note that Gardner breather solutions (8.4.1) are periodic in time, but not in space. Additionally, every Gardner breather satisfies Definition 3.1.1 with $T = \frac{2\pi}{\alpha(\gamma-\delta)} > 0$ and $L = -\gamma T$. Another interesting fact is that in the formal limit $\mu \to 0$, one recovers the standard mKdV breather (4.1.7). Moreover from (8.4.1) one has, for any $k \in \mathbb{Z}$,

$$B_{\alpha,\beta,(-1)^k\mu}\left(t, x; x_1 + \frac{k\pi}{\alpha}, x_2\right) = (-1)^k B_{\alpha,\beta,\mu}(t, x; x_1, x_2),$$

which are also solutions of (8.1.1). This identity reveals the periodic character of the first translation parameter x_1, coupled this time to the parameter $(-1)^k\mu,\ k \in \mathbb{Z}$.

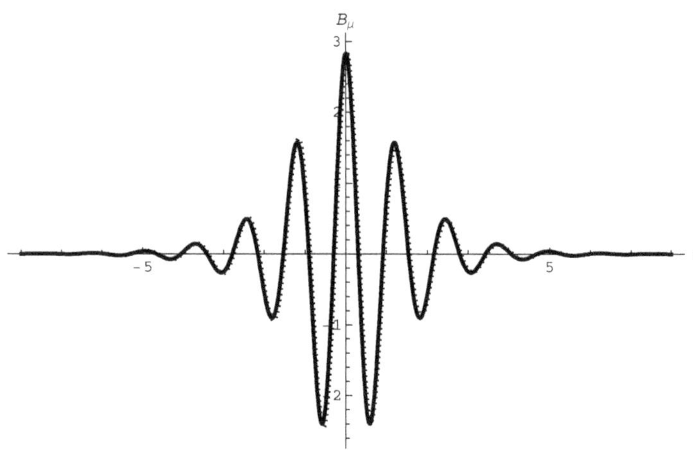

Fig. 8.2 Comparison between the mKdV breather (full line, $\beta = 1, \alpha = 5, \mu = 0$) and the Gardner breather (dotted) (8.4.1) with $\beta = 1, \alpha = 5, \mu = 0.5$.

Gardner breather solutions have also been used in [Ale12d] to prove the ill-posedness for the Gardner equation (8.1.1) in the line, showing that solutions cannot depend on a uniformly continuous form on their initial data in the Sobolev spaces $H^s(\mathbb{R})$ for $s < 1/4$.

Gardner breathers were discovered in [PeGr97]. A more detailed version can be also found in [Ale12c]. Both results use the ISM. Finally note that spatially periodic Gardner breathers can be obtained from periodic mKdV breathers with NVBC, through Lemma 8.1.1. See Chapter 11.11 for further details.

Remark 8.4.2. The Gardner breather is a four-parametric solution, with two scalings (α, β) and two shift translations (x_1, x_2). Note that we impose $\alpha^2 + \beta^2 - \frac{2}{9}\mu^2 > 0$, since we deal with real-valued solutions. Therefore $\mu \in (0, \mu_{max})$, where

$$\mu_{max} := 3\sqrt{\frac{\alpha^2+\beta^2}{2}}.$$

Remark 8.4.3. Note that we can formally take the limit when $\alpha \to 0$ in[1] (8.4.1), obtaining the so-called *double pole* solution for the Gardner equation (8.1.1), which is a natural generalization of the well-known double pole solution of mKdV:

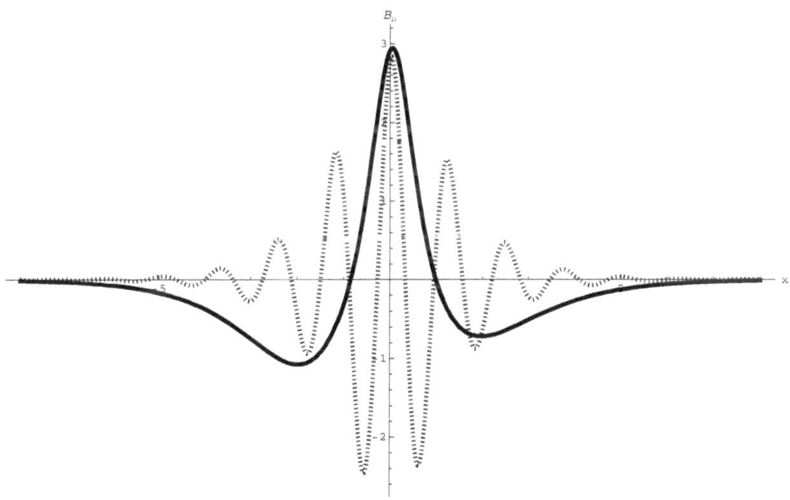

Fig. 8.3 Comparison between the Gardner breather (8.4.1) (dotted, $\beta = 1, \alpha = 5, \mu = 0.75$) and the double pole (full) (8.4.2) with $\beta = 1, \mu = 0.75$.

Definition 8.4.4. Let $\beta, \mu \in \mathbb{R}\backslash\{0\}$ such that $\Delta_0 = \beta^2 - \frac{2}{9}\mu^2 > 0$, and $x_1, x_2 \in \mathbb{R}$. The real-valued double pole solution of the Gardner equation (8.1.1) is given explicitly by the formula

$$B_{\beta,\mu}(t,x) := \partial_x \widetilde{B} := 2\sqrt{2}\partial_x\left(\arctan\left(\frac{G_{\beta,\mu}(t,x)}{F_{\beta,\mu}(t,x)}\right)\right), \tag{8.4.2}$$

[1] Actually to a phase transformed of (8.4.1), namely, with

$$G_{\alpha,\beta,\mu}(t,x) := \frac{\beta\sqrt{\alpha^2+\beta^2}\sin(\alpha y_1)}{\alpha\sqrt{\Delta}} - \frac{\sqrt{2}\mu\beta(\cosh(\beta y_2)+\sinh(\beta y_2))}{3\Delta},$$

$$F_{\alpha,\beta,\mu}(t,x) := \cosh(\beta y_2) - \frac{\sqrt{2}\mu\beta(\alpha\cos(\alpha y_1) - \beta\sin(\alpha y_1))}{3\alpha\sqrt{\alpha^2+\beta^2}\sqrt{\Delta}}.$$

with

$$G_{\beta,\mu}(t,x) := \frac{\beta^2 y_1}{\sqrt{\Delta_0}} - \frac{\sqrt{2}\mu\beta[\cosh(\beta y_2)+\sinh(\beta y_2)]}{3\Delta_0},$$

$$F_{\beta,\mu}(t,x) := \cosh(\beta y_2) - \frac{\sqrt{2}\mu[1-\beta y_1]}{3\sqrt{\Delta_0}},$$

and where now $y_1 = x - 3\beta^2 t + x_1$ and $y_2 = x - \beta^2 t + x_2$.

The Gardner double pole, as well as the mKdV double pole, are still not well-understood from a rigorous mathematical point of view. Being naturally unstable, constructing a manifold where the solution is stable is an interesting open problem.

8.5 Stability of Gardner breathers

Following the ideas from [AlMu13] (with no big changes), it is possible to prove that Gardner breathers are H^2-stable. Specifically, we present here a proof on the stability in $H^2(\mathbb{R})$ of Gardner breathers, showing that this stability is independent of the value of the parameter μ, which controls the strength of the quadratic nonlinear part or KdV term w^2, in its existence interval for *real* Gardner breathers.

Theorem 8.5.1 (**H^2-stability of Gardner breathers**). *Let $\alpha,\beta,\mu \in \mathbb{R}\backslash\{0\}$ such that $\Delta = \alpha^2 + \beta^2 - \frac{2\mu^2}{9} > 0$ and let $B_\mu \equiv B_{\alpha,\beta,\mu}$ be any Gardner breather* (8.4.1). *There exist positive parameters η_0, A_0, depending on α,β and μ, such that the following holds. Consider $w_0 \in H^2(\mathbb{R})$, and assume that there exists $\eta \in (0,\eta_0)$ such that*

$$\|w_0 - B_\mu(t=0;0,0)\|_{H^2(\mathbb{R})} \leq \eta.$$

Then there exist $x_1(t), x_2(t) \in \mathbb{R}$ such that the solution $u(t)$ of the Cauchy problem for the Gardner equation (8.1.1), *with initial data u_0, satisfies*

$$\sup_{t\in\mathbb{R}} \left\|w(t) - B_\mu(t;x_1(t),x_2(t))\right\|_{H^2(\mathbb{R})} \leq A_0\eta.$$

Precise estimates on the derivatives of the parameters $x_1(t)$ and $x_2(t)$ are easy to obtain using modulation theory. Since the proof is similar to the mKdV case, we highlight the most interesting steps in the proof.

8.6 Elliptic equations satisfied by Gardner breathers

For the Gardner case, one has the following result [Ale18]:

Theorem 8.6.1. *Let $B_\mu \equiv B_{\alpha,\beta,\mu}$ be any Gardner breather* (8.4.1). *Then, for any fixed $t \in \mathbb{R}$, B_μ satisfies the nonlinear stationary equation*

$$\begin{aligned} G_\mu[B_\mu] &:= B_{\mu,4x} - 2(\beta^2 - \alpha^2)(B_{\mu,xx} + \mu B_\mu^2 + B_\mu^3) + (\alpha^2 + \beta^2)^2 B_\mu \\ &\quad + 5B_\mu B_{\mu,x}^2 + 5B_\mu^2 B_{\mu,xx} + \frac{3}{2}B_\mu^5 + \frac{5}{3}\mu B_{\mu,x}^2 \\ &\quad + \frac{10}{3}\mu B_\mu B_{\mu,xx} + \frac{10}{9}\mu^2 B_\mu^3 + \frac{5}{2}\mu B_\mu^4 = 0. \end{aligned} \tag{8.6.1}$$

Notice that one recovers the mKdV case (see (4.2.1)) whenever μ approaches zero.

Corollary 8.6.2. *Any Gardner breather B_μ is a critical point of the functional*

$$\mathcal{H}_\mu[w](t) := F_\mu[w](t) + 2(\beta^2 - \alpha^2)E_\mu[w](t) + (\alpha^2 + \beta^2)^2 M[w](t),$$

where M is the mass (L^2 norm), E_μ is given in (8.3.1), *and F_μ as in* (8.3.2).

Note that (8.6.1) reduces to (4.2.1) when $\mu = 0$.

8.7 Linear analysis

As a consequence of Theorem 8.9.1, one is led to the study of a linear operator. In the mKdV case, this is of order four. In the Gardner case, the situation is more complicated because of new terms.

Definition 8.7.1. Let $z \in H^4(\mathbb{R})$, and B_μ be any Gardner breather. We define $\mathcal{L}_\mu$ as the linearized operator associated to B_μ, as follows:

$$\begin{aligned} \mathcal{L}_\mu z &:= z_{4x} - 2(\beta^2 - \alpha^2)z_{xx} + (\alpha^2 + \beta^2)^2 z \\ &\quad + 5B_\mu^2 z_{xx} + 10B_\mu B_{\mu,x} z_x \\ &\quad + \left(5B_{\mu,x}^2 + 10B_\mu B_{\mu,xx} + \frac{15}{2}B_\mu^4 - 6(\beta^2 - \alpha^2)B_\mu^2\right) z \\ &\quad + \frac{10}{3}\mu B_\mu z_{xx} + \frac{10}{3}\mu B_{\mu,x} z_x \\ &\quad + \mu\left(10B_\mu^3 - 4(\beta^2 - \alpha^2)B_\mu + \frac{10}{3}B_{\mu,xx} + \frac{10}{3}\mu B_\mu^2\right) z. \end{aligned} \tag{8.7.1}$$

Figure 8.4 numerically shows the spectrum of $\mathcal{L}_\mu$ (8.7.1). It naturally shows the presence of a unique negative eigenvalue, two elements in the kernel, and a suitable spectral gap. This behavior will be rigorously proved in forthcoming sections.

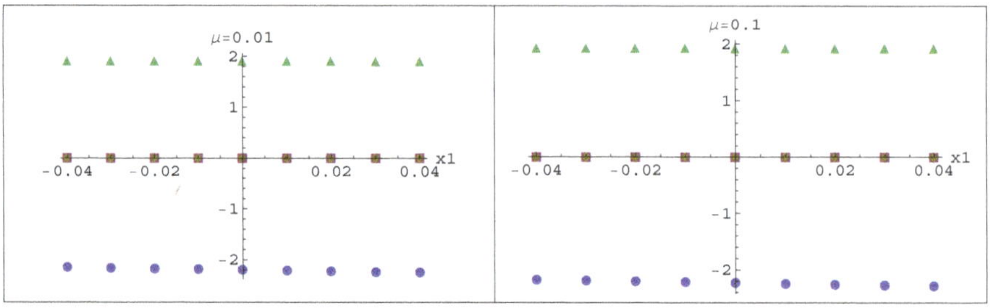

Fig. 8.4 The graph of the negative eigenvalue, the double zero kernel and the fourth eigenvalues for the linearized operator $\mathcal{L}_\mu$ (8.7.1) around a Gardner breather, for $\beta = 1$, $\alpha = 0.5$ and two differents $\mu = 0.01$ and $\mu = 0.1$, while x_1 varying between -0.4 and 0.4. Computations were made with $N = 50$ test functions.

8.8 Nonlinear identities and Weinstein conditions

The aim of this section is to get stability tests by computing generalized Weinstein conditions for any Gardner breather B_μ (8.4.1). We begin with the simpler case of the Gardner soliton (8.2.1), where the mass and the energy are given by the quantities

$$M[Q_{c,\mu}] := 2\sqrt{c} - \frac{2\sqrt{2}}{3}\mu \arctan\left(\frac{3\sqrt{c}}{\sqrt{2}\mu}\right),$$

and

$$E_\mu[Q_{c,\mu}] := -\frac{2}{3}c^{3/2} + \frac{4}{9}\mu^2\sqrt{c} - \frac{4\sqrt{2}}{27}\mu^3 \arctan\left(\frac{3\sqrt{c}}{\sqrt{2}\mu}\right),$$

which reduce, when $\mu = 0$, to the well-known mass and energy of the mKdV soliton solution [AlMu13, eqn. (2.1)]. From these expressions we see explicitly the coupling between the soliton scaling c and the parameter μ in the computed mass and the energy. Note that the Weinstein condition [Wei86] is now, for $c > 0$, $\mu \in \mathbb{R} \setminus \{0\}$,

$$\partial_c M[Q_{c,\mu}] = \frac{\sqrt{c}}{c + \frac{2}{9}\mu^2} > 0. \tag{8.8.1}$$

This inequality guarantees the nonlinear stability of the Gardner soliton (8.2.1). Moreover note that the same condition for the energy of the Gardner soliton $\partial_c E_\mu[Q_{c,\mu}] = -\frac{c^{3/2}}{c+\frac{2}{9}\mu^2}$ does not vanish either.

Lemma 8.8.1. *Let $B_\mu = B_{\alpha,\beta,\mu}$ be any Gardner breather, for α, β, μ as in definition (8.4.1). Then, the mass of B_μ is*

$$M[B_\mu](t) = 4\beta + \frac{2\sqrt{2}}{3}\mu \arctan\left(\frac{2\sqrt{2}\mu\beta}{3\Delta}\right). \tag{8.8.2}$$

Remark 8.8.2. Note that as we could expect, after the Gardner soliton case, the mass of any Gardner breather depends on the scalings α, β, and the parameter μ. This dependence slightly differs from the exclusive dependence of the mass of any mKdV breather on the second scaling parameter β (see (4.1.11)).

From the involved integral in (8.8.2), we can define the *partial* mass of any Gardner breather as follows:

Definition 8.8.3. Let $B_\mu = B_{\alpha,\beta,\mu}$ be any Gardner breather, for α, β, μ as in definition (8.4.1) (such that $\Delta = \alpha^2 + \beta^2 - \frac{2}{9}\mu^2 > 0$). Then, we define the *partial* mass associated to any Gardner breather as:

$$\begin{aligned} \mathcal{M}_\mu(t,x) &\equiv \mathcal{M}_{\alpha,\beta,\mu}(t,x) \\ &:= \frac{1}{2}\int_{-\infty}^{x} B_\mu^2(t,s;x_1,x_2)ds \\ &= 2\beta + \partial_x \log\left(G^2_{\alpha,\beta,\mu} + F^2_{\alpha,\beta,\mu}\right)(t,x) - \frac{2\sqrt{2}}{3}\mu \arctan\left(\frac{G_{\alpha,\beta,\mu}(t,x)}{F_{\alpha,\beta,\mu}(t,x)}\right). \end{aligned} \tag{8.8.3}$$

A direct consequence of the above results are the following generalized Weinstein conditions:

Corollary 8.8.4. *Let $B_\mu = B_{\alpha,\beta,\mu}$ be any Gardner breather of the form* (8.4.1). *Given $t \in \mathbb{R}$ fixed, let*

$$\Lambda_\alpha B_\mu := \partial_\alpha B_\mu, \quad \text{and} \quad \Lambda_\beta B_\mu := \partial_\beta B_\mu.$$

Then $\forall \mu \in (0, \mu_{max})$ both functions $\Lambda_\alpha B_\mu$ and $\Lambda_\beta B_\mu$ are in the Schwartz class for the spatial variable and they satisfy the identities

$$\partial_\alpha M[B_\mu] = \int_{\mathbb{R}} B_\mu \Lambda_\alpha B_\mu = -\frac{16\mu^2\beta\alpha}{9\Delta^2 + 8\mu^2\beta^2} < 0, \tag{8.8.4}$$

and

$$\partial_\beta M[B_\mu] = \int_{\mathbb{R}} B_\mu \Lambda_\beta B_\mu = 4\left(\frac{9\Delta^2 + 2\mu^2\Delta + 8\mu^2\beta^2}{9\Delta^2 + 8\mu^2\beta^2}\right) > 0, \tag{8.8.5}$$

independently of time.

Proof. By simple inspection, one can see that, given a fixed t, $\Lambda_\alpha B_\mu$ and $\Lambda_\beta B_\mu$ are well-defined Schwartz functions. The proof of (8.8.4) and (8.8.5) is consequence of (8.8.2). □

Consider now the two directions associated to spatial translations. Let $B_{\alpha,\beta,\mu}$ as introduced in (8.4.1). We define

$$\begin{aligned} B_1(t;x_1,x_2) &:= \partial_{x_1} B_{\alpha,\beta,\mu}(t;x_1,x_2),\\ B_2(t;x_1,x_2) &:= \partial_{x_2} B_{\alpha,\beta,\mu}(t;x_1,x_2). \end{aligned} \tag{8.8.6}$$

It is clear that, for all $t \in \mathbb{R}$, and α,β,μ as in definition 8.4.1 and $x_1,x_2 \in \mathbb{R}$, both B_1 and B_2 are real-valued functions in the Schwartz class, exponentially decreasing in space. Moreover, it is not difficult to see that they are *linearly independent* as functions of the x-variable, for all time t fixed.

Lemma 8.8.5. *Let $B_\mu = B_{\alpha,\beta,\mu}$ be any Gardner breather of the form* (8.4.1). *Then we have*

(1) $B_\mu = \widetilde{B}_{\mu,x}$, with $\widetilde{B}_\mu = \widetilde{B}_{\alpha,\beta,\mu}$ given by the smooth L^∞-function

$$\widetilde{B}_\mu(t,x) := 2\sqrt{2}\arctan\left(\frac{G_{\alpha,\beta,\mu}}{F_{\alpha,\beta,\mu}}\right). \tag{8.8.7}$$

(2) For any fixed $t \in \mathbb{R}$, we have $(\widetilde{B}_\mu)_t$ well-defined in the Schwartz class, satisfying

$$B_{\mu,xx} + \widetilde{B}_{\mu,t} + \mu B_\mu^2 + B_\mu^3 = 0. \tag{8.8.8}$$

(3) Let $\mathcal{M}_\mu$ be defined by (8.8.3). Then

$$B_{\mu,x}^2 + \frac{1}{2}B_\mu^4 + \frac{2}{3}\mu B_\mu^3 + 2B_\mu\widetilde{B}_{\mu,t} - 2(\mathcal{M}_\mu)_t = 0. \tag{8.8.9}$$

(4) Finally, let B_1 and B_2 as in (8.8.6) *and $\widetilde{B}_i \equiv \partial_{x_i}\widetilde{B}_\mu$, $\widetilde{B}_{ij} \equiv \partial_{x_i}\partial_{x_j}\widetilde{B}_\mu$, $i,j = 1,2$. Then*

$$\int_{-\infty}^{x} (\widetilde{B}_{12}^2 - \widetilde{B}_{11}\widetilde{B}_{22}) = -\left(\frac{\mu}{3} + B_\mu\right)\widetilde{B}_{11} + \partial_{x_1}^2\partial_x \log\left(G_{\alpha,\beta,\mu}^2 + F_{\alpha,\beta,\mu}^2\right). \tag{8.8.10}$$

Proof. The first item above is a direct consequence of the definition of $B_\mu = B_{\alpha,\beta,\mu}$ in (8.4.1). On the other hand, (8.8.8) is a consequence of (8.8.7) and integration in space (from $-\infty$ to x) of (8.1.1). To obtain (8.8.9) we multiply (8.8.8) by $B_{\mu,x}$ and integrate in space. Finally to prove (8.8.10), since we are working with smooth functions, one has $B_\mu = \widetilde{B}_1 + \widetilde{B}_2$, and also

$$B_1 = \widetilde{B}_{11} + \widetilde{B}_{12}, \qquad \text{and} \qquad B_2 = \widetilde{B}_{12} + \widetilde{B}_{22}.$$

Now, since $\widetilde{B}_{12}^2 = B_1^2 + \widetilde{B}_{11}^2 - 2B_1\widetilde{B}_{11}$ and $\widetilde{B}_{22} = B_2 - B_1 + \widetilde{B}_{11}$, we get

$$\begin{aligned} (\widetilde{B}_{12}^2 - \widetilde{B}_{11}\widetilde{B}_{22}) &= B_1^2 + \widetilde{B}_{11}^2 - 2B_1\widetilde{B}_{11} - \widetilde{B}_{11}(B_2 - B_1 + \widetilde{B}_{11})\\ &= B_1^2 - \widetilde{B}_{11}(B_1 + B_2) = B_1^2 - \widetilde{B}_{11}\partial_x(\widetilde{B}_1 + \widetilde{B}_2) = B_1^2 - \widetilde{B}_{11}B_{\mu,x}. \end{aligned} \tag{8.8.11}$$

Now integrating (8.8.11) in x, we obtain:

$$\int_{-\infty}^{x} (\widetilde{B}_{12}^2 - \widetilde{B}_{11}\widetilde{B}_{22}) = \int_{-\infty}^{x} (B_1^2 - \widetilde{B}_{11} B_{\mu,x})$$
$$= \int_{-\infty}^{x} B_1^2 - \left(B_\mu \widetilde{B}_{11}|_{-\infty}^{x} - \int_{-\infty}^{x} B_\mu B_{11} \right).$$

Using that $B_\mu B_{11} = \partial_{x_1}^2(\frac{1}{2}B_\mu^2) - B_1^2$, the last integral above simplifies as follows:

$$\int_{-\infty}^{x} B_1^2 - \left(B_\mu \widetilde{B}_{11}|_{-\infty}^{x} - \int_{-\infty}^{x} B_\mu B_{11} \right)$$
$$= \int_{-\infty}^{x} B_1^2 - \left(B_\mu \widetilde{B}_{11}|_{-\infty}^{x} - \int_{-\infty}^{x} \left(\partial_{x_1}^2 \left(\frac{1}{2} B_\mu^2 \right) - B_1^2 \right) \right)$$
$$= -B_\mu \widetilde{B}_{11}|_{-\infty}^{x} + \partial_{x_1}^2 \int_{-\infty}^{x} \frac{1}{2} B_\mu^2 = -B_\mu \widetilde{B}_{11} + \partial_{x_1}^2 \mathcal{M}_\mu[t,x].$$

Now, using (8.8.3), we get

$$\partial_{x_1}^2 \mathcal{M}_\mu = \partial_{x_1}^2 \partial_x \log\left(G_{\alpha,\beta,\mu}^2 + F_{\alpha,\beta,\mu}^2 \right) - \frac{\mu}{3} \widetilde{B}_{11}.$$

Finally, substituting this derivative in the last equation above, we obtain the desired simplification. The proof is complete. □

We compute now the energy (8.3.1) of any Gardner breather solution.

Lemma 8.8.6. *Let $B_\mu = B_{\alpha,\beta,\mu}$ be any Gardner breather, for α, β, μ as in definition (8.4.1). Then the energy of B_μ is*

$$E_\mu[B_\mu] := \frac{4}{3}\beta\gamma + \frac{8}{9}\beta\mu^2 + \frac{4\sqrt{2}}{27}\mu^3 \arctan\left(\frac{2\sqrt{2}\mu\beta}{3\Delta} \right). \tag{8.8.12}$$

Remark 8.8.7. Note that in the Gardner case, in comparison with the mKdV case ($\mu = 0$), the sign of the energy is dictated by a nonlinear balance among the velocity γ and the μ terms.

Proof. First of all, let us prove the following reduction

$$E_\mu[B_\mu](t) = \frac{1}{3} \int_{\mathbb{R}} \left((\mathcal{M}_\mu)_t(t,x) - \frac{\mu}{3} B_\mu^3(t,x) \right) dx. \tag{8.8.13}$$

Indeed, we multiply (8.8.8) by B_μ and integrate in space: we get

$$\int_{\mathbb{R}} B_{\mu,x}^2 = \int_{\mathbb{R}} B_\mu \widetilde{B}_{\mu,t} + \mu \int_{\mathbb{R}} B_\mu^3 + \int_{\mathbb{R}} B_\mu^4.$$

On the other hand, integrating (8.8.9),

$$\int_{\mathbb{R}} B_{\mu,x}^2 + \frac{1}{2} \int_{\mathbb{R}} B_\mu^4 + \frac{2}{3}\mu \int_{\mathbb{R}} B_\mu^3 + 2 \int_{\mathbb{R}} B_\mu \widetilde{B}_{\mu,t} - 2 \int_{\mathbb{R}} (\mathcal{M}_\mu)_t = 0.$$

From these two identities, we get

$$\int_{\mathbb{R}} B_\mu^4 = \frac{4}{3}\int_{\mathbb{R}} (\mathcal{M}_\mu)_t - 2\int_{\mathbb{R}} B_\mu \widetilde{B}_{\mu,t} - \frac{10}{9}\mu \int_{\mathbb{R}} B_\mu^3,$$

and therefore

$$\int_{\mathbb{R}} B_{\mu,x}^2 = \frac{4}{3}\int_{\mathbb{R}} (\mathcal{M}_\mu)_t - \int_{\mathbb{R}} B_\mu \widetilde{B}_{\mu,t} - \frac{\mu}{9}\int_{\mathbb{R}} B_\mu^3.$$

Finally, substituting the last two identities into (8.3.1), we get (8.8.13), as desired.

Now we prove (8.8.12). From (8.8.3), we have that

$$\mathcal{M}_\mu(t,x) = 2\beta + \partial_x \log\left(G_{\alpha,\beta,\mu}^2 + F_{\alpha,\beta,\mu}^2\right) - \frac{\mu}{3}\widetilde{B}_\mu,$$

and hence,

$$(\mathcal{M}_\mu)_t(t,x) = \partial_x\partial_t \log\left(G_{\alpha,\beta,\mu}^2 + F_{\alpha,\beta,\mu}^2\right) - \frac{\mu}{3}\widetilde{B}_{\mu,t}.$$

Now substituting in the energy (8.8.13), remembering the identity (8.8.8) and the explicit expression for $\mathcal{M}_\mu[B_\mu]$ in (8.8.2), we get

$$\begin{aligned}
E_\mu[B_\mu](t) &= \frac{1}{3}\int_{\mathbb{R}}\left((\mathcal{M}_\mu)_t(t,x) - \frac{\mu}{3}B_\mu^3\right)dx \\
&= \frac{1}{3}\int_{\mathbb{R}}\left(\partial_x\partial_t \log\left(G_{\alpha,\beta,\mu}^2 + F_{\alpha,\beta,\mu}^2\right) - \frac{\mu}{3}\widetilde{B}_{\mu,t} - \frac{\mu}{3}B_\mu^3\right)dx \\
&= \frac{1}{3}\int_{\mathbb{R}}\left(\partial_x\partial_t \log\left(G_{\alpha,\beta,\mu}^2 + F_{\alpha,\beta,\mu}^2\right) + \frac{\mu}{3}[B_{\mu,xx} + \mu B_\mu^2 + B_\mu^3] - \frac{\mu}{3}B_\mu^3\right)dx \\
&= \frac{1}{3}\int_{\mathbb{R}}\left(\partial_x\partial_t \log\left(G_{\alpha,\beta,\mu}^2 + F_{\alpha,\beta,\mu}^2\right) + \frac{\mu}{3}B_{\mu,xx} + \frac{\mu^2}{3}B_\mu^2\right)dx \\
&= \left(\frac{1}{3}\partial_t \log\left(G_{\alpha,\beta,\mu}^2 + F_{\alpha,\beta,\mu}^2\right) + \frac{\mu}{9}B_{\mu,x}\right)\Big|_{-\infty}^{+\infty} + \frac{\mu^2}{9}\int_{\mathbb{R}} B_\mu^2 dx \\
&= \frac{4}{3}\beta\gamma + 8\beta\frac{\mu^2}{9} + \frac{4\sqrt{2}}{27}\mu^3 \arctan\left(\frac{2\sqrt{2}\mu\beta}{3\Delta}\right).
\end{aligned}$$

□

Corollary 8.8.8. *Let $B_\mu = B_{\alpha,\beta,\mu}$ be any Gardner breather. Then*

$$\begin{aligned}
\partial_\alpha E_\mu[B_\mu] &= 8\alpha\beta\left(1 - \frac{4\mu^4}{81\Delta^2 + 72\mu^2\beta^2}\right), \\
\partial_\beta E_\mu[B_\mu] &= 4(\alpha^2 - \beta^2) + \frac{8}{9}\mu^2\left(\frac{9\Delta^2 + 2\mu^2\Delta + 4\mu^2\beta^2}{9\Delta^2 + 8\mu^2\beta^2}\right).
\end{aligned} \tag{8.8.14}$$

Remark 8.8.9. Note that the condition $\alpha = \beta$ is equivalent to the identity

$$\partial_\beta E_\mu[B_\mu] = \frac{72\mu^2\beta^4}{81\beta^4 + \mu^4},$$

which is always positive, on the contrary to the mKdV breather case ($\mu = 0$). On the other hand the similar identity (8.8.1) for the energy of Gardner solitons can not vanish for any μ.

8.9 Nonlinear stationary equation for Gardner breathers

Our aim in this section is to prove that any Gardner breather solution B_μ satisfies a suitable stationary, elliptic equation, similarly as in the mKdV case.

Theorem 8.9.1. *Let $B_\mu \equiv B_{\alpha,\beta,\mu}$ be any Gardner breather* (8.4.1). *Then, for any fixed $t \in \mathbb{R}$, B_μ satisfies the nonlinear stationary equation*

$$\begin{aligned} J_\mu[B_\mu] := {} & B_{\mu,4x} - 2(\beta^2-\alpha^2)(B_{\mu,xx} + \mu B_\mu^2 + B_\mu^3) \\ & +(\alpha^2+\beta^2)^2 B_\mu + 5B_\mu B_{\mu,x}^2 + 5B_\mu^2 B_{\mu,xx} \\ & +\frac{3}{2}B_\mu^5 + \frac{5}{3}\mu B_{\mu,x}^2 + \frac{10}{3}\mu B_\mu B_{\mu,xx} + \frac{10}{9}\mu^2 B_\mu^3 + \frac{5}{2}\mu B_\mu^4 = 0. \end{aligned} \tag{8.9.1}$$

Remark 8.9.2. This identity can be seen as the *nonlinear stationary equation* satisfied by any Gardner breather profile (8.4.1), and therefore it is independent of time and translation parameters $x_1, x_2 \in \mathbb{R}$. One can compare with the Gardner soliton solution $Q_{c,\mu}(x - ct - x_0)$, which satisfies the standard elliptic equation (2.10.2), obtained as the first variation of the H^1 Weinstein functional (2.10.4). Moreover note that (8.9.1) and (8.4.1) reduces to [AlMu13, eqn. (5.2)] and (4.2.1) (up to translations) when $\mu = 0$.

Although the shift parameters x_1, x_2 are chosen independently of time, a simple argument ensures that the previous Theorem still holds under time-dependent, translation parameters $x_1(t)$ and $x_2(t)$.

Corollary 8.9.3. *Let $B^0_{\alpha,\beta,\mu} = B^0_{\alpha,\beta,\mu}(t,x;0,0)$ be any Gardner breather as in* (8.4.1), *and $x_1(t)$, $x_2(t) \in \mathbb{R}$ two continuous functions, defined for all t in a given interval. Consider the modified breather*

$$B_{\alpha,\beta,\mu}(t,x) := B^0_{\alpha,\beta,\mu}(t,x;x_1(t),x_2(t)), \qquad (\textit{cf.}\ (8.4.1)).$$

Then $B_{\alpha,\beta,\mu}$ satisfies (8.9.1), *for all t in the considered interval.*

Proof. A direct consequence of the invariance of the equation (8.9.1) under spatial translations. □

8.10 Variational characterization of Gardner breathers

In this section we introduce a new H^2-Lyapunov functional for the Gardner equation (1.4.2). Consider $w_0 \in H^2(\mathbb{R})$ and let $w = w(t) \in H^2(\mathbb{R})$ be the associated local-in-time solution of the Cauchy problem associated to (1.4.2), with initial condition

$w(0) = w_0$. We first remember the H^2-functional introduced before (8.3.2)

$$\begin{aligned} F_\mu[w](t) := &\frac{1}{2}\int_{\mathbb{R}} w_{xx}^2 dx - \frac{5}{3}\mu\int_{\mathbb{R}} w w_x^2 dx + \frac{5}{18}\mu^2\int_{\mathbb{R}} w^4 dx \\ &- \frac{5}{2}\int_{\mathbb{R}} w^2 w_x^2 dx + \frac{1}{2}\mu\int_{\mathbb{R}} w^5 + \frac{1}{4}\int_{\mathbb{R}} w^6 dx. \end{aligned} \tag{8.10.1}$$

Lemma 8.10.1. *Let w be a local (in time) H^2-solution of the Gardner equation* (1.4.2) *with initial data w_0. Then the functional $F_\mu[w](t)$* (8.10.1) *is a conserved quantity. Moreover, w is a global (in time) H^2-solution.*

The existence of this last conserved quantity is a deep consequence of the *integrability property* for the Gardner equation. In particular, it is not present in a general, nonintegrable gKdV equation. The proof of Lemma 8.10.1 is an easy computation, that we partially leave to the reader.

Moreover, and notably, we are also able to compute (8.10.1) at B_μ, namely

Lemma 8.10.2. *Let $B_\mu = B_{\alpha,\beta,\mu}$ be any Gardner breather, for α, β, μ as in definition* (8.4.1)*. Then we have that*

$$\begin{aligned} F_\mu[B_\mu] := &\frac{4}{15}\left(3\beta(\beta^4 - 10\beta^2\alpha^2 + 5\alpha^4) - \frac{10}{9}\mu^2\beta\left(\beta^2 - 3\alpha^2 - \frac{2}{3}\mu^2\right)\right) \\ &+ \frac{8\sqrt{2}}{243}\mu^5 \arctan\left(\frac{2\sqrt{2}\mu\beta}{3\Delta}\right). \end{aligned} \tag{8.10.2}$$

Using the functional $F_\mu[w]$ (8.10.1), we build a new H^2-Lyapunov functional specifically associated to the breather solution. Let $B_\mu = B_{\alpha,\beta,\mu}$ be any Gardner breather, and $t \in \mathbb{R}$, and $M[w]$ and $E_\mu[w]$ given in (4.1.4) and (2.10.3) respectively. We define

$$\mathcal{H}_\mu[w](t) := F_\mu[w](t) + 2(\beta^2 - \alpha^2)E_\mu[w](t) + (\alpha^2 + \beta^2)^2 M[w](t). \tag{8.10.3}$$

Therefore, $\mathcal{H}_\mu[w]$ is a real-valued conserved quantity, well-defined for H^2-solutions of (1.4.2). Note additionally that the functionals $\mathcal{H}_{\mu=0}$ (4.9.24) and $\mathcal{H}_\mu$ for the mKdV and Gardner equations are surprisingly the same.

Lemma 8.10.3. *Let $B_\mu = B_{\alpha,\beta,\mu}$ be any Gardner breather, for α, β, μ as in definition* (8.4.1)*. Then we have that*

$$\mathcal{H}_\mu[B_\mu] := h_1 + h_2 \frac{2\sqrt{2}}{3}\mu \arctan\left(\frac{2\sqrt{2}\mu\beta}{3\Delta}\right),$$

with

$$
\begin{aligned}
h_1 &= \frac{8\beta}{15}\left(4\beta^4 + 20\alpha^2\beta^2 + \frac{5}{9}\mu^2\left(5\beta^2 - 3\alpha^2 + \frac{2}{3}\mu^2\right)\right),\\
h_2 &= \left((\alpha^2+\beta^2)^2 + \frac{4}{9}\mu^2\left(\beta^2 - \alpha^2 + \frac{\mu^2}{9}\right)\right).
\end{aligned} \tag{8.10.4}
$$

Proof. Collecting the mass (8.8.2), the energy (8.8.12) and F_μ at B_μ (8.10.2), and substituting at (8.10.3), we get (8.10.4). □

Let $z \in H^4(\mathbb{R})$, and B_μ be any Gardner breather, with shift parameters x_1, x_2. We define $\mathcal{L}_\mu$ as the linearized operator associated to B_μ, i.e. the bilinear operator obtained after a linearization of the Lyapunov functional (8.10.3) at the Gardner breather B_μ, as follows:

$$
\begin{aligned}
\mathcal{L}_\mu z := {}& z_{4x} - 2(\beta^2-\alpha^2)z_{xx} + (\alpha^2+\beta^2)^2 z\\
&+ 5B_\mu^2 z_{xx} + 10B_\mu B_{\mu,x}z_x\\
&+ \left(5B_{\mu,x}^2 + 10B_\mu B_{\mu,xx} + \frac{15}{2}B_\mu^4 - 6(\beta^2-\alpha^2)B_\mu^2\right)z\\
&+ \frac{10}{3}\mu B_\mu z_{xx} + \frac{10}{3}\mu B_{\mu,x}z_x\\
&+ \mu\left(10B_\mu^3 - 4(\beta^2-\alpha^2)B_\mu + \frac{10}{3}B_{\mu,xx} + \frac{10}{3}\mu B_\mu^2\right)z.
\end{aligned} \tag{8.10.5}
$$

Let us also define

$$
\mathcal{Q}_\mu := \frac{1}{2}\int z\mathcal{L}_\mu z. \tag{8.10.6}
$$

The following concept, associated to a bilinear operator like $\mathcal{Q}_\mu$ (8.10.6), is standard and it will be useful for us.

Definition 8.10.4. Any nonzero $z \in H^2$ is said to be a positive (null, negative) direction for $\mathcal{Q}_\mu$ if we have $\mathcal{Q}_\mu[z] > 0$ $(= 0, < 0)$.

One has the following

Lemma 8.10.5. *Gardner breathers B_μ (8.4.1) are critical points of the Lyapunov functional $\mathcal{H}_\mu$ (8.10.3). In fact, for any $z \in H^2(\mathbb{R})$ with sufficiently small H^2-norm, and $B_\mu = B_{\alpha,\beta,\mu}$ any Gardner breather solution, then, for all $t \in \mathbb{R}$, one has*

$$
\mathcal{H}_\mu[B_\mu + z] - \mathcal{H}_\mu[B_\mu] = \frac{1}{2}\mathcal{Q}_\mu[z] + \mathcal{N}_\mu[z],
$$

with $\mathcal{Q}_\mu$ being the quadratic form defined in (8.10.6), and $\mathcal{N}_\mu[z]$ satisfying $|\mathcal{N}_\mu[z]| \le K\|z\|^3_{H^2(\mathbb{R})}$.

Proof. Exercise. □

8.11 Spectral properties around Gardner breathers

Essential for the proof of the main result of this section, Theorem 11.7.1, is the spectral study of the associated linear operator $\mathcal{L}_\mu$ appearing from Theorem 8.9.1 and particularly, equation (8.9.1). Hence, in this section we describe the spectrum of this operator. More precisely, our main purpose is to find a suitable coercivity property, independently of the nature of scaling parameters. The main result of this section is contained in Proposition 8.11.11. Part of the analysis carried out in this section has been previously introduced for solitons by Lax [Lax68], Maddocks-Sachs [MaSa93] and for mKdV breathers by Alejo-Muñoz [AlMu13], so we follow their arguments adapted to the Gardner breather case, sketching several proofs.

Lemma 8.11.1. *$\mathcal{L}_\mu$ is a linear, unbounded operator in $L^2(\mathbb{R})$, with dense domain $H^4(\mathbb{R})$. Moreover, $\mathcal{L}_\mu$ is self-adjoint.*

A consequence of the previous result is the fact that the spectrum of $\mathcal{L}_\mu$ is real-valued. Furthermore, the following Lemma describes the continuous spectrum of $\mathcal{L}_\mu$.

Lemma 8.11.2. *Let α, β, μ as in definition* (8.4.1). *The operator $\mathcal{L}_\mu$ is a compact perturbation of the constant coefficients operator*

$$\mathcal{L}_0 z := z_{4x} - 2(\beta^2 - \alpha^2) z_{xx} + (\alpha^2 + \beta^2)^2 z.$$

In particular, the continuous spectrum of $\mathcal{L}_\mu$ is the closed interval $[(\alpha^2+\beta^2)^2, +\infty)$ in the case $\beta \geq \alpha$, and $[4\alpha^2\beta^2, +\infty)$ in the case $\beta < \alpha$. No embedded eigenvalues are contained in this region. The eigenvalue zero is isolated.

Proof. This result is a consequence of the Weyl Theorem on continuous spectrum. Let us note that the nonexistence of embedded eigenvalues is consequence of the rapidly decreasing character of the potentials involved in the definition of $\mathcal{L}_\mu$.

The isolatedness of the zero eigenvalue is a direct consequence of standard elliptic estimates for the eigenvalue problem associated to $\mathcal{L}_\mu$, corresponding uniform convergence on compact subsets of $\mathbb{R}$, and the nondegeneracy of the kernel associated to $\mathcal{L}_\mu$. □

Now, remembering the definition (4.9.4) of the two directions associated to spatial translations B_1, B_2, it is easy to see the following:

Lemma 8.11.3. *For each $t \in \mathbb{R}$, one has*

$$\ker \mathcal{L}_\mu = \operatorname{span}\{B_1(t; x_1, x_2), B_2(t; x_1, x_2)\}.$$

Proof. From Theorem 8.9.1, one has that $\partial_{x_1} J_\mu[B_\mu] = \partial_{x_2} J_\mu[B_\mu] \equiv 0$. Writing down these identities, we obtain

$$\mathcal{L}_\mu B_1(t;x_1,x_2) = \mathcal{L}_\mu B_2(t;x_1,x_2) = 0, \tag{8.11.1}$$

with $\mathcal{L}_\mu$ the linearized operator defined in (8.10.5) and B_1, B_2 defined in (4.9.4). A direct analysis involving ordinary differential equations shows that the null space of $\mathcal{L}_0$ is spawned by functions of the type

$$e^{\pm\beta x}\cos(\alpha x), \quad e^{\pm\beta x}\sin(\alpha x), \quad \alpha,\beta > 0,$$

(note that this set is linearly independent). Among these four functions, there are only two L^2-integrable ones in the semi-infinite line $[0,+\infty)$. Therefore, the null space of $\mathcal{L}_\mu|_{H^4(\mathbb{R})}$ is spanned by at most two L^2-functions. Finally, comparing with (8.11.1), we have the desired conclusion. □

We consider now the natural modes associated to the scaling parameters, which are the best candidates to generate negative directions for the related quadratic form defined by $\mathcal{L}_\mu$. Recall the definitions of $\Lambda_\alpha B_\mu$ and $\Lambda_\beta B_\mu$ introduced in (4.9.1). For these two directions, one has the following

Lemma 8.11.4. *Let $B_\mu = B_{\alpha,\beta,\mu}$ be any Gardner breather. Consider the scaling directions $\Lambda_\alpha B_\mu$ and $\Lambda_\beta B_\mu$ introduced in (4.9.1). Then, given $\alpha,\beta > 0$ and $\forall \mu \in (0,\mu_{max})$, we have*

$$\int_{\mathbb{R}} \Lambda_\alpha B_\mu \, \mathcal{L}_\mu \Lambda_\alpha B_\mu = 32\alpha^2\beta\left(1 + \frac{2\mu^2\Delta}{9\Delta^2 + 8\mu^2\beta^2}\right) > 0, \tag{8.11.2}$$

and

$$\begin{aligned}&\int_{\mathbb{R}} \Lambda_\beta B_\mu \, \mathcal{L}_\mu \Lambda_\beta B_\mu \\ &= -16\beta\left(\alpha^2 - \beta^2 + \left(\alpha^2 + \beta^2 + \frac{2}{9}\mu^2\right)\left(\frac{9\Delta^2 + 2\mu^2\Delta + 4\mu^2\beta^2}{9\Delta^2 + 8\mu^2\beta^2}\right)\right) < 0.\end{aligned} \tag{8.11.3}$$

Proof. From (8.9.1), we get after derivation with respect to α and β,

$$\begin{aligned}\mathcal{L}_\mu \Lambda_\alpha B_\mu &= -4\alpha[B_{\mu,xx} + B_\mu^3 + \mu B_\mu^2 + (\alpha^2+\beta^2)B_\mu],\\ \mathcal{L}_\mu \Lambda_\beta B_\mu &= 4\beta[B_{\mu,xx} + B_\mu^3 + \mu B_\mu^2 - (\alpha^2+\beta^2)B_\mu].\end{aligned}$$

We deal with the first identity (8.11.2). Note that from (8.8.4), (2.10.3) and (8.8.14),

$$\begin{aligned}&\int_{\mathbb{R}} \Lambda_\alpha B_\mu \, \mathcal{L}_\mu \Lambda_\alpha B_\mu \\ &= -4\alpha\int_{\mathbb{R}} [B_{\mu,xx} + B_\mu^3 + \mu B_\mu^2 + (\alpha^2+\beta^2)B_\mu]\Lambda_\alpha B_\mu \\ &= 4\alpha\partial_\alpha E_\mu[B_\mu] - 4\alpha(\alpha^2+\beta^2)\partial_\alpha M[B_\mu] \\ &= 32\alpha^2\beta\left(1 - \frac{4}{9}\frac{\mu^4}{9\Delta^2 + 8\mu^2\beta^2} + \frac{2\mu^2(\alpha^2+\beta^2)}{9\Delta^2+8\mu^2\beta^2}\right) = (8.11.2).\end{aligned}$$

Following a similar analysis, we have

$$
\begin{aligned}
&\int_{\mathbb{R}} \Lambda_\beta B_\mu \mathcal{L}_\mu \Lambda_\beta B_\mu \\
&= 4\beta \int_{\mathbb{R}} [B_{\mu,xx} + B_\mu^3 + \mu B_\mu^2 + (\alpha^2+\beta^2)B_\mu]\Lambda_\beta B_\mu \\
&= -4\beta\partial_\beta E_\mu[B_\mu] - 4\beta(\alpha^2+\beta^2)\partial_\beta M[B_\mu] \\
&= -4\beta\left[4(\alpha^2-\beta^2) + \frac{8}{9}\mu^2\left(1+\frac{2\mu^2(\Delta-2\beta^2)}{9\Delta^2+8\mu^2\beta^2}\right)\right] \\
&\quad - 4\beta(\alpha^2+\beta^2)4\left(1+\frac{2\mu^2(\Delta-2\beta^2)}{9\Delta^2+8\mu^2\beta^2}\right) = (8.11.3).
\end{aligned}
$$

□

A direct consequence of the previous identities and Corollary 8.8.4, is the following:

Corollary 8.11.5. *With the notation of Lemma 8.11.4 let*

$$
B_{0,\mu} := \frac{\alpha\Lambda_\beta B_\mu + \beta\Lambda_\alpha B_\mu}{8\alpha\beta(\alpha^2+\beta^2)}. \tag{8.11.4}
$$

Then $B_{0,\mu}$ is Schwartz, satisfies $\mathcal{L}_\mu B_{0,\mu} = -B_\mu$ and $\forall \mu \in (0, \mu_{max})$

$$
\int_{\mathbb{R}} B_{0,\mu}B_\mu = \frac{1}{2\beta(\alpha^2+\beta^2)}\left(\frac{9\Delta^2+2\mu^2\Delta+4\beta^2\mu^2}{9\Delta^2+8\mu^2\beta^2}\right) > 0. \tag{8.11.5}
$$

Moreover,

$$
\tfrac{1}{2}\textstyle\int_{\mathbb{R}} B_{0,\mu}\mathcal{L}_\mu B_{0,\mu} < 0. \tag{8.11.6}
$$

Remark 8.11.6. In other words, from Definition 8.10.4, we can see $B_{0,\mu}$ as a negative direction of Q_μ $\forall \mu \in (0, \mu_{max})$. Besides that, $B_{0,\mu}$ is not orthogonal to the breather itself. Note additionally that constants involved in (8.11.5) are independent of time.

Proof. Using (8.11.4), we are lead to understanding the sign of the function

$$
\begin{aligned}
\int_{\mathbb{R}} B_{0,\mu}B_\mu &= \frac{1}{8\alpha\beta(\alpha^2+\beta^2)}\int_{\mathbb{R}}(\alpha\Lambda_\beta B_\mu + \beta\Lambda_\alpha B_\mu)B_\mu \\
&= \frac{1}{8\alpha\beta(\alpha^2+\beta^2)}(\alpha\partial_\beta M[B_\mu] + \beta\partial_\alpha M[B_\mu]) \\
&= \frac{1}{8\alpha\beta(\alpha^2+\beta^2)}\left(4\alpha\left[1+\frac{2\mu^2\Delta}{9\Delta^2+8\mu^2\beta^2}\right] - \frac{16\mu^2\alpha\beta^2}{9\Delta^2+8\mu^2\beta^2}\right) \\
&= (8.11.5),
\end{aligned}
$$

where $\partial_\beta M[B_\mu]$ was computed in (8.8.5). □

Now, in order to prove that $\mathcal{L}_\mu$ possesses, for all time, *only one negative eigenvalue*, we follow the Greenberg and Maddocks-Sachs strategy [Gre91, MaSa93], applied this time to the linear, *oscillatory* operator $\mathcal{L}_\mu$. More specifically, we will use the following

Lemma 8.11.7 (Uniqueness criterion, see also [Gre91, MaSa93]). *Let* $B_\mu = B_{\alpha,\beta,\mu}$ *be any Gardner breather, and* B_1, B_2 *the corresponding kernel of the operator* $\mathcal{L}_\mu$. *Then* $\mathcal{L}_\mu$ *has*

$$\sum_{x\in\mathbb{R}} \dim\ker W[B_1,B_2](t;x)$$

negative eigenvalues, counting multiplicity. Here, W *is the Wronskian matrix of the functions* B_1 *and* B_2,

$$W[B_1,B_2](t;x) := \begin{bmatrix} B_1 & B_2 \\ (B_1)_x & (B_2)_x \end{bmatrix}(t,x) \tag{8.11.7}$$

Proof. This result is essentially contained in [Gre91, Theorem 2.2], where the finite interval case was considered. As shown in several articles (see e.g. [MaSa93, HPZ11]), the extension to the real line is direct and does not require additional efforts. We skip the details. □

In what follows, we compute the Wronskian (8.11.7). The surprising fact is the following greatly simplified new expression for the determinant of (8.11.7) and which generalizes the Wronskian for the mKdV's breather case ($\mu = 0$) (see [AlMu13, Lemma 4.7]):

Lemma 8.11.8. *Let* $B_\mu = B_{\alpha,\beta,\mu}$ *be any Gardner breather,* B_1, B_2 *the corresponding kernel elements defined in* (4.9.4) *and* $D_\mu = F^2_{\alpha,\beta,\mu} + G^2_{\alpha,\beta,\mu}$. *Then*

$$\begin{aligned}\det W[B_1,B_2](t;x) &:= \frac{4\beta^3(\alpha^2+\beta^2)^2[9(\alpha^2+\beta^2)^2 - 4\mu^2(\alpha^2-\mu^2)]}{9\Delta^3 D_\mu^2} \\ &\quad \Bigg(\sinh(2\beta y_2) + \frac{4\beta^2\mu^2\cosh(2\beta y_2)}{9(\alpha^2+\beta^2)^2 - 4\mu^2(\alpha^2-\mu^2)} \\ &\quad - \frac{\beta\Delta(9(\alpha^2+\beta^2)^2 - 2\mu^2(\alpha^2-\beta^2))\sin(2\alpha y_1)}{\alpha(\alpha^2+\beta^2)(9(\alpha^2+\beta^2)^2-4\mu^2(\alpha^2-\mu^2))} \\ &\quad + \frac{4\beta^2\mu^2\Delta\cos(2\alpha y_1)}{(\alpha^2+\beta^2)(9(\alpha^2+\beta^2)^2-4\mu^2(\alpha^2-\mu^2))}\Bigg).\end{aligned} \tag{8.11.8}$$

Proof. We start with a very useful simplification. We claim that

$$\begin{aligned}&\det W[B_1,B_2](x) \\ &= -2(\alpha^2+\beta^2)\Big(-\Big(\frac{\mu}{3} + B_\mu\Big)\widetilde{B}_{11} + \partial^2_{x_1}\partial_x \log\Big(G^2_{\alpha,\beta,\mu} + F^2_{\alpha,\beta,\mu}\Big)\Big),\end{aligned} \tag{8.11.9}$$

with $\widetilde{B}_\mu = \widetilde{B}_\mu(t,x;x_1,x_2)$ defined in (8.8.7), and $\widetilde{B}_j$, $\widetilde{B}_{ij}$, $i,j = 1,2$, as in (8.8.6). In order to prove the above simplification, we start from (8.8.8), and taking derivative with respect to x_1 and x_2, we get

$$(B_1)_{xx} + (\widetilde{B}_1)_t + 3B_\mu^2 B_1 + 2\mu B_\mu B_1 = 0,$$
$$(B_2)_{xx} + (\widetilde{B}_2)_t + 3B_\mu^2 B_2 + 2\mu B_\mu B_2 = 0.$$

Multiplying the first equation above by B_2 and the second by $-B_1$, and adding both equations, we obtain

$$(B_1)_{xx}B_2 - (B_2)_{xx}B_1 + (\widetilde{B}_1)_t B_2 - (\widetilde{B}_2)_t B_1 = 0,$$

that is,

$$((B_1)_x B_2 - (B_2)_x B_1)_x = (\widetilde{B}_2)_t B_1 - (\widetilde{B}_1)_t B_2. \tag{8.11.10}$$

On the other hand, since we are working with smooth functions, one has $B_\mu = \widetilde{B}_1 + \widetilde{B}_2$,

$$B_1 = \widetilde{B}_{11} + \widetilde{B}_{12}, \quad B_2 = \widetilde{B}_{12} + \widetilde{B}_{22},$$

and

$$(\widetilde{B}_1)_t = \delta\widetilde{B}_{11} + \gamma\widetilde{B}_{12}, \quad (\widetilde{B}_2)_t = \delta\widetilde{B}_{12} + \gamma\widetilde{B}_{22}.$$

Substituting into (8.11.10), we get

$$((B_1)_x B_2 - (B_2)_x B_1)_x = (\delta - \gamma)(\widetilde{B}_{12}^2 - \widetilde{B}_{11}\widetilde{B}_{22}).$$

Now, integrating in x and using the nonlinear identity (8.8.10) we get

$$\det W[B_1, B_2](x)$$
$$= -2(\alpha^2+\beta^2)\Big(-\Big(\frac{\mu}{3} + B_\mu\Big)\widetilde{B}_{11} + \partial_{x_1}^2 \partial_x \log\big(G_{\alpha,\beta,\mu}^2 + F_{\alpha,\beta,\mu}^2\big)\Big).$$

Finally to prove (4.9.9), we write explicitly the two terms involved at the r.h.s. of equation above. We will follow notation $G := G_{\alpha,\beta,\mu}$, $G_1 := G_x$, $G_2 := G_{x_1}$, $G_3 := G_{xx_1}$, $G_4 := G_{x_1x_1}$, $G_5 := G_{xx_1x_1}$ and $F := F_{\alpha,\beta,\mu}$, $F_1 := F_x$, $F_2 := F_{x_1}$, $F_3 := F_{xx_1}$, $F_4 := F_{x_1x_1}$, $F_5 := F_{xx_1x_1}$. Hence we get,

$$-(\mu/3 + B_\mu) = -\frac{(\mu/3)D_\mu - 2\sqrt{2}(-F_1G + FG_1)}{D_\mu},$$

$$\begin{aligned}\widetilde{B}_{11} &= \frac{2\sqrt{2}}{D_\mu^2} \\ &\quad \times\big[-G^2(F_4G - 2F_2G_2) - F^2(F_4G + 2F_2G_2) \\ &\qquad +F^3G_4 + FG(2F_2^2 - 2G_2^2 + GG_4)\big],\end{aligned}$$

and finally

$$-(\mu/3+B_\mu)\widetilde{B}_{11} = -2\sqrt{2}\frac{M_1}{D_\mu^3}, \tag{8.11.11}$$

with

$$\begin{aligned} M_1 := &\Big((\mu/3)D_\mu - 2\sqrt{2}(-F_1G+FG_1)\Big)\\ &\times\Big(-G^2(F_4G-2F_2G_2)-F^2(F_4G+2F_2G_2)\\ &\quad + F^3G_4 + FG(2F_2^2-2G_2^2+GG_4)\Big). \end{aligned}$$

Similarly, for the second term in (8.11.9) at the r.h.s. we have

$$\partial_{x_1}^2\partial_x \log\left(G_{\alpha,\beta,\mu}^2+F_{\alpha,\beta,\mu}^2\right) = \frac{M_2}{D_\mu^3}, \tag{8.11.12}$$

with

$$\begin{aligned} M_2 := &\,16(FF_1+GG_1)(FF_2+GG_2)^2\\ &-4D_\mu(FF_1+GG_1)(F_2^2+FF_4+G_2^2+GG_4)\\ &-8D_\mu^2(FF_2+GG_2)(F_1F_2+FF_3+G_1G_2+GG_3)\\ &+2D_\mu^2(2F_2F_3+F_1F_4+FF_5+2G_2G_3+G_1G_4+GG_5). \end{aligned}$$

Therefore putting together (8.11.11) and (8.11.12), we get

$$-(\mu/3+B_\mu)\widetilde{B}_{11} + \partial_{x_1}^2\partial_x\log\left(G_{\alpha,\beta,\mu}^2+F_{\alpha,\beta,\mu}^2\right) = \frac{-2\sqrt{2}M_1+M_2}{D_\mu^3}.$$

Indeed, it is possible to see that the above numerator reduces to

$$-2\sqrt{2}M_1 + M_2 = 2D_\mu M_3,$$

where M_3 is given by

$$\begin{aligned} M_3 := &\,F^2[-2F_2F_3-F_1F_4+2G_2G_3-3G_1G_4\\ &\quad +GG_5+\sqrt{2}(\mu/3)GF_4+2\sqrt{2}(\mu/3)F_2G_2]\\ &+F^3(F_5-\sqrt{2}(\mu/3)G_4)\\ &+F[F_5G^2+2F_4GG_1-4F_3GG_2+4F_2G_1G_2\\ &\quad -4F_2GG_3-\sqrt{2}(\mu/3)(2F_2^2G-2GG_2^2+G^2G_4)\\ &\quad +2F_1(F_2^2-G_2^2+GG_4)]\\ &+G[-3F_1F_4G-2F_2^2G_1+2G_1G_2^2-2GG_2G_3-GG_1G_4\\ &\quad +G^2G_5+\sqrt{2}(\mu/3)F_4G^2\\ &\quad +2F_2(F_3G+2F_1G_2-\sqrt{2}(\mu/3)GG_2)]. \end{aligned} \tag{8.11.13}$$

We verify, using the symbolic software *Mathematica*, that after substituting $F's$ and $G's$ terms explicitly in M_3 and lengthy rearrangements, (8.11.13) simplifies as follows:

$$\begin{aligned}2M_3 =& \frac{-2\beta^3(\alpha^2+\beta^2)(9(\alpha^2+\beta^2)^2-4\mu^2(\alpha^2-\mu^2))}{9\Delta^3}\\ &\times\left(\sinh(2\beta y_2)+\frac{4\beta^2\mu^2\cosh(2\beta y_2)}{9(\alpha^2+\beta^2)^2-4\mu^2(\alpha^2-\mu^2)}\right.\\ &\quad-\frac{\beta\Delta(9(\alpha^2+\beta^2)^2-2\mu^2(\alpha^2-\beta^2))\sin(2\alpha y_1)}{\alpha(\alpha^2+\beta^2)(9(\alpha^2+\beta^2)^2-4\mu^2(\alpha^2-\mu^2))}\\ &\quad\left.+\frac{4\beta^2\mu^2\Delta\cos(2\alpha y_1)}{(\alpha^2+\beta^2)(9(\alpha^2+\beta^2)^2-4\mu^2(\alpha^2-\mu^2))}\right).\end{aligned}$$

Finally we get

$$\begin{aligned}-(\mu+B_\mu)\widetilde{B}_{11}+\partial_{x_1}^2\partial_x\log\left(G^2_{\alpha,\beta,\mu}+F^2_{\alpha,\beta,\mu}\right) &= \frac{-2\sqrt{2}M_1+M_2}{D_\mu^3}\\ &= \frac{2D_\mu M_3}{D_\mu^3} \overset{(8.11.8)}{=} \frac{\ }{-2(\alpha^2+\beta^2)}.\end{aligned}$$

□

Proposition 8.11.9. *The operator $\mathcal{L}_\mu$ defined in* (8.10.5) *and for every $\mu \in (0,\mu_{max})$ has a unique negative eigenvalue $-\lambda_0^2 < 0$, of multiplicity one, and $\lambda_0 = \lambda_0(\alpha,\beta,\mu,x_1,x_2,t)$.*

Proof. The proof is very similar to the mKdV case. □

We consider now some standard remarks. We can reduce the spectral problem to another independent of time. Indeed, after translation and redefinition of the shift parameters x_1 and x_2 we can assume that

$$B_\mu = B_{\alpha,\beta,\mu}(0,x;x_1,0), \quad x_1 \in \left[0,\frac{2\pi}{\alpha}\right].$$

In what follows we assume that B_μ is given by the previous formula.

Let $B_{-1} \in \mathcal{S}\setminus\{0\}$ be an eigenfunction associated to the unique negative eigenvalue of the operator $\mathcal{L}_\mu$, as stated in Proposition 8.11.9. We assume that B_{-1} has unit L^2-norm, so B_{-1} is now unique. In particular, one has $\mathcal{L}_\mu B_{-1} = -\lambda_0^2 B_{-1}$. It is clear from Proposition 8.11.9 and Lemma 8.11.3 that the following result holds.

Lemma 8.11.10. *There exists a continuous function $\nu_0 = \nu_0(\alpha,\beta,\mu)$, well-defined and positive for all $\alpha,\beta > 0$, with $\mu \in (0,\mu_{max})$, and such that, for all $z_0 \in H^2(\mathbb{R})$ satisfying*

$$\int_{\mathbb{R}} z_0 B_{-1} = \int_{\mathbb{R}} z_0 B_1 = \int_{\mathbb{R}} z_0 B_2 = 0,$$

then

$$\mathcal{Q}_\mu[z_0] \geq \nu_0 ||z_0||^2_{H^2(\mathbb{R})}. \tag{8.11.14}$$

Proof. The existence of a *positive* constant $\nu_0 = \nu_0(\alpha, \beta, \mu, x_1)$ such that (8.11.14) is satisfied is similar to the mKdV case. Moreover, thanks to the periodic character of the variable x_1, and the nondegeneracy of the kernel, we obtain a uniform, positive bound independent of x_1, x_2 and t, still denoted ν_0. The proof is complete. □

Since B_{-1} is difficult to work with, we look for an easier version of the previous result. We can easily prove, as in [AlMu13, Proposition (4.11)], that the eigenfunction B_{-1} associated to the negative eigenvalue of $\mathcal{L}_\mu$ can be replaced by the breather itself, which has better behavior in terms of error controlling, unlike the first eigenfunction. This simple fact allows us to prove the nonlinear stability result as in the standard approach, without using scaling modulations. Recall that using the first eigenfunction as orthogonality condition does not guarantee a suitable control on the scaling modulation parameter, because the control given by this direction might not be good enough to close the stability estimates. However, the breather can be used as an alternative direction, and all these previous arguments remain valid, exactly as in [AlMu13], provided the Weinstein's sign condition

$$\int_{\mathbb{R}} B_{0,\mu} B_\mu > 0 \qquad \left(\text{or equivalently} \int_{\mathbb{R}} B_{0,\mu} \mathcal{L}_\mu B_{0,\mu} < 0\right), \tag{8.11.15}$$

do hold. Note that this precisely is what happens in (8.11.6) when $\mu \in (0, \mu_{max})$.

Proposition 8.11.11. *Let $B_\mu = B_{\alpha,\beta,\mu}$ be any Gardner breather, and B_1, B_2 the corresponding kernel of the associated operator $\mathcal{L}_\mu$. Let $\alpha, \beta > 0$ and $\mu \in (0, \mu_{max})$. There exists $\sigma_0 > 0$, depending on α, β, μ only, such that, for any $z \in H^2(\mathbb{R})$ satisfying*

$$\int_{\mathbb{R}} B_1 z = \int_{\mathbb{R}} B_2 z = 0,$$

one has

$$\mathcal{Q}_\mu[z] \geq \sigma_0 ||z||^2_{H^2(\mathbb{R})} - \frac{1}{\sigma_0}\left(\int_{\mathbb{R}} z B_\mu\right)^2.$$

Proof. The proof follows a similar strategy as stated in [AlMu13]. □

Theorem 8.11.12 (H^2-stability of Gardner breathers). *Let $\alpha, \beta \in \mathbb{R}\backslash\{0\}$ and $\mu \in (0, \mu_{\max})$. Moreover let $B_\mu = B_{\alpha,\beta,\mu}$ be any Gardner breather* (8.4.1). *There* *exist positive parameters η_0, A_0, depending on α, β and μ, such that the following holds. Consider $w_0 \in H^2(\mathbb{R})$, and assume that there exists $\eta \in (0, \eta_0)$ such that*

$$\|w_0 - B_\mu(t = 0; 0, 0)\|_{H^2(\mathbb{R})} \leq \eta.$$

Then there exist $x_1(t), x_2(t) \in \mathbb{R}$ such that the solution $w(t)$ of the Cauchy problem for the Gardner equation (1.4.2), with initial data w_0, satisfies

$$\sup_{t\in\mathbb{R}} \left\|w(t) - B_\mu(t; x_1(t), x_2(t))\right\|_{H^2(\mathbb{R})} \leq A_0\eta,$$

with

$$\sup_{t\in\mathbb{R}} |x_1'(t)| + |x_2'(t)| \leq KA_0\eta,$$

for a constant $K > 0$.

Proof of Theorem 8.11.12. The proof of this result is completely similar to the proof of the H^2–stability of mKdV breathers [AlMu13, Theorem (6.1)], after following the same steps, and once we guarantee coercivity of the bilinear, form Q_μ (see Proposition 8.11.11 and Lemma 8.11.10), and the nonvanishing of the denominator $\int_{\mathbb{R}} B_{0,\mu}B_\mu + Q_\mu[b_0]$ appearing in (4.9.21). □

8.12 5th order Gardner equation

In this section, we shall quickly review some recent results concerning the *focusing* 5th order Gardner equation

$$\begin{aligned} &w_t + w_{5x} + 10\mu^2 w_{3x} + \mathcal{K}_\mu(w) = 0, \quad \mu \in \mathbb{R}^+, \\ \mathcal{K}_\mu(w) :=\ &20\mu w w_{3x} + 10w^2 w_{3x} + 120\mu^3 w w_x + 180\mu^2 w^2 w_x \\ &+ 120\mu w^3 w_x + 10w_x^3 + 40\mu w_x w_{xx} + 40 w w_x w_{xx} + 30 w^4 w_x. \end{aligned} \tag{8.12.1}$$

This higher order Gardner equation can be obtained from the corresponding 5th order *focusing* modified Korteweg-de Vries equation (shortly, 5th mKdV)

$$u_t + (u_{4x} + 10uu_x^2 + 10u^2 u_{xx} + 6u^5)_x = 0, \tag{8.12.2}$$

when one formally considers mKdV solutions of the form $u(t, x) = \mu + w(t, x)$, with $\mu \in \mathbb{R}^+$ and a suitable spatial translation. When $\mu \ll 1$, equation (8.12.1) can be seen as a perturbed model of the 5th mKdV (8.12.2). See [Ka69, KiOl92, MaSm90, Olv84] and references therein for further details on physical applications.

The 5th order Gardner equation (8.12.1), as well as the 5th mKdV equation, is a well-known *completely integrable* model, with infinitely many conservation laws and well-known (long-time) asymptotic behavior of its solutions obtained with the help of the ISM [GSP10].

Concerning explicit solutions of higher order mKdV and Gardner models, Matsuno [Mat80] proved the existence and built explicitly the N-soliton solution of the focusing mKdV hierarchy of equations by using inverse scattering techniques and the bilinear Hirota decomposition. Gomes et al. [GFZ11] dealt with the defocusing mKdV with NVBC and the associated defocusing Gardner hierarchy, showing multisolitonic structures. Unfortunately, many of the solutions they obtained are singular solutions (up to the kink which is in L^∞).

The 5th order Gardner equation (8.12.1), as a completely integrable system, has an infinite set of conserved quantities. Indeed some of the (first) standard conservation laws of the (8.12.1) are the *mass*

$$M[w](t) := \frac{1}{2}\int_{\mathbb{R}} w^2(t,x)dx = M[w](0), \tag{8.12.3}$$

the *energy*

$$E_\mu[w](t) := \int_{\mathbb{R}} \left(\frac{1}{2}w_x^2 - 2\mu w^3 - \frac{1}{2}w^4\right)(t,x)dx = E_\mu[w](0), \tag{8.12.4}$$

and the *higher order energy*, defined respectively in $H^2(\mathbb{R})$

$$\begin{aligned} E_{5\mu}[w](t) :=& \int_{\mathbb{R}} \left(\frac{1}{2}w_{xx}^2 - 10\mu w w_x^2 + 10\mu^2 w^4 - 5w^2 w_x^2 + 6\mu w^5 + w^6\right)(t,x)dx \\ =& E_{5\mu}[w](0). \end{aligned} \tag{8.12.5}$$

We are interested in the regularity properties of the 5th order Gardner equation (8.12.1) and long time behavior of H^2 global solutions to (8.12.1), in particular, local and global existence of solutions for the 5th order Gardner equation (8.12.1), together with a characterization of its weak-ill-posedness properties, and the orbital stability result for new solutions discovered for this model.

8.12.0.1 *Well-posedness theory*

The nonlinearity of (8.12.1) consists of several terms which break the balance with the 5th order linear dispersive part of (8.12.1). The notion of the well-posedness, which is taken into account in this paper, is as follows:

Definition 8.12.1 (Well-posedness). We say that the 5th Gardner equation (8.12.1) is local-in-time (or locally) well-posed in $H^s(\mathbb{R})$, if for any $R > 0$ and any $w_0 \in \{f \in H^s(\mathbb{R}) : \|f\|_{H^s} \leq \mathbb{R}\}$, there exists a local time $T = T(R) > 0$ and a unique solution u to (8.12.1) in $C([0,T]; H^s(\mathbb{R})) \cap X_T$, for some auxiliary space X_T. Moreover, the solution map $w_0 \mapsto w(t)$ is continuous from $\{f \in H^s(\mathbb{R}) : \|f\|_{H^s} \leq R\}$ to $C([0,T]; H^s(\mathbb{R}))$. The local result is extended to the global one, if $T > 0$ is independent of R.

The 5th order Gardner equation (8.12.1) is locally well-posed in H^2 via the classical energy method in addition to the short time Fourier restriction norm method. We state the local well-posedness result as follows:

Theorem 8.12.2. *The 5th order Gardner equation* (8.12.1) *is locally well-posed in* $H^s(\mathbb{R})$, $s \geq 2$.

The proof of Theorem 8.12.2 uses the short-time Fourier restriction norm method in a frequency dependent time interval. This is introduced by Ionescu, Kenig and Tataru [IKT08] in the context of KP-I equation. The global well-posedness follows immediately from the above local result and the conservation of the second order energy (8.12.5).

Theorem 8.12.3. *The 5th order Gardner equation* (8.12.1) *is globally well-posed in the energy space* $H^2(\mathbb{R})$.[2]

Does the flow map from data to solutions fail to be (locally) uniformly continuous for all regularities? As an immediate answer, we have the following (weak) ill-posedness result, which extends Alejo-Cardoso's recent result [AaCa18] to all regularities:

Theorem 8.12.4. *The 5th order Gardner equation* (8.12.1) *is weakly ill-posed in* $H^s(\mathbb{R})$, *for* $s > 0$ *in the following sense: there exist* $c, C > 0$, $0 < T \leq 1$, *and two sequences* w_n *and* v_n *of solutions to* (8.12.1) *such that*

$$\sup_n \|w_n(t)\|_{H^s} + \sup_n \|v_n(t)\|_{H^s} \leq C, \qquad t \in [0, T]$$

and initially

$$\lim_{n \to \infty} \|w_n(0) - v_n(0)\|_{H^s} = 0,$$

but for every $t \in [0, T]$

$$\liminf_{n \to \infty} \|w_n(t) - v_n(t)\|_{H^s} \geq c|\sin t| \sim c|t|.$$

Theorem 8.12.4 can be expected from the classical observation in the linear *local smoothing effect*

$$\left\| \partial_x^2 e^{-t\partial_x^5} w_0 \right\|_{L_x^\infty L_t^2} \lesssim \|w_0\|_{L^2}$$

compared to the three derivatives in the quadratic nonlinearity.

Corollary 8.12.5. *The 5th order Gardner equation* (8.12.1) *is (weakly) ill-posed in* $H^s(\mathbb{R})$, *for* $s \in \mathbb{R}$, *in the sense of the statement given in Theorem 8.12.4.*

---

[2]The persistence of regularities ensures the global well-posedness in $H^s(\mathbb{R})$, $s \geq 2$.

8.12.0.2 *Orbital stability for 5th order Gardner breather solutions*

Another interesting topic in the theory of PDEs is to study the global dynamics of a solution to a given equation, in particular, the stability theory for a special solution. In this paper, we shall focus on a special type of "wave packet" solution, defined as follows:

Definition 8.12.6 (Breather solutions). We shall say that a *nontrivial* global strong solution $w = w(t, x)$ of (8.12.1) is a *breather solution* if w is *periodic* in time (up to translation) and localized in space.

In this case, we have the following example.

Definition 8.12.7 (5th order Gardner breather). Let $\alpha, \beta \in \mathbb{R}\backslash\{0\}$, $\mu \in \mathbb{R}^+\backslash\{0\}$ such that $\Delta = \alpha^2 + \beta^2 - 4\mu^2 > 0$, and $x_1, x_2 \in \mathbb{R}$. The 5th order breather solution $B_\mu \equiv B_{\mu,5}$ of the 5th order Gardner equation (8.12.1), is given explicitly by the formula

$$B_\mu \equiv B_{\alpha,\beta,\mu,5}(t, x; x_1, x_2) := 2\partial_x\left[\arctan\left(\frac{G_{\alpha,\beta,\mu,5}(t, x)}{F_{\alpha,\beta,\mu,5}(t, x)}\right)\right], \tag{8.12.6}$$

where

$$G_{\alpha,\beta,\mu,5}(t, x) := \frac{\beta\sqrt{\alpha^2+\beta^2}}{\alpha\sqrt{\Delta}}\sin(\alpha y_1) - \frac{2\mu\beta[\cosh(\beta y_2) + \sinh(\beta y_2)]}{\Delta},$$
$$F_{\alpha,\beta,\mu,5}(t, x) := \cosh(\beta y_2) - \frac{2\mu\beta[\alpha\cos(\alpha y_1) - \beta\sin(\alpha y_1)]}{\alpha\sqrt{\alpha^2+\beta^2}\sqrt{\Delta}},$$

with y_1 and y_2

$$y_1 = x + \delta_5 t + x_1, \quad y_2 = x + \gamma_5 t + x_2,$$

and with velocities

$$\delta_5 := -\alpha^4 + 10\alpha^2\beta^2 - 5\beta^4 + 10(\alpha^2 - 3\beta^2)\mu^2,$$
$$\gamma_5 := -\beta^4 + 10\alpha^2\beta^2 - 5\alpha^4 + 10(3\alpha^2 - \beta^2)\mu^2.$$

The well-posedness results in H^2 are needed in advance of the study on the breather's stability theory, as follows:

Theorem 8.12.8. *Let $\alpha, \beta \in \mathbb{R}\backslash\{0\}$ be given. Breather solutions* (8.12.6) *of the 5th order Gardner equation* (8.12.1) *are orbitally stable for H^2 perturbations, whenever the parameter $\mu \in \left(0, \frac{\sqrt{\alpha^2+\beta^2}}{2}\right)$.*

The proof of this stability result is based on two main ingredients: the first one is a particular nonlinear identity, satisfied as far as we know, only by 5th order breather solutions (8.12.6). On the other hand, the second ingredient is the *universal* fourth order ODE satisfied by (8.12.6). Using the nonlinear identity obtained from the first step, we prove that breather solutions (8.12.6) of the 5th order Gardner equation (8.12.1) satisfy a fourth order ODE (see [AlMu15, Theorem 3.5] for the same as the one satisfied by *classical* Gardner breather solutions). To complete the proof of Theorem 8.12.8, we introduce an H^2 Lyapunov functional as the linear combination of the energy (8.12.4), the mass (8.12.3) and (8.12.5) in the following form:

$$\mathcal{H}_\mu[w](t) := E_{5\mu}[w](t) + 2(\beta^2 - \alpha^2)E_\mu[w](t) + (\alpha^2 + \beta^2)^2 M[w](t).$$

The rest of the proof uses perturbation of the functional $\mathcal{H}_\mu[w](t)$ and the spectral properties of the linearized operator $\mathcal{L}_\mu$ defined as (12.7.7) corresponding to the quadratic form arising in the small perturbation of the functional $\mathcal{H}_\mu[w](t)$. The spectral properties, in particular the coercivity property of the quadratic form and existence of a unique negative eigenvalue for the linearized operator around the breather solution, ensure the H^2-stability of the 5th order Gardner breathers (8.12.6).

8.13 Exercises

(8.1) Prove Lemma 8.1.1, i.e. the connection between mKdV and Gardner solutions.

(8.2) (*Gardner Transform* [AMV13]) Check the Gardner transform
$$v = i\frac{3}{\sqrt{2}}w_x + \frac{3}{2}w^2 + \mu w.$$
Namely, prove that if we assume that w is a solution of the Gardner equation (8.1.1), then v is a solution of the KdV equation (2.7.1).

(8.3) Obtain the double pole solution in Definition 8.4.4 from the Gardner breather (8.4.1).

(8.4) Prove that $F_\mu[w](t)$ (8.3.2) is a conserved quantity, explicitly, that $F_\mu[w](t) = F_\mu[w](0)$.

(8.5) Prove by hand Lemma 8.8.1. (See [Ale18] for some help.)

(8.6) Prove by hand Lemma 8.10.5. (See [Ale18] for some help.)

(8.7) Obtain the coefficients A, B in the transformation
$$v = Aiu_x + Bu^2$$
such that if u is a solution of the 5th mKdV equation (8.12.2), then v is a solution of the 5th KdV equation
$$v_t + (v_{xxxx} + 10vv_{xx} + 5v_x^2 + 10v^3)_x = 0.$$

Chapter 9

Breathers and their stability in the sine-Gordon equation

9.1 Introduction

Recall the sine-Gordon (SG) equation, already introduced in (3.2.3):

$$\partial_t^2 u - \partial_x^2 u + \sin u = 0, \quad u = u(t,x) \in \mathbb{R}, \quad (t,x) \in \mathbb{R}^2. \tag{9.1.1}$$

This equation is one particular case of the big class of one-dimensional *scalar field* equations

$$\partial_t^2 u - \partial_x^2 u + f(u) = 0, \quad f(0) = 0,$$

appearing in Quantum Field Theory [JKW14]. Along this chapter we will use the convention that the speed of light in the model is $c = 1$. This will be clarified when talking about Lorentz boosts.

From this general class of nonlinearities $f(u)$, maybe the most important example is the ϕ^4 *model* appearing in the formulation of the Standard Model of quantum particles:

$$f(u) := -u + u^3,$$

which can be recast as the third order approximation of $-\sin u$, up to a dilation to hide the coefficient $1/3!$.

In order to make sense of the Cauchy problem, it is necessary to think about SG as a matrix system for the variables u and $\partial_t u$ included in the vector $U := (u, \partial_t u)^T$, in such a way that

$$\partial_t U = \partial_t \begin{pmatrix} u \\ \partial_t u \end{pmatrix} = \begin{pmatrix} \partial_t u \\ \partial_t^2 u \end{pmatrix} = \begin{pmatrix} \partial_t u \\ \partial_x^2 u - \sin u \end{pmatrix} = \mathcal{L}_0 U + F(U),$$

where

$$\mathcal{L}_0 := \begin{pmatrix} 0 & 1 \\ \partial_x^2 & 0 \end{pmatrix} \quad \text{and} \quad F(U) := \begin{pmatrix} 0 \\ -\sin u \end{pmatrix}.$$

Written in this form, it is clear that SG represents an order one (in time) dispersive model. Additionally, it reveals clear that, in order to put a good initial data space for solving this equation, it is enough to take $(u(0), \partial_t u(0)) \in H^s \times H^{s-1}$ for some suitable s. Even better, since the nonlinear term $\sin u$ is uniformly bounded independently of the size of u, one can find a satisfactory $H^1 \times L^2$ global well-posedness theory, see e.g. Bourgain [Bou99].

The SG equation shares many similarities with the mKdV equation (4.1.1), but it differs in many others. These differences will be very important when trying to exemplify the main results in this chapter. For the moment, let us talk about similarities. SG is also completely integrable, with a (matrix) Lax pair formulation very similar to the mKdV one.

9.2 Invariances in SG

Solutions of SG are invariant under space and time translations. Indeed, for any $t_0, x_0 \in \mathbb{R}$, $u(t - t_0, x - x_0)$ is also a solution. Furthermore, an additional (and very important) feature of the SG equation is the invariance under **Lorentz transformations**:

Definition 9.2.1 (Lorentz boost). Given any $v \in (-1, 1)$, and u solution of (9.1.1), we say that

$$u_v(t, x) := u(\gamma(t - vx), \gamma(x - vt)), \quad \gamma := (1 - v^2)^{-1/2}, \tag{9.2.1}$$

is a Lorentz boost of u by a *velocity* v. Moreover, u_v is a new solution of (9.1.1).

Notice that when considering a physically relevant speed of light c, the term $t - vx$ makes no sense unless one writes explicitly c. Naturally, this invariance of wave-type equations has to be taken into account when describing the family of breathers.

9.3 Conserved quantities

Additionally, because of the integrable character of the equation, SG has infinitely many conserved quantities. For the purposes of the exposition, we will only need the first three, or the low order ones. In order to avoid too much notation, we will use the simplification $\partial_t u =: u_t$, $\partial_x u =: u_x$ and so on. At the $H^1 \times L^2$-level of regularity, we have the *energy*

$$\begin{aligned} E[u, u_t](t) &:= \frac{1}{2} \int_{\mathbb{R}} (u_x^2 + u_t^2)(t, x) dx + \int_{\mathbb{R}} (1 - \cos u(t, x)) dx \\ &= E[u, u_t](0), \end{aligned} \tag{9.3.1}$$

and the *momentum*

$$P[u,u_t](t) := \frac{1}{2}\int_{\mathbb{R}} u_t(t,x)u_x(t,x)dx = P[u,u_t](0). \tag{9.3.2}$$

Precisely, the momentum will be zero in the case of a *static* solution (in a sense to be made precise below), and not zero for non static solutions.

Now we consider data with better regularity. Let $(u_0,u_1)\in H^2(\mathbb{R})\times H^1(\mathbb{R})$ and $(u(t),u_t(t))\in H^2(\mathbb{R})\times H^1(\mathbb{R})$ be the corresponding local-in-time solution of the Cauchy problem associated to (9.1.1), with initial condition $(u(0),u_t(0)) = (u_0,u_1)$ (cf. e.g. Bourgain [Bou99]). There is a second order conserved quantity for (9.1.1), given by

$$\begin{aligned}F[u,u_t](t) :=& \frac{1}{2}\int_{\mathbb{R}}(u_{xx}^2+u_{tx}^2)(t,x)dx - \frac{1}{32}\int_{\mathbb{R}}(u_t^4+u_x^4)(t,x)dx\\ &-\frac{3}{16}\int_{\mathbb{R}} u_t^2(t,x)u_x^2(t,x)dx\\ &+\frac{5}{8}\int_{\mathbb{R}} u_x^2(t,x)\cos u(t,x)dx\\ &+\frac{1}{8}\int_{\mathbb{R}}(\sin^2 u(t,x)+u_t^2(t,x)\cos u(t,x))dx.\end{aligned} \tag{9.3.3}$$

As one may expect, this quantity is key in the variational stability problem.

9.4 A quick review on kinks

SG equation (9.1.1) has soliton solutions usually referred as *kinks*. Indeed, given $v\in(0,1)$, $x_0\in\mathbb{R}$ and γ defined as in (9.2.1), SG (9.1.1) has kinks of the form

$$u(t,x) = \varphi(\gamma(x-vt-x_0)), \quad \varphi(s) := 4\arctan e^s. \tag{9.4.1}$$

It is not difficult to see that these solutions can be associated to a well-known functional defined in the $H^1\times L^2$ topology. Kinks solutions of (9.1.1) are orbitally stable in the energy space $H^1\times L^2$. Indeed, the energy (9.3.1) is a conserved quantity and kinks can be viewed as relative minimizers of a suitable energy functional. In this case, the situation is somehow easier and the required functional is given by

$$E[u,u_t](t) + cP[u,u_t](t),$$

for a suitable constant c, and E and P conserved quantities as in (9.3.1) and (9.3.2). For the proofs of this result for the SG case and more general equations, we refer to the works by Henry-Perez-Wreszinski [HPW82], Grillakis-Shatah-Strauss [GSS87]. An alternative proof, involving the 4-norm

$$(E[u,u_t](t))^2 - (P[u,u_t](t))^2,$$

can be found in [KMM21].

9.5 The sine-Gordon breather

As we already introduced in (3.2.4), there is an explicit breather solution for SG (9.1.1); remembering its definition:

Definition 9.5.1 ([Lam80], see also Fig. 9.1). Let $v \in (-1, 1)$, $\gamma := (1 - v^2)^{-1/2}$, $\beta \in (0, \gamma)$ and $x_1, x_2 \in \mathbb{R}$. Any breather solution of SG (9.1.1) is given by the expression

$$B := B_{\beta,v}(t, x; x_1, x_2) := 4 \arctan\left(\frac{\beta}{\alpha} \frac{\cos(\alpha y_1)}{\cosh(\beta y_2)}\right), \tag{9.5.1}$$

with

$$y_1 := t - vx + x_1, \quad y_2 := x - vt + x_2, \quad \alpha := \sqrt{\gamma^2 - \beta^2}. \tag{9.5.2}$$

The parameters β and v correspond to the scaling and velocity of the breather, and the case $v = 0$ represents a standing SG breather.

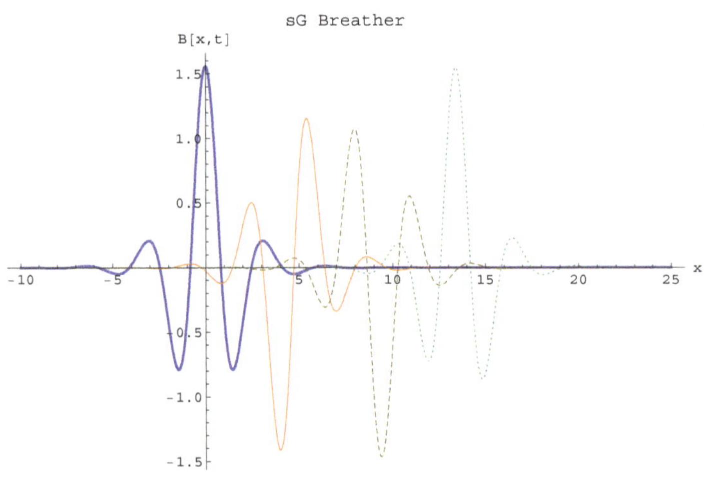

Fig. 9.1 Evolution of a sine-Gordon breather moving rightward in space as time evolves (blue color to green color).

Note that the SG breather (9.5.1) satisfies Definition 3.1.1 with $T = \frac{2\pi}{\alpha(1-v^2)} > 0$ and $L = vT \in \mathbb{R}$. Also, the Definition 9.5.1 takes into account the velocity v of the breather via a Lorentz boost, therefore it is slightly different to the one written in [Lam80]. However, after redefining the parameters in (9.5.2) and reverting the Lorentz boost, it is not difficult to check that they represent the same solution.

Additionally, note that in the SG case the two parameters α and β are not independent, unlike the mKdV case. The parameter β corresponds to the scaling of the breather. In order to make sense for a suitable Cauchy theory, our previous definition requires an additional description of the time derivative of a breather solution. Since we are going to work with several time-dependent parameters, it is certainly necessary to give a precise definition of this second nonlinear mode.

Definition 9.5.2. For any $x_1, x_2 \in \mathbb{R}$ fixed, we define the time derivative of B, denoted by $B_t = (B_{\beta,v})_t$, as follows

$$B_t(t,x;x_1,x_2) := -4\alpha\beta\left[\frac{\alpha\sin(\alpha y_1)\cosh(\beta y_2) - \beta v\cos(\alpha y_1)\sinh(\beta y_2)}{\alpha^2\cosh^2(\beta y_2) + \beta^2\cos^2(\alpha y_1)}\right]. \tag{9.5.3}$$

Note that B_t can be identically zero at some particular time. However, unless the initial data is exactly zero, the pair (B, B_t) cannot vanish.

Along these pages we will show that SG breathers satisfy the same rigidity properties as mKdV breathers. We will also describe a numerical method allowing us to understand the spectral properties of the cumbersome linearized operators appearing in the stability analysis of SG breathers.

9.6 Energy identities

In the next lines, we summarize some well-known identities for the energy and momentum of a fixed breather. For the sake of completeness, we give full proofs of these results. Recall the definition of breather (B, B_t) in (9.5.1)-(9.5.3).

Lemma 9.6.1. *Let (B, B_t) be any SG breather, with parameters for $\beta > 0$, $v \in (-1,1)$, and $x_1, x_2 \in \mathbb{R}$. Let $P[B, B_t]$ be the momentum of a breather defined in (9.3.2). Then*

$$P[B, B_t](t) = -8\beta v. \tag{9.6.1}$$

Proof. Exercise. □

We compute now the energy of a breather. See Lamb [Lam80] for a similar result in the case where $v = 0$.

Lemma 9.6.2. *Let (B, B_t) be any SG breather of parameters for $\beta > 0$, $v \in (-1,1)$, and $x_1, x_2 \in \mathbb{R}$. Then*

$$E[B, B_t](t) = 16\beta. \tag{9.6.2}$$

Moreover, this result does not change if we replace x_1 and x_2 by time-dependent parameters.[1]

Proof. This is a classical result. From (9.5.3) and (9.7.4), we have

$$\begin{aligned}&\frac{1}{2}(B_t^2 + B_x^2)\\ &= \frac{8\alpha^2\beta^2}{\left(\alpha^2\cosh^2(\beta y_2) + \beta^2\cos^2(\alpha y_1)\right)^2}\\ &\quad \Bigg((1+v^2)\alpha^2\cosh^2(\beta y_2)\sin^2(\alpha y_1) + (1+v^2)\beta^2\sinh^2(\beta y_2)\cos^2(\alpha y_1)\\ &\qquad - 4\alpha\beta v\sin(\alpha y_1)\cos(\alpha y_1)\sinh(\beta y_2)\cosh(\beta y_2)\Bigg).\end{aligned}$$

On the other hand, from (9.7.7),

$$1 - \cos B = \frac{8\alpha^2\beta^2\cosh^2(\beta y_2)\cos^2(\alpha y_1)}{\left(\alpha^2\cosh^2(\beta y_2) + \beta^2\cos^2(\alpha y_1)\right)^2}.$$

Therefore using double angle formulas, we have

$$\frac{1}{2}(B_t^2 + B_x^2) + (1 - \cos B) = \frac{8\alpha^2\beta^2\tilde{h}(t,x)}{\left(\frac{\alpha^2+\beta^2}{2} + \frac{\alpha^2}{2}\cosh(2\beta y_2) + \frac{\beta^2}{2}\cos(2\alpha y_1)\right)^2},$$

where

$$\begin{aligned}\tilde{h}(t,x) := &\frac{1}{4}((1+v^2)(\alpha^2-\beta^2)+1) + \frac{1}{4}((1+v^2)(\alpha^2+\beta^2)+1)\cosh(2\beta y_2)\\ &-\alpha\beta v\sin(2\alpha y_1)\sinh(2\beta y_2) - \frac{1}{4}((1+v^2)(\alpha^2+\beta^2)-1)\cos(2\alpha y_1)\\ &+\frac{1}{4}((1+v^2)(\beta^2-\alpha^2)+1)\cosh(2\beta y_2)\cos(2\alpha y_1).\end{aligned}$$

Now with the relation among α, β (4.1.9) a direct computation shows that

$$\tilde{h} = g(h_e)_x - h_e g_x,$$

where $h_e(t,x) := \frac{1}{2\alpha\beta}[\alpha\sinh(2\beta y_2) + \beta v\sin(2\alpha y_1)]$ and g is a given explicit function. From this identity we obtain

$$E[B, B_t] = 8\alpha^2\beta^2\int_{\mathbb{R}}\left(\frac{h_e}{g}\right)_x = 16\beta.$$

□

Remark 9.6.3. The reader may compare (9.6.2) with a similar result for the mass of the mKdV breather, see (4.1.11) and [AlMu13, eqn. (2.4)], where $M[B] = 4\beta$. In that sense, both results reveal that the mass (or energy) does not depend on the oscillatory parameter, but only on the main scaling β. This property seems inherent to aperiodic breathers.

[1] But now (B, B_t) ceases being an exact solution of SG.

9.7 Variational characterization

We introduce now some useful notation. Let $v \in (-1, 1)$, $\gamma = (1 - v^2)^{-1/2} \geq 1$, and $\beta \in (0, \gamma)$. Define the parameters

$$a := -\frac{1}{4} + \beta^2 + v^2(2\gamma^4 - \gamma^2 + \beta^2), \qquad b := 4v(\gamma^4 - \beta^2). \tag{9.7.1}$$

Note that $a + \frac{1}{4} > 0$ and $b \in \mathbb{R}$. The reader may observe that whenever $v = 0$ (the static breather), we have

$$a + \frac{1}{4} = \beta^2 \in (0, 1), \quad b = 0.$$

Theorem 9.7.1. *Let (B, B_t) be any SG breather of parameters $v \in (-1, 1)$, $\beta \in (0, \gamma)$, and a, b as in (9.7.1). Then, for any fixed $t \in \mathbb{R}$, (B, B_t) satisfy the nonlinear equations*

$$B_{txx} + \frac{1}{8}B_t^3 + \frac{3}{8}B_x^2 B_t - \frac{1}{4}B_t \cos B - aB_t - \frac{b}{2}B_x = 0, \tag{9.7.2}$$

and

$$\begin{aligned} &B_{4x} + \frac{3}{8}B_x^2 B_{xx} + \frac{3}{4}B_t B_{tx} B_x + \frac{3}{8}B_t^2 B_{xx} + \frac{5}{8}B_x^2 \sin B - \frac{5}{4}B_{xx} \cos B \\ &+ \frac{1}{4}\sin B \cos B - \frac{1}{8}B_t^2 \sin B - a(B_{xx} - \sin B) - \frac{b}{2}B_{tx} = 0. \end{aligned} \tag{9.7.3}$$

Proof of (9.7.3)**, assuming** (9.7.2)**.** Taking time derivative in equation (9.7.2), and replacing B_{tt} using (9.1.1), we have

$$\begin{aligned} 0 &= (B_{tt})_{xx} + \frac{3}{8}B_t^2 B_{tt} + \frac{3}{4}B_t B_x B_{tx} + \frac{3}{8}B_x^2 B_{tt} - \frac{1}{4}B_{tt}\cos B + \frac{1}{4}B_t^2 \sin B - aB_{tt} - \frac{b}{2}B_{tx} \\ &= (B_{xx} - \sin B)_{xx} + \frac{3}{8}B_t^2(B_{xx} - \sin B) + \frac{3}{4}B_t B_x B_{tx} + \frac{3}{8}B_x^2(B_{xx} - \sin B) \\ &\quad - \frac{1}{4}(B_{xx} - \sin B)\cos B + \frac{1}{4}B_t^2 \sin B - a(B_{xx} - \sin B) - \frac{b}{2}B_{tx} \\ &= B_{4x} - (B_x \cos B)_x + \frac{3}{8}B_t^2 B_{xx} - \frac{3}{8}B_t^2 \sin B + \frac{3}{4}B_t B_x B_{tx} + \frac{3}{8}B_x^2 B_{xx} - \frac{3}{8}B_x^2 \sin B \\ &\quad - \frac{1}{4}B_{xx}\cos B + \frac{1}{4}\sin B \cos B + \frac{1}{4}B_t^2 \sin B - a(B_{xx} - \sin B) - \frac{b}{2}B_{tx} \\ &= B_{4x} + \frac{3}{8}B_x^2 B_{xx} + \frac{3}{4}B_t B_{tx} B_x + \frac{3}{8}B_t^2 B_{xx} + \frac{5}{8}B_x^2 \sin B - \frac{5}{4}B_{xx}\cos B \\ &\quad + \frac{1}{4}\sin B \cos B - \frac{1}{8}B_t^2 \sin B - a(B_{xx} - \sin B) - \frac{b}{2}B_{tx} = \text{l.h.s. of (9.7.3)}. \end{aligned}$$

Proof of (9.7.2). We use the specific structure of the breather. From (9.5.3) and (9.5.1) we have

$$B_t = -4\alpha\beta \frac{\alpha \sin(\alpha y_1)\cosh(\beta y_2) - \beta v \cos(\alpha y_1)\sinh(\beta y_2)}{\alpha^2 \cosh^2(\beta y_2) + \beta^2 \cos^2(\alpha y_1)},$$

and

$$B_x = 4\alpha\beta\frac{v\alpha\sin(\alpha y_1)\cosh(\beta y_2) - \beta\cos(\alpha y_1)\sinh(\beta y_2)}{\alpha^2\cosh^2(\beta y_2) + \beta^2\cos^2(\alpha y_1)}. \tag{9.7.4}$$

Hence

$$\begin{aligned}
&\frac{3}{8}B_x^2 B_t \\
&= \frac{-24\alpha^3\beta^3}{(\alpha^2\cosh^2(\beta y_2) + \beta^2\cos^2(\alpha y_1))^3}[v\alpha\cosh(\beta y_2)\sin(\alpha y_1) - \beta\sinh(\beta y_2)\cos(\alpha y_1)]^2 \\
&\quad [\alpha\cosh(\beta y_2)\sin(\alpha y_1) - \beta v\sinh(\beta y_2)\cos(\alpha y_1)] \\
&= \frac{-24\alpha^3\beta^3\, h_1(t,x)}{(\alpha^2\cosh^2(\beta y_2) + \beta^2\cos^2(\alpha y_1))^3},
\end{aligned}$$

with

$$\begin{aligned}
h_1(t,x) := {}& \alpha^3 v^2\cosh^3(\beta y_2)\sin^3(\alpha y_1) - \beta^3 v\sinh^3(\beta y_2)\cos^3(\alpha y_1) \\
& - v(2+v^2)\alpha^2\beta\sinh(\beta y_2)\cosh^2(\beta y_2)\sin^2(\alpha y_1)\cos(\alpha y_1) \\
& + (1+2v^2)\alpha\beta^2\sinh^2(\beta y_2)\cosh(\beta y_2)\sin(\alpha y_1)\cos^2(\alpha y_1).
\end{aligned}$$

Similarly,

$$\frac{1}{8}B_t^3 = \frac{-8\alpha^3\beta^3\, h_2(t,x)}{(\alpha^2\cosh^2(\beta y_2) + \beta^2\cos^2(\alpha y_1))^3},$$

with

$$\begin{aligned}
&h_2(t,x) \\
&:= \alpha^3\cosh^3(\beta y_2)\sin^3(\alpha y_1) - 3\alpha^2\beta v\cosh^2(\beta y_2)\sin^2(\alpha y_1)\sinh(\beta y_2)\cos(\alpha y_1) \\
&+ 3\alpha\beta^2 v^2\cosh(\beta y_2)\sin(\alpha y_1)\sinh^2(\beta y_2)\cos^2(\alpha y_1) - \beta^3 v^3\sinh^3(\beta y_2)\cos^3(\alpha y_1).
\end{aligned}$$

We compute now the term B_{txx}. From (9.5.3) we have

$$B_{tx} = \frac{-4\alpha\beta\, h_3(t,x)}{(\alpha^2\cosh^2(\beta y_2) + \beta^2\cos^2(\alpha y_1))^2},$$

where

$$\begin{aligned}
h_3(t,x) := {}& [\alpha\beta(1-v^2)\sin(\alpha y_1)\sinh(\beta y_2) - v\gamma^2\cos(\alpha y_1)\cosh(\beta y_2)] \\
& \times[\alpha^2\cosh^2(\beta y_2) + \beta^2\cos^2(\alpha y_1)] \\
& - 2\alpha\beta[\alpha\sin(\alpha y_1)\cosh(\beta y_2) - \beta v\cos(\alpha y_1)\sinh(\beta y_2)] \\
& \times[\alpha\sinh(\beta y_2)\cosh(\beta y_2) + v\beta\sin(\alpha y_1)\cos(\alpha y_2)].
\end{aligned} \tag{9.7.5}$$

Similarly, after a long computation,

$$B_{txx} = \frac{-4\alpha\beta\, h_4(t,x)}{(\alpha^2\cosh^2(\beta y_2) + \beta^2\cos^2(\alpha y_1))^3},$$

where

$$h_4(t,x) := [\alpha^2\cosh^2(\beta y_2)+\beta^2\cos^2(\alpha y_1)](h_3)_x$$
$$-4\alpha\beta[\alpha\sinh(\beta y_2)\cosh(\beta y_2)+v\beta\sin(\alpha y_1)\cos(\alpha y_2)]h_3. \quad (9.7.6)$$

On the other hand, using the well-known formulae

$$\cos(4\theta) = 1-8\sin^2\theta+8\sin^4\theta, \quad \sin^2\theta = \frac{\tan^2\theta}{1+\tan^2\theta},$$

we have

$$\cos(4\arctan s) = \frac{(s^2-2s-1)(s^2+2s-1)}{(1+s^2)^2} = \frac{s^4-6s^2+1}{(1+s^2)^2}, \quad s\in\mathbb{R}.$$

Therefore, from (9.5.1) we obtain

$$\cos B = \frac{\beta^4\cos^4(\alpha y_1)-6\alpha^2\beta^2\cosh^2(\beta y_2)\cos^2(\alpha y_1)+\alpha^4\cosh^4(\beta y_2)}{(\alpha^2\cosh^2(\beta y_2)+\beta^2\cos^2(\alpha y_1))^2}, \quad (9.7.7)$$

and thus

$$\begin{aligned}-\frac{1}{4}B_t\cos B =& \frac{\alpha\beta}{(\alpha^2\cosh^2(\beta y_2)+\beta^2\cos^2(\alpha y_1))^3}\\ & [\alpha\cosh(\beta y_2)\sin(\alpha y_1)-\beta v\sinh(\beta y_2)\cos(\alpha y_1)]\\ & [\beta^4\cos^4(\alpha y_1)-6\alpha^2\beta^2\cosh^2(\beta y_2)\cos^2(\alpha y_1)+\alpha^4\cosh^4(\beta y_2)]\\ =& \frac{\alpha\beta\, h_5(t,x)}{(\alpha^2\cosh^2(\beta y_2)+\beta^2\cos^2(\alpha y_1))^3},\end{aligned}$$

where

$$\begin{aligned}h_5(t,x) :=\ & \alpha\beta^4\sin(\alpha y_1)\cos^4(\alpha y_1)\cosh(\beta y_2)\\ & -\alpha^4\beta v\cos(\alpha y_1)\sinh(\beta y_2)\cosh^4(\beta y_2)\\ & +\alpha^5\sin(\alpha y_1)\cosh^5(\beta y_2)-\beta^5 v\cos^5(\alpha y_1)\sinh(\beta y_2)\\ & +6\alpha^2\beta^3 v\cos^3(\alpha y_1)\sinh(\beta y_2)\cosh^2(\beta y_2)\\ & -6\alpha^3\beta^2\sin(\alpha y_1)\cos^2(\alpha y_1)\cosh^3(\beta y_2).\end{aligned}$$

Collecting the above identities, we get

$$\begin{aligned}& B_{tx}+\frac{1}{8}B_t^3+\frac{3}{8}B_x^2B_t-\frac{1}{4}B_t\cos B\\ & = \alpha\beta\frac{[-4h_4-8\alpha^2\beta^2h_2-24\alpha^2\beta^2h_1+h_5]}{(\alpha^2\cosh^2(\beta y_2)+\beta^2\cos^2(\alpha y_1))^3}.\end{aligned} \quad (9.7.8)$$

In the following lines, we simplify the numerator in the previous expression. In order to carry out this computation, the key point will be the denominator, denoted by g:

$$g(t,x) := \alpha^2\cosh^2(\beta y_2)+\beta^2\cos^2(\alpha y_1).$$

First of all, note from (9.7.5) that h_3 obeys the unique decomposition

$$h_3 =: h_{31}g - h_{32}g_x, \quad g_x = 2\alpha\beta[\alpha\sinh(\beta y_2)\cosh(\beta y_2) + \nu\beta\sin(\alpha y_1)\cos(\alpha y_1)].$$

Therefore, from (9.7.6),

$$\begin{aligned} h_4 &= g[h_{31}g - h_{32}g_x]_x - 2g_x[h_{31}g - h_{32}g_x] \\ &= g^2(h_{31})_x + g[h_{31}g_x - (h_{32}g_x)_x - 2h_{31}g_x] + 2h_{32}g_x^2 \\ &= g^2(h_{31})_x - (h_{31}g_x + (h_{32}g_x)_x)g + 2h_{32}g_x^2. \end{aligned} \tag{9.7.9}$$

On the other hand,

$$\begin{aligned} &- 8\alpha^2\beta^2(h_2 + 3h_1) \\ &= -8\alpha^5\beta^2(1 + 3\nu^2)\sin^3(\alpha y_1)\cosh^3(\beta y_2) \\ &\quad + 8\alpha^2\beta^5\nu(3 + \nu^2)\cos^3(\alpha y_1)\sinh^3(\beta y_2) \\ &\quad + 24\alpha^4\beta^3\nu(3 + \nu^2)\sin^2(\alpha y_1)\cos(\alpha y_1)\sinh(\beta y_2)\cosh^2(\beta y_2) \\ &\quad - 24\alpha^3\beta^4(3\nu^2 + 1)\sin(\alpha y_1)\cos^2(\alpha y_1)\sinh^2(\beta y_2)\cosh(\beta y_2). \end{aligned}$$

Therefore, after some simplifications,

$$\begin{aligned} &h_5 - 8\alpha^2\beta^2(h_2 + 3h_1) - 8h_{32}g_x^2 \\ &= g\Big(\alpha^3(1 - 32\beta^2)\sin(\alpha y_1)\cosh^3(\beta y_2) \\ &\qquad -\alpha^2\beta\nu(1 - 32\beta^2)\cos(\alpha y_1)\sinh(\beta y_2)\cosh^2(\beta y_2) \\ &\qquad -\beta^3\nu(1 + 32\alpha^2\nu^2)\cos^3(\alpha y_1)\sinh(\beta y_2) \\ &\qquad +\alpha\beta^2(1 + 32\alpha^2\nu^2)\sin(\alpha y_1)\cos^2(\alpha y_1)\cosh(\beta y_2) \\ &\qquad +24\alpha^3\beta^2(1 - \nu^2)\sin(\alpha y_1)\cosh(\beta y_2) \\ &\qquad -24\alpha^2\beta^3\nu(1 - \nu^2)\cos(\alpha y_1)\sinh(\beta y_2)\Big). \end{aligned} \tag{9.7.10}$$

On the other hand, we consider the second term in (9.7.9). We have

$$\begin{aligned} &4(h_{31}g_x + (h_{32}g_x)_x)g \\ &= 8\alpha\beta g\Big(2\alpha^2\beta(2 - \nu^2)\sin(\alpha y_1)\cosh^3(\beta y_2) \\ &\qquad -3\alpha^2\beta(1 - \nu^2)\sin(\alpha y_1)\cosh(\beta y_2) \\ &\qquad -2\alpha\nu(\beta^2 + \gamma^2)\cos(\alpha y_1)\sinh(\beta y_2)\cosh^2(\beta y_2) \\ &\qquad +3\alpha\beta^2\nu(1 - \nu^2)\cos(\alpha y_1)\sinh(\beta y_2) \\ &\qquad +2\alpha\beta^2\nu(2\nu^2 - 1)\cos^3(\alpha y_1)\sinh(\beta y_2) \\ &\qquad -2\beta\nu^2(\alpha^2 + \gamma^2)\sin(\alpha y_1)\cos^2(\alpha y_1)\cosh(\beta y_2)\Big). \end{aligned} \tag{9.7.11}$$

Collecting the last two estimates, we obtain

$$\begin{aligned}(9.7.10) + (9.7.11) = g^2[&\alpha(1 - 16\beta^2 v^2)\sin(\alpha y_1)\cosh(\beta y_2) \\ &- \beta v(1 + 16\alpha^2)\cos(\alpha y_1)\sinh(\beta y_2)].\end{aligned} \tag{9.7.12}$$

We finally add the first term in (9.7.9), such that

$$\begin{aligned}-4g^2(h_{31})_x + (9.7.12) = \alpha\beta g^2[&\alpha(1 + 4\alpha^2 v^2 - 8\beta^2 v^2 - 4\beta^2)\sin(\alpha y_1)\cosh(\beta y_2) \\ &-v\beta(1 + 8\alpha^2 - 4\beta^2 + 4\alpha^2 v^2)\cos(\alpha y_1)\sinh(\beta y_2)].\end{aligned}$$

Comparing with (9.5.3) and (9.7.4), in order to obtain (9.7.2) we have

$$(9.7.8) = aB_t + \frac{b}{2}B_x,$$

provided a and b are chosen as in (9.7.1). The proof is complete. □

Remark 9.7.2. The structure of equations (9.7.2)-(9.7.3) is inherent to solutions of SG (9.1.1). For example, the static kink solution $(\varphi(x), 0)$ defined in (9.4.1) is also a solution of the same set of equations, no matter what a and b are. In some sense, the proof of (9.7.2) is independent of the breather itself, but the proof of (9.7.3) explicitly involves the structure of the breather.

9.8 Linear operators and stability tests

Some essential consequences of the identities (9.7.2)-(9.7.3) are the following stability conditions, which will be useful when dealing with coercivity estimates in Subsection 9.10.

Corollary 9.8.1. *Let (B, B_t) be any SG breather of the form* (9.5.1)-(9.5.3). *For $t \in \mathbb{R}$ fixed, let*

$$\Lambda B := \partial_\beta B, \quad \Lambda B_t := \partial_\beta B_t. \tag{9.8.1}$$

Then $(\Lambda B, \Lambda B_t)$ are Schwartz in the space variable, and the following identities are satisfied

$$\partial_\beta E[B, B_t] = 16 > 0, \quad \partial_\beta P[B, B_t] = -8v, \tag{9.8.2}$$

independently of time.

Proof. By simple inspection one can see that for each time t one has that ΛB and ΛB_t are well-defined Schwartz functions. The proof of (9.8.2) is consequence of (9.6.2) and (9.6.1). □

Now, we introduce the following two directions associated to spatial translations. Let (B, B_t) as defined in (9.5.1)-(9.5.3), with main parameters $\beta > 0$ and $v \in (0, 1)$. We define

$$\begin{cases} B_1(t; x_1, x_2) := \partial_{x_1} B(t; x_1, x_2), \\ B_2(t; x_1, x_2) := \partial_{x_2} B(t; x_1, x_2). \end{cases} \tag{9.8.3}$$

Similarly, we introduce the terms involving time derivatives

$$\begin{cases} (B_t)_1(t; x_1, x_2) := \partial_{x_1} B_t(t; x_1, x_2) \\ (B_t)_2(t; x_1, x_2) := \partial_{x_2} B_t(t; x_1, x_2). \end{cases} \tag{9.8.4}$$

It is clear that, for all $t \in \mathbb{R}$, $\beta > 0$ and $x_1, x_2 \in \mathbb{R}$, both B_1 and B_2 are real-valued, Schwartz functions.[2] More explicitly, (9.8.3) read

$$\begin{aligned} B_1 &= \frac{-4\alpha^2\beta \sin(\alpha y_1)\cosh(\beta y_2)}{\alpha^2 \cosh^2(\beta y_2) + \beta^2 \cos^2(\alpha y_1)}, \\ B_2 &= \frac{-4\alpha\beta^2 \cos(\alpha y_1)\sinh(\beta y_2)}{\alpha^2 \cosh^2(\beta y_2) + \beta^2 \cos^2(\alpha y_1)}. \end{aligned} \tag{9.8.5}$$

Now we introduce the main linear operator associated to the breather (B, B_t). As it will be clear below, this matrix operator will be of fourth order in one component $\mathcal{L}_1$, and of second order in the second one $\mathcal{L}_2$. Two nontrivial nondiagonal terms $\mathcal{B}_1$ and $\mathcal{B}_2$ couple the two previous components in a nontrivial fashion, which is never zero in general, even in the case $v = 0$.

Definition 9.8.2. Let $\mathcal{L}$ be matrix linear operator

$$\mathcal{L}[z, w] := \begin{pmatrix} \mathcal{L}_1 & \mathcal{B}_1 \\ \mathcal{B}_2 & \mathcal{L}_2 \end{pmatrix} \begin{pmatrix} z \\ w \end{pmatrix}, \tag{9.8.6}$$

where

$$\begin{aligned} &\mathcal{L}_1 z \\ &:= z_{4x} - \left(a - \frac{3}{8}(B_x^2 + B_t^2) + \frac{5}{4}\cos B\right) z_{xx} \\ &\quad + \left(\frac{3}{4} B_x B_{xx} + \frac{3}{4} B_t B_{tx} + \frac{5}{4}\sin B B_x\right) z_x \\ &\quad + \left(a\cos B + \frac{5}{8} B_x^2 \cos B + \frac{5}{4} B_{xx}\sin B + \frac{1}{4}(\cos^2 B - \sin^2 B) - \frac{1}{8} B_t^2 \cos B\right) z, \end{aligned} \tag{9.8.7}$$

[2] Additionally, we can express B_t and B_x in terms of B_1 and B_2: we have

$$B_t = B_1 - vB_2, \qquad B_x = -vB_1 + B_2.$$

$$\mathcal{L}_2 w := -w_{xx} + \frac{1}{4}\left(4a + \cos B - \frac{3}{2}(B_t^2 + B_x^2)\right)w, \tag{9.8.8}$$

$$\mathcal{B}_1 w := \frac{1}{4}\left(3B_{tx}B_x + 3B_tB_{xx} - B_t \sin B\right)w + \frac{1}{4}\left(3B_tB_x - 2b\right)w_x, \tag{9.8.9}$$

and finally

$$\mathcal{B}_2 z := \frac{1}{2}\left(b - \frac{3}{2}B_tB_x\right)z_x - \frac{1}{4}B_t \sin B z. \tag{9.8.10}$$

In order to ensure a correct definition for $\mathcal{L}$ (9.8.6)-(9.8.7)-(9.8.8)-(9.8.9)-(9.8.10), we will set $\mathcal{L}$ on the Hilbert space $L^2(\mathbb{R})^2$, with dense domain $H^4(\mathbb{R}) \times H^2(\mathbb{R})$. Note additionally that $\mathcal{B}_1$ and $\mathcal{B}_2$ in (9.8.9)-(9.8.10) do not vanish even if the speed v is zero.

A simple consequence of Theorem 9.7.1 is the following partial kernel description.

Corollary 9.8.3. *Let a, b be defined as in (9.7.1), and B_1, B_2 given in (9.8.5). We have*

$$\mathcal{L}[B_1, (B_t)_1] = \mathcal{L}[B_2, (B_t)_2] = \begin{pmatrix} 0 \\ 0 \end{pmatrix}.$$

Moreover, $(B_1, (B_t)_1)^T$ and $(B_2, (B_t)_2)^T$ are linearly independent in $\mathbb{R}$.

Now we consider the natural directions associated to the scaling parameters. First of all, we define the quadratic form associated to $\mathcal{L}$, namely

$$Q[z, w] := \frac{1}{2}\int_{\mathbb{R}} (z, w)\mathcal{L}[z, w]. \tag{9.8.11}$$

A more detailed version of Q is the following expression, obtained after integration by parts:

$$\begin{aligned}
&Q[z, w] \\
&= \int_{\mathbb{R}} \{z\mathcal{L}_1 z + z\mathcal{B}_1 w + w\mathcal{B}_2 z + w\mathcal{L}_2 w\} \\
&= \int_{\mathbb{R}} z_{xx}^2 + \int_{\mathbb{R}} w_x^2 + \int_{\mathbb{R}} \left(a - \frac{3}{8}(B_x^2 + B_t^2) + \frac{5}{4}\cos B\right)z_x^2 \\
&\quad + \int_{\mathbb{R}} \left(\frac{5}{8}B_x^2 \cos B + \frac{5}{4}B_{xx}\sin B + \frac{1}{4}(\cos^2 B - \sin^2 B) - \frac{1}{8}B_t^2 \cos B + a\cos B\right)z^2 \\
&\quad + \int_{\mathbb{R}} \left(\left(a + \frac{1}{4}\right)\cos B - \frac{3}{8}(B_t^2 + B_x^2) + \frac{3}{4}B_{tx}B_x + \frac{3}{4}B_tB_{xx} - \frac{1}{4}B_t \sin B\right)w^2 \\
&\quad + \int_{\mathbb{R}} \left(\left(b - \frac{3}{2}B_tB_x\right)z_x w - \frac{1}{2}B_t \sin B z w\right).
\end{aligned} \tag{9.8.12}$$

Recall the following definition:

Definition 9.8.4. Any nonzero pair $(z, w) \in H^4 \times H^2$ is said to be a positive (null, negative) direction for Q as in (9.8.12) if we have $Q[z, w] > 0$ $(= 0, < 0)$.

Recall the definitions of ΛB and ΛB_t introduced in (9.8.1). For this direction, one has the result below.

Corollary 9.8.5. *The following are satisfied:*

(1) Let (B, B_t) be any SG breather. Consider the scaling *direction $(\Lambda B, \Lambda B_t)$ introduced in (9.8.1). Then $(\Lambda B, \Lambda B_t)$ is a negative direction for Q. Moreover,*

$$Q[\Lambda B, \Lambda B_t] = -32(1 + 3v^2)\beta < 0. \tag{9.8.13}$$

(2) The quadratic functional Q is bounded below, namely

$$Q[z, w] \geq -c_{\beta,v}\|(z, w)\|^2_{H^2(\mathbb{R})\times H^1(\mathbb{R})}, \tag{9.8.14}$$

for some nonnegative constant $c_{\beta,v}$. Moreover, $\mathcal{L}$ has at least *one negative eigenvalue, and therefore a minimal one, which is simple.*

Proof. The proof of (9.8.14) is standard. On the other hand, from (9.7.3), we get after derivation with respect to β,

$$\mathcal{L}_1\Lambda B + \mathcal{B}_2\Lambda B_t = a'(\beta)(B_{xx} - \sin B) + \frac{1}{2}b'(\beta)B_{tx}, \tag{9.8.15}$$

and from (9.7.2),

$$\mathcal{L}_2\Lambda B_t + \mathcal{B}_1\Lambda B = -a'(\beta)B_t - \frac{1}{2}b'(\beta)B_x. \tag{9.8.16}$$

Integrating against ΛB and ΛB_t respectively, we get from (4.9.3),

$$\begin{aligned} Q[\Lambda B, \Lambda B_t] &= -\frac{1}{2}a'(\beta)\partial_\beta \int_{\mathbb{R}} (B_x^2 + B_t^2 + 1 - \cos B) - \frac{1}{2}b'(\beta)\partial_\beta \int_{\mathbb{R}} B_t B_x \\ &= -a'(\beta)\partial_\beta E[B, B_t] - b'(\beta)\partial_\beta P[B, B_t] \\ &= -32(1 + v^2)\beta - 64v^2\beta \\ &= -32(1 + 3v^2)\beta < 0, \end{aligned}$$

where we have used (9.7.1) to obtain $a'(\beta) = 2(1 + v^2)\beta$ and $b'(\beta) = -8v\beta$. This last identity proves (9.8.13). □

From the previous result we conclude the following useful identity:

Corollary 9.8.6. *Let $(B_0, \widetilde{B}_0)$ denote the following direction*

$$(B_0, \widetilde{B}_0) := -\frac{1}{2\beta}(\Lambda B, \Lambda B_t).$$

Then

$$\mathcal{L}[B_0, \widetilde{B}_0] = (1 + v^2)\begin{pmatrix} \sin B - B_{xx} \\ B_t \end{pmatrix} + 2v\begin{pmatrix} B_{tx} \\ -B_x \end{pmatrix}, \qquad (9.8.17)$$

and

$$-\int_{\mathbb{R}} (B_0, \widetilde{B}_0) \cdot \mathcal{L}[B_0, \widetilde{B}_0] = -\frac{1}{4\beta^2} Q[\Lambda B, \Lambda B_t] = \frac{8}{\beta}(1 + 3v^2) > 0. \qquad (9.8.18)$$

Proof. A direct consequence of (9.8.15) and (9.8.16). □

Remark 9.8.7. The previous result states that $(B_0, \widetilde{B}_0)$ is a very good candidate to replace the first eigenfunction of $\mathcal{L}$ coming from Assumption 2 below. This idea, originally coming from Weinstein [Wei85], has been recently used in several works (see [AlMu13] for example) where no knowledge of the ground state is at hand.

9.9 Lyapunov functional in the SG case

Now we present some important consequences of Theorem 9.7.1. These results imply that there is a suitable Lyapunov functional that describes the local $H^2 \times H^1$ dynamics of SG breathers. Indeed, define the functional $\mathcal{H}$ as the following linear combination of F in (9.3.3), the energy E in (9.3.1) and the momentum P from (9.3.2):

$$\mathcal{H}[u, u_t] = F[u, u_t] + aE[u, u_t] + bP[u, u_t],$$

where a and b are the constants previously introduced in (9.7.1). Clearly the functional $\mathcal{H}$ is conserved along the flow, thanks to

Lemma 9.9.1. *Given $(u, u_t)(t)$, local $H^2 \times H^1$-solution of (9.1.1) with initial data (u_0, u_1) in the same vector space, the functional $F[u, u_t](t)$ is a conserved quantity. In particular, (u, u_t) is a global-in-time $H^2 \times H^1$-solution.*

At this point we make use of the integrability of the equation, since a quantity such as F is not present in a general, nonintegrable scalar field equation. The verification of Lemma 9.9.1 is a tedious but direct computation.

Lemma 9.9.2. *Let $(z, w) \in H^2(\mathbb{R}) \times H^1(\mathbb{R})$ be any pair of functions with sufficiently small $H^2 \times H^1$-norm, and (B, B_t) be any breather solution, with corresponding parameters $\beta > 0$ and $v \in (-1, 1)$. Then, for all $t \in \mathbb{R}$, one has*

$$\mathcal{H}[B + z, B_t + w](t) - \mathcal{H}[B, B_t](t) = \frac{1}{2} Q[z, w] + \mathcal{N}[z, w],$$

with Q being the quadratic form defined in (9.8.11), and $\mathcal{N}[z,w]$ a small, nonlinear term satisfying

$$|\mathcal{N}[z,w]| \lesssim p(\|z\|_{H^2(\mathbb{R})}, \|w\|_{H^1(\mathbb{R})}),$$

where p is a positive, third order monomial on its corresponding variables.

Proof. We compute:

$$\begin{aligned}
&\mathcal{H}[B+z, B_t+w] \\
&= \frac{1}{2}\int_{\mathbb{R}}[(B+z)_{xx}^2 + (B_t+w)_x^2] - \frac{1}{32}\int_{\mathbb{R}}[(B_t+w)^4 + (B+z)_x^4] \\
&\quad - \frac{3}{16}\int_{\mathbb{R}}(B_t+w)^2(B+z)_x^2 + \frac{5}{8}\int_{\mathbb{R}}(B+z)_x^2\cos(B+z) \\
&\quad + \frac{1}{8}\int_{\mathbb{R}}[\sin^2(B+z) + (B_t+w)^2\cos(B+z)] \\
&\quad + \frac{a}{2}\int_{\mathbb{R}}[(B+z)_x^2 + (B_t+w)^2] + a\int_{\mathbb{R}}[1-\cos(B+z)] + \frac{b}{2}\int_{\mathbb{R}}(B_t+w)(B+z)_x.
\end{aligned}$$

Expanding every term above, as is done in [AlMu13], we get

$$\mathcal{H}[B+z, B_t+w] = \mathcal{H}[B, B_t] + \int_{\mathbb{R}}(9.7.3)z - \int_{\mathbb{R}}(9.7.2)w + \frac{1}{2}Q[z,w] + \mathcal{N}[z,w],$$

where Q is defined in (9.8.11). The second and third term in the r.h.s of the above equation vanish. Finally, the term $N[z,w]$ is given by

$$\begin{aligned}
\mathcal{N}[z,w] := -\frac{1}{8}\int_{\mathbb{R}}\Big(& B_t w^3 + B_x z_x^3 + \frac{1}{4}(w^4 + z_x^4) + 3B_t w z_x^2 + 3B_x z_x w^2 \\
&+ \frac{3}{2}w^2 z_x^2 + \frac{5}{2}\cos B z^2 z_x^2 + 5\sin B z z_x^2 - \frac{1}{4}\sin^2 B z^4 \\
&+ \sin B\cos B z^3 + \frac{1}{2}\cos B w^2 z^2 + B_t w z^2 + \sin B z w^2\Big).
\end{aligned}$$

Therefore, from direct estimates one has $\mathcal{N}[z,w] \lesssim p(\|z\|_{H^2(\mathbb{R})}, \|w\|_{H^1(\mathbb{R})})$, where p is a positive, third order polynomial in its variables, and where the constant is independent of the size of (z,w) and the time, provided the former are chosen small. □

Corollary 9.9.3. *For initial data in $H^2 \times H^1$, the functional $\mathcal{H}$ is conserved along the flow. Moreover, breathers are critical points of $\mathcal{H}$.*

The main consequence of previous results is the possibility of proving stability, independent of how complicated breather dynamics can be. First of all, we assume two conditions:

Assumption 1: (*Nondegeneracy of the kernel*) For each $v \in (-1, 1)$, $x_1 \in \mathbb{R}$ and $\beta \in (0, \gamma)$, $\ker \mathcal{L}$ is spanned by the two elements $(B_1, (B_t)_1)^T$ and $(B_2, (B_t)_2)^T$; and there is a (uniform in x_1) gap between the kernel and the bottom of the positive spectrum.

Assumption 2: (*Unique, simple negative eigenvalue*) For each $\beta \in (0, \gamma)$, $v \in (-1, 1)$ and $x_1 \in [0, 2\pi\alpha^{-1}]$, the operator $\mathcal{L}$ has a unique simple, negative eigenvalue $\lambda_1 = \lambda_1(\beta, v, x_1) < 0$ associated to the unit $L^2 \times L^2$-norm eigenfunction $(\widetilde{B}, \hat{B})^T$. Moreover, there is $\lambda_1^0 < 0$ depending on β and v only, such that $\lambda_1 \leq \lambda_1^0$ for all x_1.

Theorem 9.9.4. *Under Assumptions 1 and 2, SG breathers are stable under small $H^2 \times H^1$ perturbations. More precisely, there are $\eta_0 > 0$ and $K_0 > 0$, only depending on β, such that if $0 < \eta < \eta_0$ and*

$$\left\|(u_0, u_1) - (B_{\beta,v}, (B_{\beta,v})_t)(t = 0, \cdot; 0, 0)\right\|_{H^2 \times H^1} < \eta, \tag{9.9.1}$$

then there are real-valued parameters $x_1(t)$ and $x_2(t)$ for which the global $H^2 \times H^1$-solution $(u, u_t)(t)$ of (9.1.1) with initial data (u_0, u_1) satisfies

$$\sup_{t \in \mathbb{R}} \left\|(u, u_t)(t) - (B_{\beta,v}, (B_{\beta,v})_t)(t, \cdot; x_1(t), x_2(t))\right\|_{H^2 \times H^1} < K_0 \eta,$$

with similar estimates for the derivatives of the shift parameters x_1, x_2.

Below we present compelling numerical evidence revealing that Assumptions 1 and 2 do hold for any breather solution, regardless of its size, initial velocity or position, see Figs. 9.2, 9.4 and 9.1.

Remark 9.9.5. Looking for previous contributions to the stability problem we have found the work by Ercolani, Forest and McLaughlin [EFM85], where a sketch of the stability proof is presented. To be more precise, their argument states that, by using the inverse scattering theory applied to a small perturbation of the breather, the corresponding solution to (9.1.1) must remain uniformly close in time to a modulated breather solution. However, it is important to stress that any rigorous argument involving the inverse scattering method requires a *nontrivial amount of decay* on the initial data, an assumption that is not needed in our case.

The purpose of the next Section is to give the main ingredients for the proof of Theorem 9.9.4. In a first subsection, we prove some energy and momentum identities, allowing us to show that, at least formally, there is only one negative direction, and moreover, it is possible to control such a bad direction by using only the breather (B, B_t) as replacement.

9.10 Spectral analysis

Let $z = z(x)$ and $w = w(x)$ be two functions, to be specified below, and let $(B, B_t)(t, x; x_1, x_2)$ be any breather solution, with parameters x_1, x_2 possibly depending on time. In this section we describe part of the spectrum of the operator $\mathcal{L}$. We start with the following result.

Lemma 9.10.1. *$\mathcal{L}$ is a linear, unbounded operator in $L^2(\mathbb{R})^2$, with dense domain $H^4(\mathbb{R}) \times H^2(\mathbb{R})$. Moreover, $\mathcal{L}$ is self-adjoint.*

Proof. Exercise. □

A consequence of the above result is the fact that the spectrum of $\mathcal{L}$ is real-valued. Furthermore, $\mathcal{L}$ is a compact perturbation of a constant coefficients operator. Now we look at the kernel of $\mathcal{L}$. Let us fix $t \in \mathbb{R}$, $v \in (-1, 1)$ and $\beta \in (0, \gamma)$. It is not difficult to check that any breather (B, B_t) can be seen as a function independent of time, by considering the profiles

$$\begin{aligned}(B, B_t)(t, x; x_1, x_2) &= (B, B_t)(0, x; \tilde{x}_1, \tilde{x}_2), \quad \tilde{x}_1 := t + x_1, \ \tilde{x}_2 := -vt + x_2, \\ &=: (B, B_t)(x; \tilde{x}_1, \tilde{x}_2).\end{aligned} \tag{9.10.1}$$

Moreover, it turns out that if[3]

$$(z(x; \tilde{x}_1, \tilde{x}_2), w(x; \tilde{x}_1, \tilde{x}_2), \lambda(\tilde{x}_1, \tilde{x}_2)) \in H^4 \times H^2 \times \mathbb{R}$$

satisfy the eigenvalue-eigenfunction problem $\mathcal{L}[z, w] = \lambda(z, w)^T$ with (B, B_t) as in (9.10.1), then the new function

$$(\tilde{z}, \tilde{w})(x; \tilde{x}_1, \tilde{x}_2) := (z, w)(x - \tilde{x}_2; \tilde{x}_1, \tilde{x}_2)$$

will satisfy the same spectral equation with (B, B_t) replaced by $(B, B_t)(x; \tilde{x}_1 - \tilde{x}_2, 0)$. After redefining $\tilde{x}_1$, we can assume without loss of generality that (B, B_t) is only depending on x, β, v and $\tilde{x}_1$, and therefore $(\tilde{z}, \tilde{w})$ will also depend on the same variables. We will have $\lambda = \lambda(\beta, v, \tilde{x}_1)$ only. Finally, since the breather was periodic in the variable x_1, with period $2\pi/\alpha$, we can also assume

$$\tilde{x}_1 \in \left(0, \frac{2\pi}{\alpha}\right), \quad \alpha = \sqrt{\gamma^2 - \beta^2}.$$

In conclusion, with no loss of generality we set through this section $(B, B_t) = (B, B_t)(0, x; x_1, 0)$, with x_1 lying in the interval $[0, 2\pi/\alpha]$ and each breather-like potential periodic on that variable. The same property applies for the functions $(B_1, (B_t)_1)$ and $(B_2, (B_t)_2)$.

[3]Of course, we also have a dependence on β and v, but it is not needed at this moment.

In Section 9.12 we perform numerical computations that reveals that both conditions are naturally correct for every set of parameters that we numerically tested. The main consequence of the two preceding assumptions is the following direct coercivity property (recall the quadratic form Q, associated to $\mathcal{L}$, and defined in (9.8.11)):

Lemma 9.10.2. *Let $(z,w) \in H^2(\mathbb{R}) \times H^1(\mathbb{R})$, and (B, B_t) be any SG breather, and let $(\widetilde{B}, \hat{B})^T$ be the first eigenfunction from Assumption 2. If (z,w) are such that they satisfy the orthogonality conditions*

$$\int_{\mathbb{R}} (z,w)(B_1,(B_1)_t)^T = \int_{\mathbb{R}} (z,w)(B_2,(B_2)_t)^T = 0, \tag{9.10.2}$$

then there is $\mu_0 = \mu_0(\beta, v) > 0$ such that

$$Q[z,w] \geq \mu_0 \|(z,w)\|^2_{H^2\times H^1} - \frac{1}{\mu_0}\left|\int_{\mathbb{R}} (z,w)(\widetilde{B},\dot{B})^T\right|^2. \tag{9.10.3}$$

Following a similar strategy as in [AlMu13], we must use a different orthogonality condition in order to ensure a good control on the scaling parameter β, in such a form that we can run a stability argument without using modulations, which are very difficult to estimate for breather solutions.[4] Indeed, consider the new direction

$$\begin{pmatrix} A \\ \tilde{A} \end{pmatrix} := (1+v^2)\begin{pmatrix} \sin B - B_{xx} \\ B_t \end{pmatrix} + 2v\begin{pmatrix} B_{tx} \\ -B_x \end{pmatrix} \in H^2 \times H^1. \tag{9.10.4}$$

Note that from (9.8.17), we have

$$\mathcal{L}[B_0, \widetilde{B_0}] = \begin{pmatrix} A \\ \tilde{A} \end{pmatrix}.$$

We must have in mind that $(A, \tilde{A})^T$ is a sort of *generalized negative direction* (see (9.8.18)), that will replace $(\widetilde{B}, \hat{B})$ in (9.10.3). The advantage of taking this new set of orthogonality conditions comes from the fact that $(A, \tilde{A})^T$ are **naturally associated to the energy and momentum conservation laws**, which are only $H^1 \times L^2$ based, and therefore, low-regularity conserved quantities compared with the $H^2 \times H^1$ global dynamics. Indeed, we prove the following

Lemma 9.10.3. *Let $(z,w) \in H^2 \times H^1$ satisfying the orthogonality conditions (9.10.2), and $(A, \tilde{A})$ the direction defined in* (9.10.4). *Then there is $\nu_0 = \nu_0(\beta, v) > 0$ such that*

$$Q[z,w] \geq \nu_0 \|(z,w)\|^2_{H^2\times H^1} - \frac{1}{\nu_0}\left|\int_{\mathbb{R}} (z,w)(A,\tilde{A})^T\right|^2.$$

[4]In other words, we control the variations of the scaling parameter β at least at the second order in the error term (z,w).

Proof. We follow a similar strategy as stated in [AlMu13]. It is enough to prove that, under the orthogonality conditions (9.10.2), and the additional constraint

$$\int_{\mathbb{R}} \begin{pmatrix} z \\ w \end{pmatrix} \cdot \begin{pmatrix} A \\ \tilde{A} \end{pmatrix} = 0,$$

we have

$$\mathcal{Q}[z,w] \geq \nu_0 \|(z,w)\|^2_{H^2\times H^1}.$$

We write

$$(z,w) = (\tilde{z},\tilde{w}) + \delta_0(\widetilde{B},\hat{B}), \quad (B_0,\widetilde{B}_0) = (b_0,\widetilde{b}_0) + \gamma_0(\widetilde{B},\hat{B}).$$

Note that

$$\int_{\mathbb{R}} (B_0,\widetilde{B}_0)(B_1,(B_t)_1)^T = \int_{\mathbb{R}} (B_0,\widetilde{B}_0)(B_2,(B_t)_2)^T = 0,$$

and we can assume

$$\int_{\mathbb{R}} (\tilde{z},\tilde{w})(\widetilde{B},\hat{B})^T = \int_{\mathbb{R}} (b_0,\widetilde{b}_0)(\widetilde{B},\hat{B})^T = 0.$$

Therefore,

$$\begin{aligned} \mathcal{Q}[z,w] &= \mathcal{Q}[\tilde{z}+\delta_0\widetilde{B},\tilde{w}+\delta_0\hat{B}] \\ &= \mathcal{Q}[\tilde{z},\tilde{w}] + \delta_0^2\mathcal{Q}[\widetilde{B},\hat{B}] = \mathcal{Q}[\tilde{z},\tilde{w}] - \delta_0^2\lambda_0^2. \end{aligned} \tag{9.10.5}$$

Now we must replace δ_0 and λ_0 by suitable expressions. First of all,

$$0 = \int_{\mathbb{R}} (z,w)(A,\tilde{A})^T = \int_{\mathbb{R}} (z,w)\mathcal{L}[B_0,\widetilde{B}_0] = \int_{\mathbb{R}} (B_0,\widetilde{B}_0)\mathcal{L}[z,w].$$

Since $\mathcal{L}[z,w] = \mathcal{L}[\tilde{z},\tilde{w}] - \delta_0\lambda_0^2(\widetilde{B},\hat{B})^T$ and $(B_0,\widetilde{B}_0) = (b_0,\widetilde{b}_0) + \gamma_0(\widetilde{B},\hat{B})$, we obtain

$$\int_{\mathbb{R}} (b_0,\widetilde{b}_0)\mathcal{L}[\tilde{z},\tilde{w}] = \gamma_0\delta_0\lambda_0^2. \tag{9.10.6}$$

This last expression involves γ_0, δ_0 and λ_0. We want now an expression for γ_0. We have

$$\begin{aligned} \int_{\mathbb{R}} (A,\tilde{A})(B_0,\widetilde{B}_0)^T &= \int_{\mathbb{R}} (B_0,\widetilde{B}_0)\mathcal{L}[B_0,\widetilde{B}_0] \\ &= \int_{\mathbb{R}} \Big((b_0,\widetilde{b}_0) + \gamma_0(\widetilde{B},\hat{B})\Big)\Big(\mathcal{L}[b_0,\widetilde{b}_0] - \gamma_0\lambda_0^2(\widetilde{B},\hat{B})^T\Big) \\ &= \mathcal{Q}[b_0,\widetilde{b}_0] - \gamma_0^2\lambda_0^2. \end{aligned}$$

Note in addition that from (9.8.17) and (9.8.18),

$$\int_{\mathbb{R}} (A,\tilde{A})(B_0,\widetilde{B}_0)^T = \int_{\mathbb{R}} (B_0,\widetilde{B}_0)\mathcal{L}[B_0,\widetilde{B}_0] = -\frac{8}{\beta}(1+3v^2).$$

Therefore,

$$\gamma_0^2\lambda_0^2 = Q[b_0,\widetilde{b_0}] + \frac{8}{\beta}(1+3v^2).$$

Replacing this last expression in (9.10.5), and using (9.10.6), we get

$$\begin{aligned}Q[z,w] &= Q[\tilde{z},\tilde{w}] - \frac{(\gamma_0\delta_0\lambda_0^2)^2}{\gamma_0^2\lambda_0^2}\\ &= Q[\tilde{z},\tilde{w}] - \frac{\left(\displaystyle\int_{\mathbb{R}}(b_0,\widetilde{b_0})\mathcal{L}[\tilde{z},\tilde{w}]\right)^2}{Q[b_0,\widetilde{b_0}] + \frac{8}{\beta}(1+3v^2)}.\end{aligned} \tag{9.10.7}$$

Note that the denominator above is **always positive**. If now $(\tilde{z},\tilde{w}) = \lambda(b_0,\widetilde{b_0})$, for some $\lambda \neq 0$, we have

$$\begin{aligned}\left(\int_{\mathbb{R}}(b_0,\widetilde{b_0})\mathcal{L}[\tilde{z},\tilde{w}]\right)^2 &= \left(\int_{\mathbb{R}}(\tilde{z},\tilde{w})\mathcal{L}[\tilde{z},\tilde{w}]\right)\left(\int_{\mathbb{R}}(b_0,\widetilde{b_0})\mathcal{L}[b_0,\widetilde{b_0}]\right)\\ &= Q[\tilde{z},\tilde{w}]Q[b_0,\widetilde{b_0}],\end{aligned}$$

and therefore,

$$\frac{\left(\displaystyle\int_{\mathbb{R}}(b_0,\widetilde{b_0})\mathcal{L}[\tilde{z},\tilde{w}]\right)^2}{Q[b_0,\widetilde{b_0}] + \frac{8}{\beta}(1+3v^2)} = \frac{Q[\tilde{z},\tilde{w}]Q[b_0,\widetilde{b_0}]}{Q[b_0,\widetilde{b_0}] + \frac{8}{\beta}(1+3v^2)} = \rho Q[\tilde{z},\tilde{w}],$$

with $\rho \in (0,1)$ independent of (z,w). We have then,

$$Q[z,w] = (1-\rho)Q[\tilde{z},\tilde{w}].$$

Now if $(\tilde{z},\tilde{w})$ lies in the orthogonal vector space spanned by $(b_0,\widetilde{b_0})$, $(\widetilde{B},\hat{B})$, and the kernel $(B_1,(B_1)_t)$, $(B_2,(B_2)_t)$, we have that $Q[\tilde{z},\tilde{w}]$ defines an internal product, for which the Cauchy-Schwarz's inequality holds:

$$\left(\int_{\mathbb{R}}(b_0,\widetilde{b_0})\mathcal{L}[\tilde{z},\tilde{w}]\right)^2 < Q[\tilde{z},\tilde{w}]Q[b_0,\widetilde{b_0}],$$

(note that there is no equality since $(b_0,\tilde{b}_0)$ and $(\tilde{z},\tilde{w})$ are not orthogonal). We conclude now that

$$\frac{\left(\displaystyle\int_{\mathbb{R}}(b_0,\widetilde{b_0})\mathcal{L}[\tilde{z},\tilde{w}]\right)^2}{Q[b_0,\widetilde{b_0}] + \frac{8}{\beta}(1+3v^2)} < \frac{Q[\tilde{z},\tilde{w}]Q[b_0,\widetilde{b_0}]}{Q[b_0,\widetilde{b_0}] + \frac{8}{\beta}(1+3v^2)} = \rho Q[\tilde{z},\tilde{w}],$$

with $\rho \in (0,1)$. We conclude that

$$Q[z,w] > (1-\rho)Q[\tilde{z},\tilde{w}] \geq 0.$$

Therefore, in (9.10.5) we get

$$Q[\tilde{z},\tilde{w}] \geq \delta_0^2\lambda_0^2.$$

Finally, we have

$$\begin{aligned}
Q[z,w] &> (1-\rho)Q[\tilde{z},\tilde{w}] \\
&= \frac{1}{2}(1-\rho)Q[\tilde{z},\tilde{w}] + \frac{1}{2}(1-\rho)Q[\tilde{z},\tilde{w}] \\
&\geq \frac{1}{2}(1-\rho)\mu_0\|(\tilde{z},\tilde{w})\|^2_{H^1\times L^2} + \frac{1}{2}(1-\rho)\delta_0^2\lambda_0^2 \\
&\gtrsim \|(\tilde{z},\tilde{w})\|^2_{H^2\times H^1} + \|(\widetilde{B},\hat{B})\|^2_{H^2\times H^1} \gtrsim \|(z,w)\|^2_{H^2\times H^1}.
\end{aligned}$$

□

9.11 Proof of stability

In this subsection we prove Theorem 9.9.4. This proof follows similar lines as in [AlMu13]. Assume (u_0,u_1) satisfy the hypothesis (9.9.1), for some $\eta < \eta_0$ small. Let (u,u_t) be the associated $H^2\times H^1$ solution to (9.1.1) with initial data (u_0,u_1). Given $K^* > 1$ to be chosen later, we denote $T^* = T^*(K^*) > 0$ as the maximal time for which, for all time $T \in (0,T^*]$,

$$\sup_{t\in[0,T]} \|(u,u_t)(t) - (B,B_t)_{\beta,v}(0,\cdot;\tilde{x}_1(t),\tilde{x}_2(t))\|_{H^2\times H^1} < K^*\eta,$$

is satisfied for some choice of modulation parameters $\tilde{x}_1(t)$ and $\tilde{x}_2(t)$, not necessarily unique. For the sake of simplicity, and if no confusion arises, we denote

$$(B,B_t) := (B,B_t)_{\beta,v}.$$

If we assume that $T^*(K^*)$ is finite, we can choose (assuming η_0 smaller if necessary), using the Implicit Function Theorem, special parameters $x_1(t)$ and $x_2(t)$ such that

$$\int_{\mathbb{R}} (z,w)(B_1,(B_1)_t)^T = \int_{\mathbb{R}} (z,w)(B_1,(B_1)_t)^T = 0,$$

where

$$z(t,x) := u(t,x) - B(t,x;x_1(t),x_2(t)), \tag{9.11.1}$$

$$w(t,x) := u_t(t,x) - B_t(t,x;x_1(t),x_2(t)), \tag{9.11.2}$$

and B_1, B_2 are given in (9.8.3)-(9.8.4). These conditions are well-defined since the matrix with coefficients

$$\begin{pmatrix} \int_{\mathbb{R}} \left(B_1^2 + ((B_1)_t)^2\right) & \int_{\mathbb{R}} B_1 B_2 + (B_1)_t (B_2)_t \\ \int_{\mathbb{R}} B_1 B_2 + (B_1)_t (B_2)_t & \int_{\mathbb{R}} B_2^2 + ((B_2)_t)^2 \end{pmatrix}$$

has nonzero determinant everywhere (cf. Corollary 9.8.3). Moreover, thanks to (9.9.1), we have

$$\|(z, w)(0)\|_{H^2 \times H^1} \lesssim \eta,$$

with constant independent of K^*.

Consider now the decomposition (9.11.1)-(9.11.2), with (B, B_t) depending on the modulation parameters $x_1(t)$ and $x_2(t)$. A simple argument reveals that $\mathcal{H}[B, B_t](t)$ is still independent of time (see [AlMu13] for a similar proof). Therefore we obtain

$$Q[z, w](t) \leq Q[z, w](0) + \sup_{t \in [0,T^*]} \mathcal{N}[z, w](t) \lesssim \eta^2 + (K^*)^3 \eta^3,$$

Now, from Proposition 9.10.3 we have

$$\|(z, w)(t)\|_{H^2 \times H^1}^2 \lesssim \eta^2 + (K^*)^3 \eta_0^3 + \left| \int_{\mathbb{R}} (z, w)(A, \tilde{A})^T (t) \right|^2, \tag{9.11.3}$$

Recall that $(A, \tilde{A})$ was defined in (9.10.4). Finally, using the energy and momentum conservation laws (9.3.1)-(9.3.2), evaluated at two different times, we obtain a good control on the term

$$\left| \int_{\mathbb{R}} (z, w)(A, \tilde{A})^T (t) \right|^2 .$$

Indeed, we have

$$\frac{1}{2} \int_{\mathbb{R}} ((B_x + z_x)^2 + (B_t + w)^2)(t) + \int_{\mathbb{R}} (1 - \cos(B + z))(t) =$$
$$= \frac{1}{2} \int_{\mathbb{R}} ((B_x + z_x)^2 + (B_t + w)^2)(0) + \int_{\mathbb{R}} (1 - \cos(B + z))(0),$$

from which

$$\int_{\mathbb{R}} z(\sin B - B_{xx})(t) + \int_{\mathbb{R}} B_t w(t) = \int_{\mathbb{R}} z(\sin B - B_{xx})(0) + \int_{\mathbb{R}} B_t w(0)$$
$$+ O\left(\sup_{t \in [0,T^*]} \|(z, w)(t)\|_{H^1 \times L^2}^2 \right).$$

Similarly,

$$\int_{\mathbb{R}} (B_x + z_x)(B_t + w)(t) = \int_{\mathbb{R}} (B_x + z_x)(B_t + w)(0),$$

so that

$$\int_{\mathbb{R}} (B_{tx}z - B_x w)(t) = \int_{\mathbb{R}} (B_{tx}z - B_x w)(0) + O\left(\sup_{t\in[0,T^*]} \|(z,w)(t)\|^2_{H^1\times L^2} \right).$$

Consequently, reconstructing $(A, \tilde{A})$ as it was defined in (9.10.4),

$$\begin{aligned}\left| \int_{\mathbb{R}} (z,w)(A,\tilde{A})^T(t) \right| &\lesssim \left| \int_{\mathbb{R}} (z,w)(A,\tilde{A})^T(0) \right| + \sup_{t\in[0,T^*]} \|(z,w)(t)\|^2_{H^2\times H^1} \\ &\lesssim \eta + (K^*)^2\eta^2.\end{aligned}$$

Finally, substituting in (9.11.3), we get

$$\|(z,w)(t)\|^2_{H^2\times H^1} \lesssim \eta^2 + (K^*)^3\eta^3.$$

Taking K^* large enough, and then $\eta(K^*) > 0$ small, we have

$$\|(z,w)(t)\|^2_{H^2\times H^1} \leq \frac{1}{4}(K^*)^2\eta^2,$$

so we improve the original estimate on (z, w), which contradicts the finiteness of T^*. Therefore, for all K^* large enough, $T^* = +\infty$.

9.12 Numerical analysis

The purpose of this paragraph is to give strong evidence of the fact that Assumptions 1 and 2 in page 145 do hold. We have run a similar numerical scheme as in the case of the mKdV and Gardner breathers, but now it is important to mention that the resulting linear system is no longer a scalar valued one; instead, we have to deal with a vector valued linear system, which formally implies that we need twice as many test functions as usual, compared to the mKdV case.

First test. Recall that we can always assume $x_2 = 0$ and $x_1 \in [0, 2\pi/\alpha]$. We have run a primary test with $v = 0.7$, different values of β, and x_1 varying between -0.4 and 0.3. It seems that the SG case is numerically simpler than the mKdV one, because of the fact that the SG breather requires one derivative less for its definition. Therefore, with $N = 25$ (that is to say, 50 test functions), we have obtained the results summarized in Fig. 9.2, in agreement with Assumptions 1 and 2 up to two significant digits. We also check the expected oscillatory behavior of the minimal eigenvalue of the linearized operator $\mathcal{L}$ around a SG breather for variations of the shift x_1. We show this result in Fig. 9.3 where in that example, we consider $\beta = 0.5$ and $v = 0.7$ and shift parameter x_1 varying from -3.0 to 3.0.

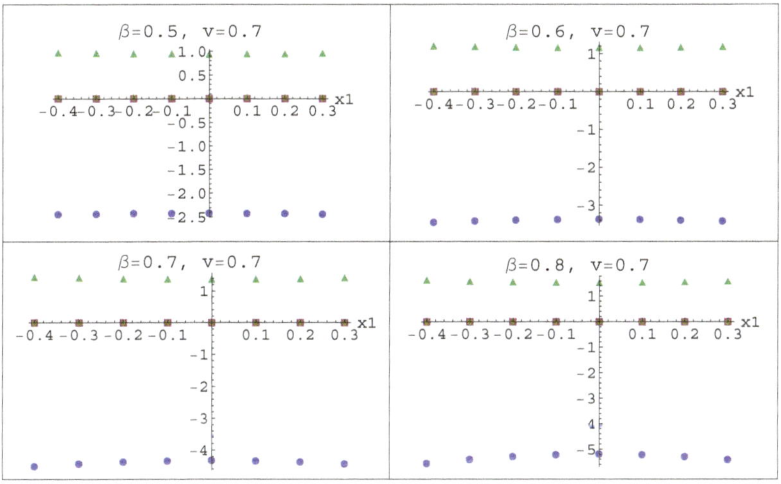

Fig. 9.2 The graph of the four minimal eigenvalues (from the first negative eigenvalue (blue circle), to the double zero kernel (purple box and brown diamond) and the first positive eigenvalue (green triangle)) of the SG breather with $\beta = 0.5, 0.6, 0.7$ and 0.8, with x_1 varying between -0.4 and 0.3. The velocity of the relativistic breather is taken as $v = 0.7$. Note that two numerical eigenvalues are placed near zero, see also Fig. 9.1.

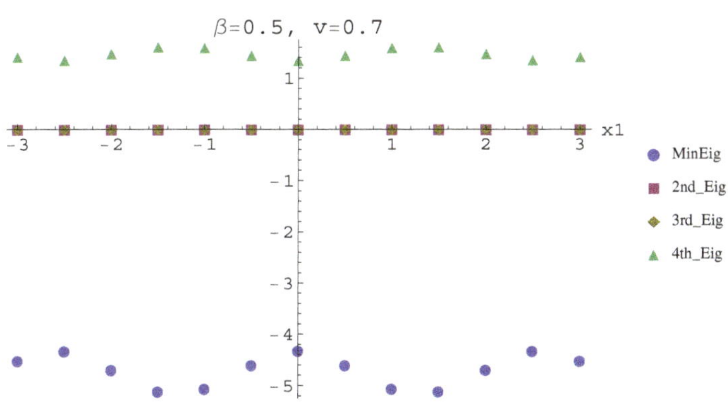

Fig. 9.3 The graph of the four minimal eigenvalues of the SG breather case, with $\beta = 0.5$, $v = 0.7$ and with x_1 varying between -3.0 and 3.0.

Second test. When β is fixed (e.g. $\beta = 0.5$), and we move the velocity v of the relativistic SG breather from 0.0 to 0.7, we obtain the results written in Fig. 9.1 (table of eigenvalues). The corresponding graphics are shown in Fig. 9.4.

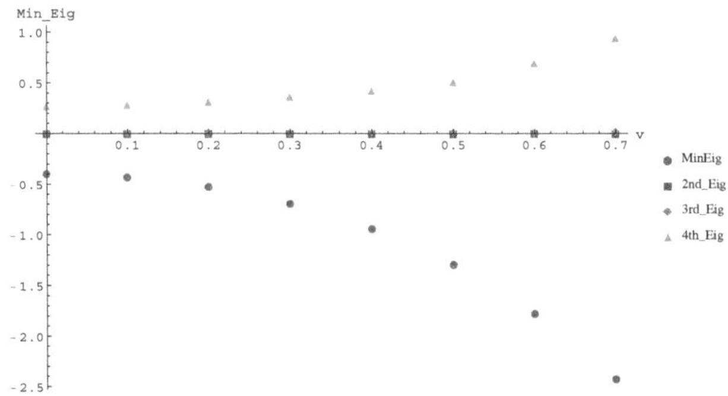

Fig. 9.4 The graph of the four minimal eigenvalues of the SG breather with $\beta = 0.5$ and the velocity v of the relativistic breather varying between 0.0 to 0.7. The shifts $x_1 = 0.1$, $x_2 = 0$ and we take $N = 50$ eigenfunctions (see also Figs. 9.5 and 9.1, the table below left).

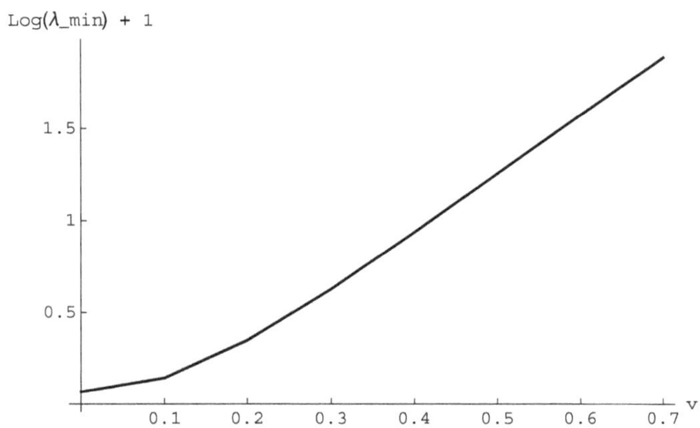

Fig. 9.5 The graph of the logarithm of the *absolute* value of the minimal eigenvalue, plus one, for the SG breather with $\beta = 0.5$ and the velocity v of the relativistic breather varying between 0.0 to 0.7. The shifts are $x_1 = 0.1$, $x_2 = 0$. The continuous curve is the set of straight lines that join each pair of points, and the approximate slope among two adjacent points varies from 3.1 to 3.2.

Additionally, from the numerical tests we have found the following

(1) For v fixed, as long as β approaches zero, we see that the negative eigenvalue converges to zero, a phenomenon that is in concordance with the convergence to zero in L^∞ norm of the breather as $\beta \to 0$.

(2) As v increases to the speed of light 1, the quality and accuracy of the resulting numerical results strongly decreases, implying that one needs to take N even larger to recover the desired spectral stability.

Table 9.1 The first four eigenvalues of $\mathcal{L}$: (left), for $\beta = 0.5$, $x_1 = 0.1$, $x_2 = 0$, and the velocity v of the relativistic SG breather varying from 0.0 to 0.7, as corresponding to Fig. 9.4 (right), for $\beta = 0.8$, $v = 0.7$, $x_2 = 0$, and x_1 varying from -0.4 to 0.3, as corresponding to bottom right Fig. 9.2. All computations were made with $N = 50$. The third and fourth columns (left) and eighth and ninth columns (right) represent the approximate kernel of $\mathcal{L}$ respectively.

v	1st eig.	2nd	3rd	4th	x_1	1st eig.	2nd	3rd	4th
0.0	-0.3932	0.0020	0.0124	0.2783	-0.4	-5.555	0.0004	0.0011	1.658
0.1	-0.4246	0.0023	0.0107	0.2883	-0.3	-5.404	0.0003	0.0012	1.612
0.2	-0.5206	0.0040	0.0069	0.3191	-0.2	-5.290	0.0002	0.0013	1.576
0.3	-0.6871	0.0037	0.0085	0.3698	-0.1	-5.218	0.0001	0.0014	1.553
0.4	-0.9367	0.0023	0.0162	0.4292	0.0	-5.194	0.0001	0.0014	1.545
0.5	-1.2892	0.0026	0.0180	0.5143	0.1	-5.218	0.0001	0.0014	1.553
0.6	-1.7724	0.0067	0.0094	0.6994	0.2	-5.290	0.0002	0.0013	1.576
0.7	-2.4203	0.0036	0.0254	0.9489	0.3	-5.404	0.0003	0.0012	1.612

In conclusion, our numerical computations reveal that, in any of parameters regime considered, there is only one negative eigenvalue, as well as a well-defined two dimensional kernel, and a spectral gap between the kernel and the continuum spectrum, supporting Assumptions 1 and 2 in page 145.

9.13 Stability in the energy space

It turns out that the previous result can be improved to the level of allowing $H^1 \times L^2$ perturbations or namely, in the energy space for the sine-Gordon equation (9.1.1). However, now the proof is not variational, since the $H^1 \times L^2$-stability will be a consequence of a *dynamical rigidity* for small perturbations of breather solutions associated to the integrability of the equation.

Theorem 9.13.1 (Stability in the energy space). *The SG breather is nonlinearly stable under perturbations in the energy space $H^1 \times L^2$. More precisely, there exist $C_0 > 0$ and $\eta_0 > 0$ such that the following holds. Let (ϕ, ϕ_t) be a solution of* (9.1.1), *with initial data (ϕ_0, ϕ_1) such that*

$$\|(\phi_0, \phi_1) - (B, B_t)(0, \cdot\, ; \beta, 0, 0)\|_{H^1 \times L^2} < \eta, \tag{9.13.1}$$

for some $0 < \eta < \eta_0$ sufficiently small. Then, there are shifts $x_1(t), x_2(t) \in \mathbb{R}$ well-defined and differentiable such that

$$\sup_{t \in \mathbb{R}} \|(\phi(t), \phi_t(t)) - (B, B_t)(t, \cdot\, ; \beta, x_1(t), x_2(t))\|_{H^1 \times L^2} < C_0 \eta. \tag{9.13.2}$$

Moreover, we have

$$\sup_{t\in\mathbb{R}} |x_1'(t)| + |x_2'(t)| \lesssim C_0\eta.$$

The proof of this result uses the Bäcklund Transformations (BT) between two solutions (ϕ, φ) for SG, and fixed parameter a:

$$\begin{aligned}
\varphi_x - \phi_t &= \frac{1}{a}\sin\left(\frac{\varphi+\phi}{2}\right) + a\sin\left(\frac{\varphi-\phi}{2}\right),\\
\varphi_t - \phi_x &= \frac{1}{a}\sin\left(\frac{\varphi+\phi}{2}\right) - a\sin\left(\frac{\varphi-\phi}{2}\right).
\end{aligned}$$

These two equations allow to describe the dynamics of breathers as in (9.13.2) using (9.13.1), and the reduction of complexity induced by the BT. These ideas have been successfully implemented in several contexts: Hoffman and Wayne [HoWa13] used BT to show a first abstract result for stability, and applied this result to show in particular orbital stability of the SG kink solution, as well as briefly discussing the case of n-kinks. Next, Mizumachi and Pelinovsky [MiPe12] showed L^2 stability of the NLS soliton using this approach. The case in [AlMu15] was the first where a BT was used in the case of breathers. In this paper, we will follow in spirit the Hoffman and Wayne ideas for the SG case.

In order to fix ideas, let us consider the case of the SG breather. First of all, we will need to work with complex-valued solutions. We will introduce the kink function (K, K_t):

$$(K, K_t)(t,x) := \left(4\arctan\left(e^{\beta x + i\alpha t}\right), \frac{4i\alpha e^{\beta x+i\alpha t}}{1+e^{2(\beta x+i\alpha t)}}\right).$$

This complex-valued SG solution is connected to zero via a BT of parameter $\beta - i\alpha$. We have

$$\begin{aligned}
K_x &= \frac{1}{\beta - i\alpha}\sin\left(\frac{K}{2}\right) + (\beta - i\alpha)\sin\left(\frac{K}{2}\right),\\
K_t &= \frac{1}{\beta - i\alpha}\sin\left(\frac{K}{2}\right) - (\beta - i\alpha)\sin\left(\frac{K}{2}\right).
\end{aligned} \tag{9.13.3}$$

On the other hand, the complex-valued kink is a singular solution to SG, in the sense that it blows up (in L^∞ norm) in a sequence of times t_k, without accumulation point. The advantage of introducing the profiles (K, K_t) in Theorem 9.13.1 is the following: this profile is connected to the breather (B, B_t) via a new BT of parameter $\beta + i\alpha$:

$$\begin{aligned}
B_x - K_t &= \frac{1}{\beta + i\alpha}\sin\left(\frac{B+K}{2}\right) + (\beta + i\alpha)\sin\left(\frac{B-K}{2}\right),\\
B_t - K_x &= \frac{1}{\beta + i\alpha}\sin\left(\frac{B+K}{2}\right) - (\beta + i\alpha)\sin\left(\frac{B-K}{2}\right).
\end{aligned} \tag{9.13.4}$$

The identities (9.13.3) and (9.13.4) are then perturbed using the Implicit Function Theorem. For details, see [MuPa19].

9.14 Sine-Gordon breathers and rigorous integrability techniques

Recently, Chen-Liu-Lu [CLL20] presented *full* asymptotic stability results for solitons and breathers of SG by using the nonlinear steepest descent approach to oscillatory Riemann-Hilbert problems and under the condition of weighted initial data strictly included in the energy space.

By full asymptotic stability we mean convergence to zero with explicit decay rates of the remainder in the solution, with norms taken on the complete real line, but with particular weights. These weights precisely avoid the existence of breathers in the dynamics. This is contrary to the so-called local asymptotic stability, which is proved in the energy space only, and not in the full real line, because of the existence of possible breathers in the energy space.

In fact, Chen-Liu-Lu proved the *soliton resolution* for the SG case, giving detailed asymptotic analysis for the Cauchy problem of SG. They proved that the solution of the Cauchy problem for any generic data of SG can be written as a superposition of solitons, breathers and radiation in the following explicit way:

$$u(t,x) = \sum_{l=1}^{N_1} Q_l + \sum_{l=1}^{N_2} B_l + R(t,x).$$

Here Q_l are soliton-type solutions (in this case, kinks and anti-kinks), B_l are breathers, and finally R is the remainder. Consider the space of Sobolev type (but with associated weights)

$$H^{2,s}_{\sin}(\mathbb{R}) \times H^{1,s}(\mathbb{R}) := \left\{ \vec{f} = (f_1, f_2) \ : f_1 \in H^{2,s}_{\sin}(\mathbb{R}), f_2 \in H^{1,s}(\mathbb{R}) \right\},$$

where

$$H^{k,s}_{\sin}(\mathbb{R}) := \{ f \in L^1_{loc}(\mathbb{R}) : \partial_x^j f \in L^{2,s}(\mathbb{R}), \ 1 \le j \le k, \quad \text{and} \quad \sin(f/2) \in L^{2,s} \}$$

$$H^{k,s}(\mathbb{R}) := \{ f \in L^1_{loc}(\mathbb{R}) : \partial_x^j f \in L^{2,s}(\mathbb{R}), \ 0 \le j \le k \}.$$

The precise statement for the *soliton resolution for the sine-Gordon equation with generic data* is the following:

Theorem 9.14.1. *Let $\vec{f}(0) = (f_1, f_2) \in H^{2,s}_{sin}(\mathbb{R}) \times H^{1,s}(\mathbb{R})$ with $s > \frac{1}{2}$ be given generic initial data in the sense of the Definition 2.6 ([CLL20, p.21]). Let f be the corresponding solution of SG, ensured by local and global well-posedness.*

Then the solution can be written as the superposition of kinks, anti-kinks, breathers and radiation as follows

$$\sin(f(t,x)) = \sum_{l=1}^{N_1} \sin(Q_l(t,x)) + \sum_{l=1}^{N_2} \sin(B_l(t,x)) + f_{r,sin}(t,x)$$

and

$$1-\cos(f(t,x)) = \sum_{l=1}^{N_1}(1-\cos(Q_l(t,x))) + \sum_{l=1}^{N_2}(1-\cos(B_l(t,x))) + f_{r,cos}(t,x),$$

where $1-\cos(B_l(t,x))$, $\sin(B_l(t,x))$ *and* $1-\cos(Q_l(t,x))$ *are given by explicit formulae (see [CLL20]).*

As for the radiation terms, $f_{r,sin}(t,x)$ *and* $f_{r,cos}(t,x)$ *by setting* $x = \mathrm{v}t$, *one obtains explicit formulae in [CLL20].*

As a consequence of Theorem 9.14.1, a *full asymptotic stability result* for the sine-Gordon equation can be obtained. To study the asymptotic stability, they first constructed general reflectionless solutions, namely, solutions without radiation terms. Starting with the following scattering data

$$S_D = \left\{0,\ \{z^K_{0,k}, c^K_{0,k}\}_{k=1}^{N_1},\ \{z^B_{0,j}, c^B_{0,j}\}_{j=1}^{N_2}\right\} \in H^{1,1} \bigoplus \mathbb{C}^{2N_1} \bigoplus \mathbb{C}^{2N_2},$$

and assuming $z^K_{0,k} = \rho^K_{0,k} i$ and $z^B_{0,k} = \rho^B_{0,k} e^{i\omega_{0,j}}$, denote

$$v^l_{0,j} = \frac{1-\rho^l_{0,j}}{1+\rho^l_{0,j}},\ l = K, B,$$

with all $v^l_{0,j}$ different. If $N_l = 0$, $l = 1,2$ then the corresponding set will be the empty set. Therefore, one can construct a reflectionless solution $N(t,x)$ using the discrete scattering data (9.1) and Proposition 2.9 (see [CLL20]). Moreover, this solution has the following asymptotics as $t \to \infty$:

$$\sin(N(t,x)) = \sum_{l=1}^{N_1} \sin(Q_l(t,x)) + \sum_{l=1}^{N_2} \sin(B_l(t,x)) + O(e^{-\varepsilon t}),$$

and

$$1-\cos(N(t,x)) = \sum_{j=1}^{N_1}(1-\cos(Q_j(t,x))) + \sum_{j=1}^{N_2}(1-\cos(B_j(t,x))) + O(e^{-\varepsilon t}),$$

where $\sin(Q_j(t,x))$, $1-\cos(Q_j(t,x))$ and $\sin(B_j(t,x))$, $1-\cos(Q_j(t,x))$ can be reconstructed. Therefore by the continuity of the scattering data, we can also consider the stability of the sum

$$\sum_{j=1}^{N_1} \sin(Q_j(t,x)) + \sum_{j=1}^{N_2} \sin(B_j(t,x)),$$

and

$$\sum_{j=1}^{N_1}(1-\cos(Q_j(t,x))) + \sum_{j=1}^{N_2}(1-\cos(B_j(t,x))),$$

provided the kinks, anti-kinks and breathers in these sums are sufficiently separated.

Corollary 9.14.2. *Consider the reflectionless solution $N(t,x)$ to SG. Assume that*

$$\|\vec{f}(0) - (N(0), \partial_t N(0))\|_{H^{2,1}\times H^{1,1}} < \varepsilon,$$

for $0 \leq \varepsilon \ll 1$ small enough. Let $\vec{f}$ be the solution of (9.1.1) *with the initial data $\vec{f}(0)$, then there exist scattering data*

$$S = \left\{ r(z),\ \{z_{1,k}, c_{1,k}\}_{k=1}^{N_1},\ \{z_{1,j}, c_{1,j}\}_{j=1}^{N_2} \right\} \in H^{1,1}(\mathbb{R}) \bigoplus \mathbb{C}^{2N_1} \bigoplus \mathbb{C}^{2N_2},$$

computed in terms of $\vec{f}(0)$ such that

$$\|r\|_{H^{1,1}} + \sum_{l=1}^{N_1} (|z_{0,l} - z_{1,l}| + |c_{0,l} - c_{1,l}|) + \sum_{l=1}^{N_2} (|z_{0,l} - z_{1,l}| + |c_{0,l} - c_{1,l}|) \lesssim \varepsilon.$$

Moreover, with the scattering data S one can write the solution f as follows

$$\sin(f(t,x)) = \sum_{l=1}^{N_1} \sin(\tilde{Q}_l(t,x)) + \sum_{l=1}^{N_2} \sin(\tilde{B}_l(t,x)) + f_{r,sin}(t,x)$$

and

$$1 - \cos(f(t,x)) = \sum_{l=1}^{N_1} \left(1 - \cos(\tilde{Q}_l(t,x))\right) + \sum_{l=1}^{N_2} \left(1 - \cos(\tilde{B}_l(t,x))\right) + f_{r,cos}(t,x),$$

where $\sin(Q_j(t,x))$, $1 - \cos(Q_j(t,x))$ and $\sin(B_j(t,x))$, $1 - \cos(Q_j(t,x))$ are reconstructed using scattering data (9.4) respectively. Here, the radiation terms $f_{r,sin}(t,x)$ and $f_{r,cos}(t,x)$ have the asymptotics given in Theorem 9.14.1 using the scattering data (9.4). The interested reader can consult [CLL20, p.21].

9.15 Exercises

(9.1) Prove by hand Lemma 9.9.1.

(9.2) Prove that SG breathers (B, B_t) are actually critical points of the functional $\mathcal{H}$.

(9.3) Prove by hand Lemma 9.6.1, namely that

$$P[B, B_t](t) = -8\beta v.$$

(9.4) Prove by hand Lemma 9.10.1.

(9.5) Prove that B and K are related by a BT, namely prove (9.13.4).

Chapter 10

Advanced topics in sine-Gordon and related breathers

10.1 Uniqueness of sine-Gordon breathers

An interesting open question remains. Are SG breathers all of the form (9.5.1)? This is not an easy question, and still unsolved. However, some interesting new results have appeared during past years. For the sake of brevity, we shall review a nice result here, and let the full domain of research to a chapter of our third book series.

Recently, Mandel [Man21] proved that the SG breather (9.5.1) is the only *quasimonochromatic* breather in the context of nonlinear wave equations in $\mathbb{R}^N$. Indeed, let us consider the more general nonlinear wave equations of the form (3.3.1), where the space dimension $N \in \mathbb{N}$ and the nonlinearity $g : \mathbb{R} \to \mathbb{R}$ are arbitrary.

Definition 10.1.1. Let $N \geq 1$. We call the function $u : \mathbb{R}^N \times \mathbb{R} \to \mathbb{R}$ a *quasi-monochromatic breather* if

$$u(t, x) = F(\sin(\omega t)p(x)), \quad x \in \mathbb{R}^N, \quad t \in \mathbb{R}, \tag{10.1.1}$$

for some $\omega \in \mathbb{R} \setminus \{0\}$ and nontrivial functions $F \in C^2(\mathbb{R}),\ \ p \in C^2(\mathbb{R}^N)$ such that $F(0) = 0$ and $p(x) \to 0$ as $|x| \to \infty$.

Mandel showed that in one spatial dimension the SG breather is, up to translation and dilation, the only one for (3.3.1) and that no such breathers exist in higher dimensions as long as g does not act like a linear function.

In fact, to rule out L^∞−small solutions of linear wave equations, we assume that $g : \mathbb{R} \to \mathbb{R}$ is not a linear function near zero, i.e., that there is a nontrivial interval $I \subset \mathbb{R}$ containing 0 with the property that there is no $\beta \in \mathbb{R}$ such that $g(z) = \beta z$ for all $z \in I$.

Theorem 10.1.2. *Assume $N \in \mathbb{N}$ and that $g : \mathbb{R} \to \mathbb{R}$ is not a linear function near zero.*

(i) In the case $N \geq 2$ there is no quasimonochromatic breather solution of (3.3.1).
(ii) In the case $N = 1$ each quasimonochromatic breather solution of (3.3.1) *is of the form $u(x,t) = \kappa u^*(x - x_0, t)$ for $x_0 \in \mathbb{R}, m, \omega, \kappa \in \mathbb{R} \setminus \{0\}$ and u^* as in* (9.5.1). *The nonlinearity then satisfies $g(z) = -(m^2 + \omega^2)\kappa \sin(\kappa^{-1} z)$ whenever $|z| < 2\pi|\kappa|$.*

This result holds regardless of any smoothness assumption on g nor any kind of growth condition at 0 or infinity. Moreover, our considerations are not limited to small perturbations of u^* or small breathers in whatever sense. Following the proof of Theorem 10.1.2 one also finds that quasimonochromatic breathers of wave equations on any open set $\Omega \subset \mathbb{R}^N$ with homogeneous Dirichlet conditions

$$\partial_{tt} u - \Delta u = g(u), \qquad \text{in } \Omega \times \mathbb{R}, \qquad u = 0 \text{ on } \partial\Omega \times \mathbb{R}$$

with profile functions $p \in C^2(\Omega)$ do not exist either (even if $N = 1$) provided that g is not a linear function near zero. As a consequence, we find that $C^2([0,1] \times \mathbb{R})$-solutions of the 1D wave equation from [Rab78, Theorem 1.6] are not of quasimonochromatic type. This might be true as well for the solutions from [BCN80, Cor82], but here our argument does not apply in a direct way since the solutions are not known to be twice continuously differentiable up to the boundary.

For completeness we briefly comment on the linear case $g(z) = \beta z, \ \beta \in \mathbb{R}$. Then the profile function p of any given quasimonochromatic breather of (3.3.1) satisfies the linear elliptic PDE

$$-\Delta p - (\omega^2 + \beta) p = 0, \quad \text{in } \mathbb{R}^N.$$

For $\beta < -\omega^2$ there are positive, radially symmetric and exponentially decaying solutions p, see [GNN81, Theorem 2].

In the case $\beta > -\omega^2, N \geq 2$ one can find radial as well as nonradial solutions of the associated Helmholtz equation all of which have infinitely many nodal domains and satisfy $|p(x)| + |\nabla p(x)| \gtrsim |x|^{\frac{1-N}{2}}$ in a suitable integrated sense, see [Swa79, Theorem 1] respectively [Kat59, Theorem 1a].

For $\beta > -\omega^2, N = 1$ all solutions are linear combinations of sin and cos so that breather solutions do not exist. So we see that the picture is already quite complete in the case of linear wave equations.

In the following let $u(x,t) = F(sin(\omega t)p(x))$ be a solution of (3.3.1) as in (10.1.1) with g as in the Theorem. Plugging in this ansatz we get for all $x \in \mathbb{R}^N$

such that $p(x) \neq 0$,

$$\begin{aligned}\partial_{tt}u(x,t) &= -\,\omega^2 \sin(\omega t)p(x)F'(\sin(\omega t)p(x))\\ &\quad + \omega^2 \cos(\omega t)^2 p(x)^2 F''(\sin(\omega t)p(x))\\ &= -\,\omega^2 zF'(z) + \omega^2(p(x)^2 - z^2)F''(z),\end{aligned}$$

and

$$\begin{aligned}\Delta u(x,t) &= \sin(\omega t)\Delta p(x)F'(\sin(\omega t)p(x))\\ &\quad + \sin(\omega t)2|\nabla p(x)|^2 F''(\sin(\omega t)p(x))\\ &= \frac{\Delta p(x)}{p(x)} zF'(z) + \frac{|\nabla p(x)|^2}{p(x)^2} z^2 F''(z),\end{aligned}$$

where $z = \sin(\omega t)p(x) \in [-||p||_\infty, +||p||_\infty]$. This and (3.3.1) imply for $x \in \mathbb{R}^N, z \in \mathbb{R}$ such that $p(x) \neq 0$, and $z \in [-||p||_\infty, +||p||_\infty]$,

$$\begin{aligned}&g(F(z)) + \omega^2 zF'(z) + \omega^2 z^2 F''(z)\\ &= p(x)^2\omega^2 F''(z) - \frac{\Delta p(x)}{p(x)} zF'(z) - \frac{|\nabla p(x)|^2}{p(x)^2} z^2 F''(z).\end{aligned} \tag{10.1.2}$$

If F was linear on $[-||p||_\infty, +||p||_\infty]$, then g would have to be linear on the nontrivial interval $I := \{F(z) : |z| \leq ||p||_\infty\}$ as well. Since the latter is not the case by assumption, we know that $z \to z^2 F''(z)$ does not vanish identically on that interval.

Multiplying (10.1.2) with $p(x)$ and choosing z according to $z^2 F''(z) \neq 0$ we find that p does not change sign. Indeed, if $p(x^*) \neq 0$ and $R > 0$ is the smallest radius such that p has a fixed sign in the open ball $B_R(x^*)$, then Hopf's Lemma [GilTru98, Lemma 3.4] implies $|\nabla p| > 0$ on $\partial B_R(x^*)$. But then (10.1.2) implies that Δp is unbounded on $\partial B_R(x^*)$, which contradicts $p \in C^2(\mathbb{R}^N)$.

Hence, p does not change sign and we will without loss of generality assume that p is positive. So (10.1.2) holds for all $x \in \mathbb{R}^N$ and all $z \in [-||p||_\infty, +||p||_\infty]$ and standard elliptic regularity theory gives $p \in C^\infty(\mathbb{R}^N)$.

Differentiating (10.1.2) with respect to x_i we get

$$\partial_i(p(x)^2)\omega^2 F''(z) - \partial_i\left(\frac{\Delta p(x)}{p(x)}\right) zF'(z) - \partial_i\left(\frac{|\nabla p(x)|^2}{p(x)^2}\right) z^2 F''(z) = 0.$$

Since p^2 is non-constant, we infer that F satisfies an ODE of the form

$$F''(z) = \frac{-\mu_2 z}{\omega^2 + \mu_1 z^2} F'(z), \quad |z| \leq ||p||_\infty, \quad \mu_1 \in \mathbb{R}, \quad \mu_2 \in \mathbb{R}\setminus\{0\}. \tag{10.1.3}$$

Here, $\mu_2 \neq 0$ is due to the fact that F is not a linear function. Each nontrivial solution of such an ODE satisfies $F'(z) \neq 0$ for almost all $z \in [-||p||_\infty, +||p||_\infty]$. Combining (10.1.2) and (10.1.3) we thus infer

$$-\partial_i(p(x)^2)\frac{\mu_2\omega^2 z}{\omega^2 + \mu_1 z^2} - \partial_i\left(\frac{\Delta p(x)}{p(x)}\right) z + \partial_i\left(\frac{|\nabla p(x)|^2}{p(x)^2}\right)\frac{\mu_2 z^3}{\omega^2 + \mu_1 z^2} = 0.$$

Since (10.1.3) holds for all $i \in \{1, \ldots, N\}$ and $z \in [-\|p\|_\infty, +\|p\|_\infty]$, we get

$$-\mu_1 \partial_i\left(\frac{\Delta p(x)}{p(x)}\right) + \mu_2 \partial_i\left(\frac{|\nabla p(x)|^2}{p(x)^2}\right) = 0,$$

$$-\mu_2 \partial_i(p(x)^2) - \partial_i\left(\frac{\Delta p(x)}{p(x)}\right) = 0.$$

Since $\mu_2 \neq 0$ we can find $\lambda_1, \lambda_2 \in \mathbb{R}$ such that

$$-\mu_1 \frac{\Delta p}{p} + \mu_2 \frac{|\nabla p|^2}{p^2} = -\lambda_2 \mu_1 + \lambda_1 \mu_2, \quad -\mu_2 p^2 - \frac{\Delta p}{p} = -\lambda_2.$$

This implies

$$|\nabla p|^2 = \lambda_1 p^2 - \mu_1 p^4, \quad -\Delta p + \lambda_2 p = \mu_2 p^3. \tag{10.1.4}$$

We now use (10.1.4) and the positivity of p to show that p is radially symmetric about its maximum point $p_0 \in \mathbb{R}^N$. We concentrate on the case $N \geq 2$ since the claim for $N \geq 1$ follows from the fact that $x \to u(x_0 + x)$ and $x \to u(x_0 - x)$ solve the same initial value problem. Since p vanishes at infinity, we must have $\lambda_1 \geq 0$ and since p does not change sign, $\lambda_2 \geq 0$, see [Swa79, Th.1]. Moreover, p attains its maximum at some point $x_0 \in \mathbb{R}^N$ with $p(x_0) > 0, |\nabla p(x_0)| = 0, \Delta p(x_0) \leq 0$. This and (10.1.4) implies $\lambda_1, \mu_1 > 0$ as well as $\mu_2 \geq 0$. So we know that (10.1.4) holds for

$$\lambda_1, \mu_1 > 0, \quad \lambda_2, \mu_2 \geq 0.$$

In the case $\lambda_2 > 0$ from [GNN81, Theorem 2] implies the radial symmetry about x_0, so we are left with the case $\lambda_2 = 0$. So let us assume $\lambda_2 = 0$. Liouville's theorem implies that $\mu_2 = 0$ is impossible, so we have $\mu_2 > 0$ in this case.

Define $\alpha := 1 - \mu_1 \mu_2 \in (-\infty, 1)$. In the case $\alpha \in (0, 1)$ the function $\psi(x) := p(x)\alpha$ satisfies

$$-\psi = -\alpha(\Delta p)p^{\alpha-1} - \alpha(\alpha - 1)|\nabla p|^2 p^{\alpha-2} = \alpha(\alpha - 1)\lambda_1 \psi.$$

In view of $\alpha(\alpha - 1)\lambda_1 > 0$ Theorem 1 from [Swa79] implies that ψ has infinitely many nodal domains, which contradicts the positivity of ψ. So this case cannot occur. In the case $\alpha \in (-\infty, 0)$ radial symmetry about x_0 follows once more from [GNN81, Theorem 2], so it remains to discuss the case $\alpha = 0$, i.e., $\mu_1 = \mu_2$. Then $\psi(x) := \log(p(x))$ satisfies

$$-\Delta\psi = -(\Delta p)p^{-1} + |\nabla p|^2 p^{-2} = \lambda_1 \psi,$$

and we find as above that ψ has to change sign, which is a contradiction. So we have shown that p is radially symmetric about x_0 also in the case $\lambda_2 = 0$. So we have

$$p(x) = p_0(|x - x_0|), \quad \text{where } p_0'(r)^2 = \lambda_1 p_0(r)^2 - \mu_1 p_0(r)^4, \;\; p_0'(0) = 0.$$

Solving this ODE gives

$$p_0(r) = A\cosh(mr) \quad \text{where } \lambda_1 = m^2, \mu_1 = m^2A^{-2}.$$

for some $A > 0, m \neq 0$. So $-\nabla p + \lambda_2 p = \mu_2 p^3$ can only hold for $N = 1$ as well as $\lambda_2 = m^2, \mu_2 = 2m^2A^{-2}$. Plugging these values into (10.1.3) and solving the ODE we get from $F(0) = 0, F \neq 0$

$$F(z) = 4\kappa \arctan\left(\frac{mz}{A\omega}\right), \qquad \text{for some } \kappa \in \mathbb{R}\setminus\{0\}.$$

This implies that the breather solution is given by

$$u(x,t) = F(\sin(\omega t)p(x)) = F(\sin(\omega t)p_0(|x - x_0|)) = \kappa u^*(x - x_0, t)$$

for u^* as in (10.1.1). So we have proved the nonexistence of such breathers for $N \geq 2$ from claim (i) and the uniqueness statement from claim (ii). To see that this solution formula determines the nonlinearity g, we combine (10.1.3) and (10.1.4) to get

$$\begin{aligned} &p(x)^2\omega^2F''(z) - \frac{\Delta p(x)}{p(x)}zF'(z) - \frac{|\nabla p(x)|^2}{p(x)^2}z^2F''(z) \\ &= \frac{m^2(m^2z^2 - A^2\omega^2)}{m^2z^2 + A^2\omega^2}F'(z)z. \end{aligned}$$

Therefore,

$$\begin{aligned} g(F(z)) &= -\omega^2zF'(z) - \omega^2z^2F''(z) + \frac{m^2(m^2z^2 - A^2\omega^2)}{m^2z^2 + A^2\omega^2}F'(z)z \\ &= \frac{(m^2+\omega^2)(m^2z^2 - A^2\omega^2)}{m^2z^2 + A^2\omega^2}F'(z)z \\ &= \frac{4Am\kappa\omega(m^2+\omega^2)(m^2z^2 - A^2\omega^2)z}{(m^2z^2 + A^2\omega^2)^2}. \end{aligned}$$

Plugging in $z = \frac{A\omega}{m}\tan(\frac{y}{4\kappa})$ for $|y| < 2\pi|\kappa|$ we get $F(z) = y$ and hence

$$\begin{aligned} g(y) &= \frac{4A^2\omega^2\kappa(m^2+\omega^2)(A^2\omega^2\tan(\frac{y}{4\kappa})^2 - A^2\omega^2)\tan(\frac{y}{4\kappa})}{\left(A^2\omega^2\tan(\frac{y}{4\kappa})^2 + A^2\omega^2 + A^2\omega^2\right)^2} \\ &= \frac{4\kappa(m^2+\omega^2)(\tan(\frac{y}{4\kappa})^2 - 1)\tan(\frac{y}{4\kappa})}{(\tan(\frac{y}{4\kappa})^2 + 1)^2}F'(z)z \\ &= 4\kappa(m^2+\omega^2)\left(\sin\left(\frac{y}{4\kappa}\right)^2 - \cos\left(\frac{y}{4\kappa}\right)^2\right)\sin\left(\frac{y}{4\kappa}\right)\cos\left(\frac{y}{4\kappa}\right) \\ &= -2\kappa(m^2+\omega^2)\sin\left(\frac{y}{2\kappa}\right)\cos\left(\frac{y}{2\kappa}\right) \\ &= -\kappa(m^2+\omega^2)\sin\left(\frac{y}{\kappa}\right). \end{aligned}$$

Mandel [Man21] also proposed two interesting conjectures, deeply related with de Giorgi conjectures. These are:

Conjecture 1. Let $N \in \mathbb{N}, N \geq 2$ and let $p \in C^2(\mathbb{R}^N)$ be a solution of (10.1.4) for some $\lambda_1, \lambda_2, \mu_1, \mu_2 \in \mathbb{R}$ that satisfies (8). Then there are $\gamma, m, z \in \mathbb{R}$ such that

$$p(x) = \frac{\gamma}{\cosh(m(x_1 - z))}.$$

Conjecture 2. Let $N \in \mathbb{N}, N \geq 2$ and let $p \in C^2(\mathbb{R}^N)$ be a solution of (10.1.4) for some $\lambda_1, \lambda_2, \mu_1, \mu_2 \in \mathbb{R}$ that satisfies $\partial_1 p(x) x_1 < 0$ for all $x \in \mathbb{R}^N$ such that $x_1 \neq 0$. Then there are $\gamma, m \in \mathbb{R}$ such that

$$p(x) = \frac{\gamma}{\cosh(m x_1)}.$$

Recently, Mandel's idea has been extended to the case of generalized Korteweg-de Vries, Kadomtsev-Petviashvilii and Zakharov-Kuznetsov models. See [FFM24, ACM24] for details.

10.2 The parabolic sine-Gordon model

There is an interesting problem concerning the parabolic version of SG (9.1.1), and recently addressed by Gao and Xu [GaXu23]. Consider the following one-dimensional parabolic SG equation posed on the domain $\mathbb{T} = [-\pi, \pi]$:

$$\begin{cases} \partial_t u = \kappa^2 \Delta u - f(u), & (t, x) \in (0, \infty) \times \mathbb{T}, \\ u|_{t=0} = u_0, \end{cases} \tag{10.2.1}$$

where κ is the diffusion constant, $f(u) = -\sin u = F'(u)$, and $F(u) = \cos u + 1$ is the usual double-well potential. The unknown function $u : \mathbb{T} \to \mathbb{R}$ represents the concentration difference in the phase field context and typically takes the value in $[-\pi, \pi]$.

Concerning the SG equation (9.1.1), G.L. Lamb [Lam71] has shown that it admits a solution of the following form

$$\phi = 4 \arctan\big(X(x)/T(t)\big).$$

Based on the expression, we can easily find out that $X(x)$ and $T(t)$ must satisfy

$$\begin{aligned} (X'(x))^2 &= kX^4(x) + mX^2(x) + n, \\ (T'(t))^2 &= -kT^4(t) + (m-1)T^2(t) - n, \end{aligned}$$

where k, m, n are arbitrary constants. For a special case, $k = 0$ and $n = m$, we can get

$$X(x) = \sinh(mx), \quad T(t) = \frac{m}{m-1} \cosh\Big(\sqrt{(m-1)}t\Big),$$

and

$$\phi = 4\arctan\left(\frac{\gamma\sinh\left(\frac{x}{\sqrt{1-\gamma^2}}\right)}{\cosh\left(\frac{t}{\sqrt{1-\gamma^2}}\right)}\right),$$

where

$$\gamma = \sqrt{1 - 1/m}.$$

This solution has been already discovered, which was used to describe a collision between two kinks in a centre-of-mass coordinate system, see [PeSky62] and reference therein. It is natural to extend the study of SG (9.1.1) in high spatial dimensions. Particularly, for two spatial dimensions, the equation reads as

$$\phi_{xx} + \phi_{yy} - \phi_{tt} = \sin(\phi) \tag{10.2.2}$$

which is invariant under the Lorentz transformation. Barone-Esposito-Magee [BEM71] concluded that particle-like solutions of (10.2.2) occur only if the boundary conditions at the origin are able to provide the gradient of

$$\nabla\phi \approx -\alpha\frac{\bar{u}_r}{r},$$

where $\bar{u}_r$ is a unit vector in a radial direction. Similar conclusions also hold for high dimensions. We refer the readers to [CKS03] for more background. The parabolic SG equation (10.2.1) can be viewed as the parabolic version of the usual wave-type SG (9.1.1) in the phase-field context. Recently, in the work of Li-Quan- Tang-Yang [LQT22], they observe the symmetry breaking phenomenon in the typical high-precision computation of the Allen-Cahn equation with very smooth initial data. It happens due to the manifestation of the gradual accumulation of non-negligible machine round off errors over time. In fact, this phenomenon also happens for the other equations with double-well potential. In order to solve this issue, they proposed a new Fourier filter method which worked successfully for a class of initial data with certain symmetry and band-gap properties. Furthermore, they obtained several interesting results concerning the one-dimensional Allen-Cahn equation with periodic boundary conditions, including the complete classification of steady states, the monotonicity property of the associated energy to the diffusion constant and the asymptotic behavior of the profiles in [LQT21]. We will consider the same problem for SG, where the nonlinearity term behaves similarly as the Allen-Cahn equation. Concerning the steady state of the SG, Cheng-Li-Quan-Yang [CLQY21] established the maximum principle on the torus for all dimensions and classified all bounded steady states in the one-dimensional for (10.2.1).

To get a better understanding of the steady state, we shall consider the following elliptic equation on the real line, i.e.

$$\kappa^2 \Delta u + \sin(u) = 0, \quad x \in \mathbb{R}. \tag{10.2.3}$$

Theorem 10.2.1. *Let $0 < \kappa < 1$ and m_κ be the largest positive integer such that $m_\kappa \kappa < 1$, then equation (10.2.3) admits exactly m_κ non-constant 2π periodic solutions up to some translation and odd reflection.*

In [CLQY21], Cheng-Li-Quan-Yang showed that any periodic solution of (10.2.3) is bounded, i.e. $|u| \le \pi$. Expect for the trivial solutions $u = \pm\pi$, all the other nontrivial solutions satisfy $|u| < \pi$ and our interests are focused on these solutions. For equation (10.2.3), the associated energy is

$$E_\kappa(u) = \int_{\mathbb{R}} \left(\frac{\kappa^2}{2} |\nabla u|^2 + \cos u + 1 \right) dx. \tag{10.2.4}$$

Based on the (10.2.4), it is possible to define the energy of the periodic functions

$$E_\kappa(u) = \int_{\mathbb{T}} \left(\frac{\kappa^2}{2} |\nabla u|^2 + \cos u + 1 \right) dx.$$

Let

$$E_\kappa = \inf_{u \in \mathcal{F}} E_\kappa(u), \tag{10.2.5}$$

where

$$\mathcal{F} := \left\{ \phi \in H^1(\mathbb{T}) \middle| \phi(x) = \phi(x + 2\pi) \text{ and } |\phi| < \pi, \ x \in \mathbb{R} \right\}.$$

Usually, the 2π-periodic solution u_κ of (10.2.3) is called ground state if u_κ is odd and achieves E_κ. For a fixed $\kappa \in (0, 1)$, it can be obtained from [CLQY21, Proposition 3.2] that the ground state solution is unique after translation and reflection if necessary. For convenience, we introduce the notion of odd zero-up ground states U_κ, which is odd and the derivative at zero is positive (see Definition 3.1). For any $0 < \kappa < 1$, defining m_κ to be the unique integer such that

$$\frac{1}{1 + m_\kappa} \le \kappa < \frac{1}{m_\kappa},$$

and each $j = 1, \ldots, m_\kappa$,

$$\tilde{u}_{\kappa,j}(x) = U_{j\kappa}(jx),$$

then $\{\tilde{u}_{\kappa,j}\}_{j=1}^{m_\kappa}$ are all the possible odd zero-up solutions to (9.1.1). In fact, the energy of $\tilde{u}_{\kappa,j}(x)$ satisfies

$$E_\kappa(\tilde{u}_{\kappa,j}) = \int_{\mathbb{T}} \left(\frac{\kappa^2}{2} (\partial_x \tilde{u}_{\kappa,j})^2 + \cos \tilde{u}_{\kappa,j} + 1 \right) dx = E_{j\kappa}(U_{j\kappa}).$$

Now we state the following theorem on the energy functional $E_\kappa(u)$ of the 2π-periodic solutions.

Theorem 10.2.2. *Let $E_\kappa(u)$ be defined in* (10.2.5). *Then it can be achieved for any $\kappa > 0$. Furthermore the following hold:*

(a) $E_\kappa = 4\pi$ for $\kappa > 1$ and it is only achieved by trivial solution $u = 0$.
(b) E_κ is achieved by U_κ whenever $\kappa \in (0, 1)$.
(c) If $0 < \kappa_1 < \kappa_2 < 1$, then there is strict monotonicity property $E_{\kappa_1} < E_{\kappa_2}$.
(d) The odd zero-up ground state U_κ satisfies

$$\left| U_\kappa(x) - 2 \arcsin \tanh \left(\frac{x}{\kappa} \right) \right| \leq Ce^{-\frac{d}{\kappa}}$$

where $C > 0, d > 0$, and

$$\lim_{\kappa \to 0} \frac{E_\kappa}{\kappa} = 16 > 0.$$

(e) The 2π-periodic solutions of (10.2.3) *have the following property: any 2π-periodic solution of* (10.2.3) *which is not identically $\pm\pi$ or 0 must coincide with $\tilde{u}_{\kappa,j}$, where $j < 1/\kappa$. Furthermore $E_\kappa(u) = E_{m\kappa}(U_{m\kappa})$.*

Moreover, it is possible to rigorously prove the convergence results of the solution and identify the explicit profiles. In particular, if $\kappa > 1, u(x, t)$ converges to 0 with the rate $O(e^{-(\kappa^2-1)t})$, while $\kappa = 1$, the convergence rate becomes $O(t^{-2})$. Furthermore, they obtain the explicit profiles for $\kappa \geq 1$ (see [GaXu23, Theorem 4.2]). Summarizing the above discussion, they get the following precise and definite convergence results for $\kappa \geq 1$ and $0 < \kappa < 1$, respectively.

Theorem 10.2.3. *Let $0 < \kappa < 1$. Assume the initial data $u_0 : \mathbb{R} \to \mathbb{R}$ is 2π-periodic, odd and nonnegative in $[0, \pi]$, then we have $u(x, t) \to U_\kappa$ or 0 as $t \to \infty$. Moreover, if $u_0 \neq 0$ and $E_\kappa(u_0) \geq 4\pi$, then $u(x, t) \to U_\kappa$ as $t \to \infty$ and the rate of convergence is exponential in time.*

Theorem 10.2.4. *Let $\kappa \geq 1$. Assume u_0 is 2π-periodic, odd and bounded. Suppose u is the solution to* (10.2.1) *corresponding to the initial data u_0. If $\kappa > 1$, we have*

$$u(t, x) = e^{-(\kappa^2-1)t} \alpha_* \sin(x) + r(t), \quad \forall t \geq 1,$$

where the constant α_* depends on (u_0, κ), and $\|r(t)\|_{H^2(\mathbb{T})} = o(e^{-3(\kappa^2-1)t})$ as $t \to +\infty$. For $\kappa = 1$, we have

$$u(t, x) = t^{-\frac{1}{2}} \beta_* \sin(x) + r_1(t),$$

where the constant β_* depends on u_0, and $\|r_1(t)\|_{H^2(\mathbb{T})} = o(t^{-\frac{1}{2}})$ as $t \to +\infty$.

10.3 On Boussinesq breathers

In this section we deal with the *bad* Boussinesq equation [39]:

$$u_{tt} - u_{xx} - \left(u_{xx} + \frac{3}{2}u^2\right)_{xx} = 0. \tag{10.3.1}$$

Here $u = u(t,x)$ is a real-valued function, and $(t,x) \in \mathbb{R}^2$. Bad Boussinesq is a completely integrable equation, and it was first derived by Boussinesq itself [39] as an approximate model for shallow water waves in the one-wave regime and is linearly ill-posed, as revealed by a standard frequency analysis.[1]

For the rest of the section, we write the bad Boussinesq equation (10.3.1) as a wave system:

$$\begin{cases} u_t = v_x, \\ v_t = u_x + \left(u_{xx} + \frac{3}{2}u^2\right)_x. \end{cases} \tag{10.3.2}$$

Some lower order conservation laws for the bad Boussinesq equation (10.3.1) are

$$\int_{\mathbb{R}} u, \qquad \int_{\mathbb{R}} v, \tag{10.3.3}$$

the momentum

$$P[u,v] := \int_{\mathbb{R}} uv, \tag{10.3.4}$$

the energy

$$E[u,v] := \int_{\mathbb{R}} \left(u_x^2 - u^2 - v^2 - u^3\right), \tag{10.3.5}$$

and the H^2 conservation law

$$F[u,v] := \int_{\mathbb{R}} \left(\frac{2}{3}u_{xx}^2 - \frac{2}{3}v_x^2 - \frac{2}{3}u_x^2 - 3uu_x^2 + uv^2 + \frac{1}{2}u^4 + \frac{1}{3}u^3\right). \tag{10.3.6}$$

Following [196, eqn. (4)], the exact soliton solution of (10.3.2) is,

$$u(t,x) := Q_c(x - ct - x_0), \tag{10.3.7}$$

where

$$Q_c(x) := (c^2-1)Q\left(\sqrt{c^2-1}x\right), \quad Q(x) := \operatorname{sech}^2\left(\frac{x}{2}\right), \tag{10.3.8}$$

[1] Note that a more general model can be derived from (10.3.1), namely with $p, q \in \{-1, 1\}$,

$$u_{tt} - pu_{xx} - q\left(u_{xx} + \frac{3}{2}u^2\right)_{xx} = 0.$$

For $p = 1$, depending on the sign of q, above equations are referred with the prefix *bad* ($q = +1$) Boussinesq or *good* ($q = -1$) Boussinesq.

with $x_0 \in \mathbb{R}$ and $c > 0$ fixed parameters. Moreover in this case, even if the equation (10.3.1) is linearly ill-posed, it still does have breather-like solutions. Indeed, let $\alpha, \beta > 0$ and $x_1, x_2 \in \mathbb{R}$ be fixed parameters. From [196, eqn. (4)] we have that the breather $(u, v) := (B, \widetilde{B})$ solves the bad Boussinesq equation (10.3.2). Here, B and $\widetilde{B}$ are given by the simplified expressions

$$B(t,x) := 4\partial_x^2 \log \phi(t,x), \qquad \widetilde{B}(t,x) := 4\partial_{tx} \log \phi(t,x), \tag{10.3.9}$$

and where ϕ is the phase function

$$\phi(t,x) := 1 + e^{y_2}(2\cos y_1 + \rho e^{y_2}). \tag{10.3.10}$$

Additionally, the variables y_1, y_2 and the parameter ρ have the form

$$y_1 := \alpha x - \delta t + x_1, \quad y_2 := \beta x - \gamma t + x_2, \quad \rho := \frac{\delta^2 - \alpha^2 + 4\alpha^4}{\gamma^2 - \beta^2 - 4\beta^4}, \tag{10.3.11}$$

where the speeds γ, δ are given by

$$\delta := \frac{b}{2\gamma}, \quad \gamma := \left(\frac{1}{2}\left(a + \sqrt{a^2 + b^2}\right)\right)^{1/2}, \tag{10.3.12}$$

and finally

$$a := (\beta^2 - \alpha^2) + (\beta^2 - \alpha^2)^2 - 4\alpha^2\beta^2, \quad b := 2\alpha\beta(1 + 2(\beta^2 - \alpha^2)). \tag{10.3.13}$$

For the sake of completeness, the explicit expression for the breather B (10.3.9) is the following:

$$B(t,x) := 4\frac{\rho\beta^2 - \alpha^2 + \sqrt{\rho}(\beta^2 - \alpha^2)\cosh(\tilde{y}_2)\cos(y_1) + 2\alpha\beta\sinh(\tilde{y}_2)\sin(y_1)}{\left(\cos(y_1) + \sqrt{\rho}\cosh(\tilde{y}_2)\right)^2}, \tag{10.3.14}$$

where $\psi = \log(\sqrt{\rho})$ and $\tilde{y}_2 = y_2 + \psi$.

Remark 10.3.1 (Periodicity of the bad Boussinesq breather). Note that from (10.3.10) we have that generically B is time periodic, up to a shift in space. Indeed, if $b \neq 0$, then after a time $T := 2\pi\delta^{-1}$ the breather becomes the same, but translated a distance $\gamma T\beta^{-1}$. Assume $\gamma = 0$, that is to say, the speed of the breather is zero. Therefore, from (10.3.12) we have $b = 0$ and $a < 0$. If $b = 0$, then necessarily

$$\alpha^2 - \beta^2 = \frac{1}{2}.$$

This condition leads to $a = -\frac{1}{4} - 4\alpha^2\beta^2 < 0$. Therefore, zero speed breathers, time periodic, are naturally allowed in the bad Boussinesq model.

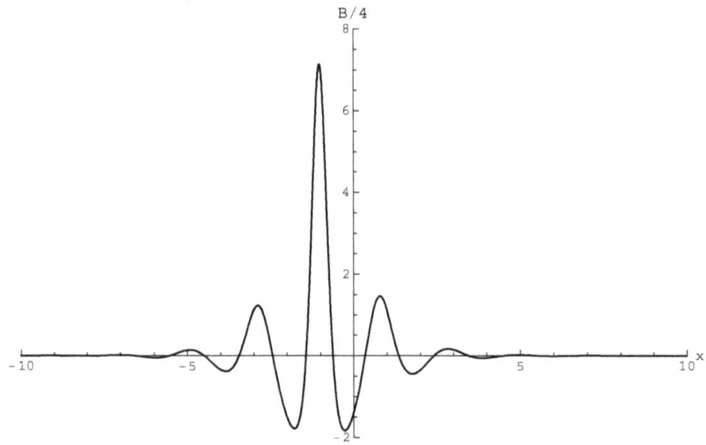

Fig. 10.1 The bad Boussinesq breather (10.3.14) $\frac{B}{4}$, with $\alpha = 3$ and $\beta = 1$ at $t = 0$.

The breather (10.3.9)-(10.3.14) was explicitly found in [196], except for some minor misprints in the formula. Here we recover the correct expression, which has only some minor modifications. In [196], it was also argued that the good Boussinesq equation ($q = -1$) cannot have *regular* breather solutions.

In this section, we prove that, contrary to what is expected, bad Boussinesq breathers *still preserve* its first order variational structure, independently of the linearly unstable character of the equation. The main result is an unpublished result worked by M. F. Cortez and the authors:

Theorem 10.3.2 (Variational structure of breathers in bad Boussinesq). *Let* $(B, \widetilde{B})(t, x)$ *be any breather solution for the bad Boussinesq equation* (10.3.2), *with parameters* $\alpha, \beta > 0$, $x_1, x_2 \in \mathbb{R}$. *Then there are constants* $\lambda, \sigma \in \mathbb{R}$, *only depending on* α, β, *such that for all* $t \in \mathbb{R}$

$$\begin{aligned} B_{xxxx} + B_{xx} + \frac{9}{4}B_x^2 + \frac{9}{2}BB_{xx} + \frac{3}{4}B^2 + \frac{3}{4}\widetilde{B}^2 + \frac{3}{2}B^3 \\ - \frac{3}{2}\lambda\left(B_{xx} + B + \frac{3}{2}B^2\right) + \frac{3}{4}\sigma\widetilde{B} = 0, \end{aligned} \tag{10.3.15}$$

and

$$\widetilde{B}_{xx} + \frac{3}{2}B\widetilde{B} - \frac{3}{2}\lambda\widetilde{B} + \frac{3}{4}\sigma B = 0. \tag{10.3.16}$$

Moreover, each breather $(B, \widetilde{B})$ *is a critical point of the Lyapunov functional*

$$\mathcal{H}[u, v] := F[u, v] + \lambda E[u, v] + \sigma P[u, v], \tag{10.3.17}$$

where the energy $E[u, v]$ and momentum $P[u, v]$ are defined in (10.3.5) *and* (10.3.4) *respectively, and $F[u, v]$* (10.3.6) *is the third conserved quantity for the bad Boussinesq equation.*

Remark 10.3.3. Theorem 10.3.2 reveals that no matter the ill-posed character of the bad Boussinesq equation, its breather follows a suitable variational structure. Of course, this structure is still unstable, however, it corroborates the principle that every breather has a variational structure.

Remark 10.3.4. In the previous Theorem 10.3.2, the parameters λ, σ are not easy to compute, and have some undesirable form. For the sake of completeness, we include them in (10.3.38), at the end of this section.

Let us first define the following auxiliary function:

$$\begin{aligned}\hat{B}(t,x) &:= 4\partial_t \log(\phi(t,x))\\ &= -4\frac{\sqrt{\rho}\gamma(\cosh(y_2+\psi)+\sinh(y_2+\psi))+\gamma\cos(\alpha y_1)-\delta\sin(\alpha y_1)}{\sqrt{\rho}\cosh(y_2+\psi)+\cos(\alpha y_1)},\end{aligned} \tag{10.3.18}$$

which satisfies that $\widetilde{B} = \partial_x \hat{B}$, as easily checked. The following result is an intermediary but key step to the proof of Theorem 10.3.2. Its proof is painful and complicated, for this reason we prove it at the end of this section.

Lemma 10.3.5. *Let $(B, \widetilde{B})$ be a breather solution of* (10.3.1) *as in* (10.3.9). *Then*

$$\widetilde{B}_{tx} - \frac{3}{4}B_x^2 + \frac{3}{2}BB_{xx} + \frac{3}{4}B^2 + \frac{3}{4}\widetilde{B}^2 + \frac{3}{2}B^3 - \frac{3}{2}\lambda\hat{B}_t + \frac{3}{4}\sigma\widetilde{B} = 0, \tag{10.3.19}$$

and

$$\widetilde{B}_{xx} + \frac{3}{2}B\widetilde{B} - \frac{3}{2}\lambda\widetilde{B} + \frac{3}{4}\sigma B = 0. \tag{10.3.20}$$

Proof. See the end of this section for a detailed proof of this system of nonlinear identities. □

Proof of Theorem 10.3.2, assuming Lemma 10.3.5. First we prove (10.3.15). Note that (10.3.16) is precisely the identity (10.3.20).

We have from (10.3.1) and the fact that $B_t = \widetilde{B}_x$,

$$B_{xxxx} = B_{tt} - \left(B + \frac{3}{2}B^2\right)_{xx} = \widetilde{B}_{tx} - (B_{xx} + 3B_x^2 + 3BB_{xx}).$$

Hence, substituting this relation in (10.3.15) and having in mind the bad Boussinesq equation (10.3.1), we rewrite (10.3.15) as follows

$$\begin{aligned}
&\widetilde{B}_{tx} - (B_{xx} + 3B_x^2 + 3BB_{xx}) + B_{xx} + \frac{9}{4}B_x^2 + \frac{9}{2}BB_{xx}\\
&\quad + \frac{3}{4}B^2 + \frac{3}{4}\widetilde{B}^2 + \frac{3}{2}B^3 - \frac{3}{2}\lambda(B_{xx} + B + \frac{3}{2}B^2) + \frac{3}{4}\sigma\widetilde{B}\\
&= \widetilde{B}_{tx} - \frac{3}{4}B_x^2 + \frac{3}{2}BB_{xx} + \frac{3}{4}B^2 + \frac{3}{4}\widetilde{B}^2 + \frac{3}{2}B^3 - \frac{3}{2}\lambda\hat{B}_t + \frac{3}{4}\sigma\widetilde{B} = 0,
\end{aligned}$$

where we have used the identity (10.3.19). We conclude.

Now, in order to see that the breather solution $(B, \widetilde{B})$ is actually a critical point of the Lyapunov functional (10.3.17), we consider $\tilde{u} = B + z$, $\tilde{v} = \widetilde{B} + w$, with $(B, \widetilde{B})$ any type of breather solution and (z, w) small $H^2 \times H^1$ functions respectively. In what follows, we compute the linear contribution of each conserved quantity above, expanded around a breather solution $(B, \widetilde{B})$. We have

$$\begin{aligned}
\mathcal{H}[B + z, \widetilde{B} + w] &= F[B + z, \widetilde{B} + w] + \lambda E[B + z, \widetilde{B} + w] + \sigma P[B + z, \widetilde{B} + w]\\
&= \mathcal{H}[B, \widetilde{B}]\\
&\quad + \int_{\mathbb{R}} z\Big[\frac{4}{3}B_{xxxx} + \frac{4}{3}B_{xx} + 3B_x^2 + 6BB_{xx} + B^2 + \widetilde{B}^2\\
&\qquad + 2B^3 - 2\lambda(B_{xx} + B + \frac{3}{2}B^2) + \sigma\widetilde{B}\Big]\\
&\quad + \int w\left[\frac{4}{3}\widetilde{B}_{xx} + 2B\widetilde{B} - 2\lambda\widetilde{B} + \sigma B\right] + O(z^2, w^2, zw).
\end{aligned} \tag{10.3.21}$$

Here $O(z^2, w^2, zw)$ denote quadratic terms in z, w and its lower order derivatives. Therefore the first variation of the Lyapunov functional give us the system of ODEs satisfied by the bad Boussinesq breather, previously proved. □

Proof of Lemma 10.3.5. Recall the breather functions $(B, \widetilde{B}, \hat{B})$ involved in (10.3.9) and (10.3.18). We write them in the following short form:

$$B(t, x) := 4\frac{F}{N^2}, \quad \widetilde{B}(t, x) := -4\frac{G}{N^2}, \quad \hat{B}(t, x) := -4\frac{H}{N}. \tag{10.3.22}$$

Here

$$\begin{aligned}
F(t, x) \equiv F &:= (\rho\beta^2 - \alpha^2) + \sqrt{\rho}(\beta^2 - \alpha^2)\cosh(y_2 + \psi)\cos(\alpha y_1)\\
&\quad + 2\alpha\beta\sqrt{\rho}\sinh(y_2 + \psi)\sin(\alpha y_1),\\
G(t, x) \equiv G &:= (\rho\beta\gamma - \alpha\delta) + \sqrt{\rho}(\beta\gamma - \alpha\delta)\cosh(y_2 + \psi)\cos(\alpha y_1)\\
&\quad + (\alpha\gamma + \beta\delta)\sqrt{\rho}\sinh(y_2 + \psi)\sin(\alpha y_1),\\
H(t, x) \equiv H &:= \sqrt{\rho}\gamma(\cosh(y_2 + \psi) + \sinh(y_2 + \psi)) + \gamma\cos(\alpha y_1) - \delta\sin(\alpha y_1),\\
N(t, x) \equiv N &:= \sqrt{\rho}\cosh(y_2 + \psi) + \cos(\alpha y_1).
\end{aligned} \tag{10.3.23}$$

Then, with this notation, we can compute each of the terms in (10.3.19) and (10.3.20). Specifically, in the case of the identity (10.3.19) we have:

$$\begin{aligned} \widetilde{B}_{tx} &= \frac{M_1}{N^6}, \\ M_1 &= N^2\Big[6N_xN_tG + N^2G_{tx} - 2N(N_xG_t + G_xN_t + GN_{tx})\Big], \end{aligned} \tag{10.3.24}$$

$$\begin{aligned} -\frac{3}{4}B_x^2 + \frac{3}{2}BB_{xx} &= \frac{M_2}{N^6}, \\ M_2 &= -\frac{3}{4}\Big[4NF(F_xN_x + N_{xx}F) + N^2(F_x^2 - 2FF_{xx}) - 8N_x^2F^2\Big], \end{aligned} \tag{10.3.25}$$

$$\begin{aligned} \frac{3}{4}B^2 + \frac{3}{4}\widetilde{B}^2 + \frac{3}{2}B^3 &= \frac{M_3}{N^6}, \\ M_3 &= \frac{3}{4}\Big[N^2F^2 + 2F^3 + N^2G^2\Big], \end{aligned} \tag{10.3.26}$$

and

$$-\frac{3}{2}\lambda\hat{B}_t + \frac{3}{4}\sigma\widetilde{B} = \frac{M_4}{N^6}, \quad M_4 = \frac{3}{4}N^4\Big[\sigma G - 2\lambda(NH_t - HN_t)\Big]. \tag{10.3.27}$$

Similarly, with the identity (10.3.20), we have

$$\widetilde{B}_{xx} = \frac{R_1}{N^4}, \quad R_1 = 6GN_x^2 - N(4N_xG_x + 2GN_{xx} - NG_{xx}), \tag{10.3.28}$$

$$+\frac{3}{2}B\widetilde{B} - \frac{3}{2}\lambda\widetilde{B} + \frac{3}{4}\sigma B = \frac{R_2}{N^4}, \quad R_2 = \frac{3}{4}\Big(2FG + \sigma FN^2 - 2\lambda GN^2\Big). \tag{10.3.29}$$

Therefore, we rewrite the identities (10.3.19)-(10.3.20) in the following simplified way

$$(10.3.19) = \frac{1}{N^6}\sum_{i=1}^{4} M_i, \tag{10.3.30}$$

$$(10.3.20) = \frac{1}{N^4}\sum_{i=1}^{2} R_i. \tag{10.3.31}$$

Now, substituting the explicit functions F, G, H, N in M_i, $i = 1, \ldots, 4$ and R_i, $i = 1, 2$ and collecting terms, we get (with notation $\tilde{y}_2 = y_2 + \psi)$)

$$
\begin{aligned}
\sum_{i=1}^{4} M_i = {} & d_1 \cosh(\tilde{y}_2)^2 + d_2 \cosh(\tilde{y}_2)\cos(y_1) + d_3 \cosh(\tilde{y}_2)^3 \cos(y_1) \\
& + d_4 \cos(y_1)^2 + d_5 \cosh(\tilde{y}_2)^2 \cos(y_1)^2 + d_6 \cosh(\tilde{y}_2)^4 \cos(y_1)^2 \\
& + d_7 \cosh(\tilde{y}_2)\cos(y_1)^3 + d_8 \cosh(\tilde{y}_2)^3 \cos(y_1)^3 + d_9 \cosh(\tilde{y}_2)^2 \cos(y_1)^4 \\
& + d_{10} \cosh(\tilde{y}_2)^4 + d_{11} \cosh(\tilde{y}_2)^2 \sinh(\tilde{y}_2)\sin(y_1) \\
& + d_{12} \cosh(\tilde{y}_2)\sinh(\tilde{y}_2)\cos(y_1)\sin(y_1) \\
& + d_{13} \cosh(\tilde{y}_2)^3 \sinh(\tilde{y}_2)\cos(y_1)\sin(y_1) \\
& + d_{14} \sinh(\tilde{y}_2)\cos(y_1)^2 \sin(y_1) \\
& + d_{15} \cosh(\tilde{y}_2)^2 \sinh(\tilde{y}_2)\cos(y_1)^2 \sin(y_1) \\
& + d_{16} \cosh(\tilde{y}_2)\sinh(\tilde{y}_2)\cos(y_1)^3 \sin(y_1) + d_{17} \cosh(\tilde{y}_2)^5 \cos(y_1) \\
& + d_{18} \cosh(\tilde{y}_2)\cos(y_1)^5 + d_{19} \cos(y_1)^4 + d_{20} \sinh(\tilde{y}_2)\cos(y_1)^4 \sin(y_1) \\
& + d_{21} \cosh(\tilde{y}_2)^4 \sinh(\tilde{y}_2)\sin(y_1),
\end{aligned}
\tag{10.3.32}
$$

and

$$
\begin{aligned}
\sum_{i=1}^{2} R_i = {} & e_1 \cosh(\tilde{y}_2)^2 \sinh(\tilde{y}_2)\sin(y_1) + e_2 \cosh(\tilde{y}_2)\sinh(\tilde{y}_2)\cos(y_1)\sin(y_1) \\
& + e_3 \sinh(\tilde{y}_2)\cos(y_1)^2 \sin(y_1) + e_4 \cosh(\tilde{y}_2)^3 \cos(y_1) \\
& + e_5 \cosh(\tilde{y}_2)^2 \cos(y_1)^2 + e_6 \cosh(\tilde{y}_2)\cos(y_1)^3 + e_7 \cosh(\tilde{y}_2)^2 \\
& + e_8 \cosh(\tilde{y}_2)\cos(y_1) + e_9 \cos(y_1)^2,
\end{aligned}
\tag{10.3.33}
$$

with coefficients $d_i,\ i = 1, \ldots, 21$, and $e_i,\ i = 1, \ldots 9$, as follows

$$
\begin{aligned}
d_1 &= 12\rho\Big(-4\alpha^6 - \alpha^2\delta^2 + \alpha^4\left(12\rho\beta^2 + 1\right) + \alpha^2\Big(\rho\Big(-12\beta^4 - 6\beta^2 + 4\beta\gamma\frac{\delta}{\alpha} + \gamma^2\Big)\Big) \\
&\quad + \rho\beta^2\delta^2 + \rho\beta^2\left(\rho\left(4\beta^4 + \beta^2 - \gamma^2\right)\right)\Big), \\
d_2 &= 2\rho^{-1/2} d_1, \\
\frac{d_3}{4} &= \rho^{3/2}\Big(- 24\alpha^6 + 4\alpha^2\delta^2 + 6\alpha^4\left((\rho - 6)\beta^2 + 1\right) + 2\alpha^2[-6(2\rho - 3)\beta^4 - 3(\rho - 3)\beta^2 \\
&\quad + 2(\rho - 3)\beta\gamma\frac{\delta}{\alpha} + \rho\gamma^2 + 3\delta^2] + \delta\left(2(\rho - 3)\beta^2\delta - 2(\rho + 3)\alpha\beta\gamma + 3(2\delta\lambda + \alpha\sigma)\right) \\
&\quad + \rho\left(18\beta^6 + 6\beta^4 + 10\beta^2\gamma^2 - 3\beta\gamma\sigma - 6\gamma^2\lambda\right) + 6\beta^2\delta^2\Big), \\
d_4 &= \rho^{-1} d_1,
\end{aligned}
\tag{10.3.34}
$$

$$
\begin{aligned}
d_5 =& -6\rho\Big(26\alpha^6 + 2\alpha^4\left(3(\rho-2)\beta^2 - 4\right) + 2\alpha^2\left((6\rho-3)\beta^4 - (\rho+1)\gamma^2 - 4\delta^2\right) \\
&+ \alpha\delta[4(\rho+1)\beta\gamma - 3(2\frac{\delta}{\alpha}\lambda + \sigma)] \\
&- \rho[26\beta^6 + 8\beta^4 + 2\beta^2\left(4\gamma^2 + \delta^2\right) - 3\beta\gamma\sigma - 6\gamma^2\lambda] - 2\beta^2\delta^2\Big), \\
\frac{d_6}{4\rho^2} =& -\Big(3\alpha^6 - 3\alpha^4\left(15\beta^2 + 1\right) + \alpha^2\left(45\beta^4 + 18\beta^2 - 4\beta\gamma\frac{\delta}{\alpha} + \gamma^2 - \delta^2\right) \\
&- \delta\left(2\beta^2\delta - 8\beta\alpha\gamma + 6\delta\lambda + 3\sigma\right) - 3\beta^6 - 3\beta^4 - \beta^2\left(\gamma^2 - 3\delta^2\right) + 3\beta\gamma\sigma \\
&+ 6\gamma^2\lambda\Big), \\
\frac{d_7}{4\sqrt{\rho}} =& -\Big(18\alpha^6 - 4\alpha^2\delta^2 + 6\alpha^4\left((6\rho-4)\beta^2 - 1\right) - 2\alpha^2[3(6\rho-1)\beta^4 + (9\rho-3)\beta^2 \\
&+ 2(1-3\rho)\beta\gamma\frac{\delta}{\alpha} + \gamma^2 + 3\delta^2] + \delta(2(3\rho-1)\beta^2\delta + 2\beta\alpha\gamma(3\rho+1) \\
&- 3(2\delta\lambda + \alpha\sigma)) - \rho\left(24\beta^6 + 6\beta^4 + 2\beta^2\left(5\gamma^2 + 3\delta^2\right) - 3\beta\gamma\sigma - 6\gamma^2\lambda\right)\Big), \\
d_8 =& -6\rho^{3/2}\Big(4\alpha^6 - 4\alpha^4\left(15\beta^2 + 1\right) + 4\alpha^2\left(15\beta^4 + 6\beta^2 - 2\beta\gamma\frac{\delta}{\alpha}\right) \\
&- \delta\left(4\beta^2\delta - 8\beta\alpha\gamma + 6\delta\lambda + 3\sigma\alpha\right) - 4\beta^6 - 4\beta^4 + 4\beta^2\delta^2 + 3\beta\gamma\sigma + 6\gamma^2\lambda\Big), \\
d_9 =& \rho^{-1}d_6, \\
d_{10} =& \rho^2\Big(-12\alpha^6 - 120\alpha^4\beta^2 + 16\alpha^2\delta^2 + 4\alpha^2\left(21\beta^4 + 12\beta^2 - 8\beta\gamma\frac{\delta}{\alpha} - \gamma^2\right) \\
&+ \delta\left(-16\beta^2\delta - 8\beta\alpha\gamma + 6\delta\lambda + 3\alpha\sigma\right) + \rho\gamma\left(16\beta^2\gamma - 3\beta\sigma - 6\gamma\lambda\right) + 12\beta^2\delta^2\Big), \\
d_{11} =& 24\rho^{3/2}\Big(-6\alpha^4\beta - \beta\delta^2 + \left(2(\rho+1)\alpha^2\beta^3 + 2\alpha^2\beta - \alpha\gamma\delta\right) \\
&+ \beta\left(\rho\left(-6\beta^4 - 2\beta^2 + \beta\gamma\frac{\delta}{\alpha} + \gamma^2\right)\right)\Big), \\
d_{12} =& 2\rho^{-1/2}d_{11}, \\
d_{13} =& 4\rho^2\Big(18\alpha^5\beta - 2\alpha^3\left(30\beta^3 + 6\beta - \gamma\frac{\delta}{\alpha}\right) - 4\delta(\alpha^2\gamma - \beta\alpha\delta) + \alpha[18\beta^5 + 12\beta^3 \\
&- 2\beta^2\gamma\frac{\delta}{\alpha} + 2\beta\left(\gamma^2 - 3\delta^2\right) - 3\gamma(2\frac{\delta}{\alpha}\lambda + \sigma)] + \delta\left(4\beta^2\gamma - 3\beta\sigma - 6\gamma\lambda\right)\Big), \\
d_{14} =& 24\rho^{1/2}\Big(-6\alpha^4\beta - 2\beta\delta^2 + \left(2(\rho+1)\beta^3\alpha^2 + 2\beta\alpha^2 - \gamma\alpha\delta\right) \\
&+ \beta\left(\rho\left(-6\beta^4 - 2\beta^2 + \beta\gamma\frac{\delta}{\alpha} + \gamma^2\right) + \delta^2\right)\Big), \\
d_{15} =& 6\rho^{3/2}\Big(24\alpha^5\beta - 4\alpha^3\left(20\beta^3 + 4\beta - \gamma\frac{\delta}{\alpha}\right) - 4\alpha^2\delta(\gamma - 2\beta\frac{\delta}{\alpha}) + \alpha[24\beta^5 \\
&+ 16\beta^3 - 4\beta^2\gamma\frac{\delta}{\alpha} - 8\beta\delta^2 - 3\gamma(2\frac{\delta}{\alpha}\lambda + \sigma)] + \delta\left(4\beta^2\gamma - 3\beta\sigma - 6\gamma\lambda\right)\Big),
\end{aligned}
\tag{10.3.35}
$$

$$
\begin{aligned}
d_{16} &= \rho d_{13},\\
d_{17} &= \rho^{5/2}(4\alpha^2\delta^2 - 4\gamma(2\beta\alpha\delta + \alpha^2\gamma) + \delta\left(-4\beta^2\delta - 8\beta\alpha\gamma + 6\delta\lambda + 3\alpha\sigma\right)\\
&\quad + \gamma\left(4\beta^2\gamma - 3\beta\sigma - 6\gamma\lambda\right)),\\
d_{18} &= \rho^{-2}d_{17},\\
d_{19} &= \left(\delta\left(16\alpha^2\delta + 6\delta\lambda + 3\alpha\sigma\right) - \rho[84\alpha^4\beta^2 - 4\alpha^2\left(30\beta^4 + 12\beta^2 - 8\beta\gamma\frac{\delta}{\alpha} - \gamma^2\right)\right.\\
&\quad \left. + 8\beta\delta(2\beta\delta + \alpha\gamma) - 12\beta^6 - 4\beta^2\left(4\gamma^2 + 3\delta^2\right) + 3\beta\gamma\sigma + 6\gamma^2\lambda]\right),\\
d_{20} &= \rho^{1/2}\left(8\alpha^2\gamma\delta + 8\alpha\beta\delta^2 + \gamma\left(-8\beta^2\delta - 8\alpha\beta\gamma + 12\delta\lambda + 3\alpha\sigma\right) + 3\delta\beta\sigma\right),\\
d_{21} &= \rho^2 d_{20},
\end{aligned}
\tag{10.3.36}
$$

and

$$
\begin{aligned}
e_1 &= 2\rho^{3/2}\left(2\alpha^3\gamma + 6\alpha^2\beta\delta + \alpha\left(-6\beta^2\gamma + 3\beta\sigma + 3\gamma\lambda\right) - 2\beta^3\delta + 3\beta\delta\lambda\right),\\
e_2 &= 4\rho\left(2\alpha^3\gamma + 6\alpha^2\beta\delta + \alpha\left(-6\beta^2\gamma + 3\beta\sigma + 3\gamma\lambda\right) - 2\beta^3\delta + 3\beta\delta\lambda\right),\\
e_3 &= 2\sqrt{\rho}\left(2\alpha^3\gamma + 6\alpha^2\beta\delta + \alpha\left(-6\beta^2\gamma + 3\beta\sigma + 3\gamma\lambda\right) - 2\beta^3\delta + 3\beta\delta\lambda\right),\\
e_4 &= \rho^{3/2}\left(-4\alpha^3\delta + 3\alpha^2(4\beta\gamma - \sigma) + 6\alpha\delta\left(2\beta^2 - \lambda\right) + \beta\left(-4\beta^2\gamma + 3\beta\sigma + 6\gamma\lambda\right)\right),\\
e_5 &= 2\rho\left(-4\alpha^3\delta + 3\alpha^2(4\beta\gamma - \sigma) + 6\alpha\delta\left(2\beta^2 - \lambda\right) + \beta\left(-4\beta^2\gamma + 3\beta\sigma + 6\gamma\lambda\right)\right),\\
e_6 &= \sqrt{\rho}\left(-4\alpha^3\delta + 3\alpha^2(4\beta\gamma - \sigma) + 6\alpha\delta\left(2\beta^2 - \lambda\right) + \beta\left(-4\beta^2\gamma + 3\beta\sigma + 6\gamma\lambda\right)\right),\\
e_7 &= -\rho\left(\alpha\left(16\alpha^2\delta + 3\alpha\sigma + 6\delta\lambda\right) + \rho\beta\left(16\beta^2\gamma - 3\beta\sigma - 6\gamma\lambda\right)\right),\\
e_8 &= -2\sqrt{\rho}\left(\alpha\left(16\alpha^2\delta + 3\alpha\sigma + 6\delta\lambda\right) + \rho\beta\left(16\beta^2\gamma - 3\beta\sigma - 6\gamma\lambda\right)\right),\\
e_9 &= -\left(\alpha\left(16\alpha^2\delta + 3\alpha\sigma + 6\delta\lambda\right) + \rho\beta\left(16\beta^2\gamma - 3\beta\sigma - 6\gamma\lambda\right)\right).
\end{aligned}
\tag{10.3.37}
$$

Now, with δ, γ, ρ as in definition (10.3.11)-(10.3.12) and λ, σ as in (10.3.38), we get, with the help of the symbolic software Mathematica and after lengthy but direct computations that all $d_i = 0,\ i = 1, \ldots, 21$, and all $e_i = 0,\ i = 1, \ldots, 9$, and therefore $\sum_{i=1}^4 M_i = 0$ and $\sum_{i=1}^2 R_i = 0$, and we conclude. □

Explicit expression for parameters λ and σ. The *variational* parameters λ and σ are defined as follows:

$$
\begin{aligned}
\lambda &:= \frac{a_0 + a_1\rho + a_2\rho^2 + a_3\rho^3 + a_4\rho^4}{b_0 + b_1\rho + b_2\rho^2 + b_3\rho^3 + b_4\rho^4},\\
\sigma &:= \frac{c_0 + c_1\rho + c_2\rho^2 + c_3\rho^3 + c_4\rho^4}{b_0 + b_1\rho + b_2\rho^2 + b_3\rho^3 + b_4\rho^4},
\end{aligned}
\tag{10.3.38}
$$

with $a_i, b_i, c_i,\ i = 0, 1, 2, 3, 4$ given by:

$$a_0 = -6\left(3\alpha^2 - \beta^2\right)\left(27\alpha^6 - 9\alpha^4\left(3\beta^2 + 1\right) + \alpha^2\left(9\beta^4 + 6\beta^2 + 9\delta^2\right) - 6\alpha\beta\gamma\delta - \beta^2\left(\beta^4 + \beta^2 - \gamma^2\right)\right),$$

$$a_1 = -12\Big[45\alpha^8 - 3\alpha^6\left(13\beta^2 + 6\right) + \alpha^4\left(65\beta^4 + 6\beta^2 + 18\delta^2\right) - 12\alpha^3\beta\gamma\delta + \alpha^2\beta^2\left(-85\beta^4 - 30\beta^2 + 2\gamma^2 + 6\delta^2\right) + 28\alpha\beta^3\gamma\delta + 2\beta^4\left(11\beta^4 + 5\beta^2 - 5\gamma^2\right)\Big],$$

$$a_2 = -4\Big[54\alpha^8 - 9\alpha^6\left(17\beta^2 + 3\right) + \alpha^4\left(-27\beta^4 + 39\beta^2 + 27\delta^2\right) - 42\alpha^3\beta\gamma\delta - 3\alpha^2\beta^2\left(49\beta^4 - 5\beta^2 + 7\gamma^2 - \delta^2\right) + 6\alpha\beta^3\gamma\delta + 185\beta^8 + 77\beta^6 - 77\beta^4\gamma^2\Big],$$

$$a_3 = -4\Big[9\alpha^8 - \alpha^6\left(71\beta^2 + 6\right) + \alpha^4\left(105\beta^4 + 26\beta^2 + 6\delta^2\right) - 20\alpha^3\beta\gamma\delta \quad \alpha^2\beta^2\left(53\beta^4 + 26\beta^2 - 6\gamma^2 + 6\delta^2\right) + 20\alpha\beta^3\gamma\delta + 6\left(3\beta^8 + \beta^6 - \beta^4\gamma^2\right)\Big],$$

$$a_4 = -2\left(\alpha^2 - 3\beta^2\right)\left(\alpha^6 - \alpha^4\left(15\beta^2 + 1\right) + \alpha^2\left(15\beta^4 + 6\beta^2 + \delta^2\right) - 6\alpha\beta\gamma\delta - 33\beta^6 - 9\beta^4 + 9\beta^2\gamma^2\right),$$

$$\frac{b_0}{27} = \left(27\alpha^6 - 9\alpha^4\left(3\beta^2 + 1\right) + \alpha^2\left(9\beta^4 + 6\beta^2 + 9\delta^2\right) - 6\alpha\beta\gamma\delta - \beta^2\left(\beta^4 + \beta^2 - \gamma^2\right)\right),$$

$$\frac{b_1}{18} = \left(45\alpha^6 - 3\alpha^4\left(35\beta^2 + 6\right) + 3\alpha^2\left(29\beta^4 + 12\beta^2 + 6\delta^2\right) - 36\alpha\beta\gamma\delta - 19\beta^6 - 10\beta^4 + 10\beta^2\gamma^2\right),$$

$$\frac{b_2}{6} = \left(54\alpha^6 - 3\alpha^4\left(70\beta^2 + 9\right) + 3\alpha^2\left(94\beta^4 + 30\beta^2 + 9\delta^2\right) - 90\alpha\beta\gamma\delta - 158\beta^6 - 59\beta^4 + 59\beta^2\gamma^2\right),$$

$$b_3 = 6\left(9\alpha^6 - \alpha^4\left(53\beta^2 + 6\right) + \alpha^2\left(107\beta^4 + 28\beta^2 + 6\delta^2\right) - 28\alpha\beta\gamma\delta - 87\beta^6 - 30\beta^4 + 30\beta^2\gamma^2\right),$$

$$b_4 = 3\left(\alpha^6 - \alpha^4\left(9\beta^2 + 1\right) + \alpha^2\left(27\beta^4 + 6\beta^2 + \delta^2\right) - 6\alpha\beta\gamma\delta - 9\left(3\beta^6 + \beta^4 - \beta^2\gamma^2\right)\right),$$

and

$$c_0 = -24\Big[81\alpha^7\delta - 27\alpha^6\beta\gamma - 27\alpha^5\left(3\beta^2\delta + \delta\right) + 9\alpha^4\left(3\beta^3 + \beta\right)\gamma + 9\alpha^3\delta(3\beta^4 + 2\beta^2 + 3\delta^2) - 3\alpha^2\beta\gamma\left(3\beta^4 + 2\beta^2 + 9\delta^2\right) - 3\alpha\beta^2\delta\left(\beta^4 + \beta^2 - 3\gamma^2\right) + \beta^3\gamma(\beta^4 + \beta^2 - \gamma^2)\Big],$$

$$c_1 = -24\Big[63\alpha^7\delta - 21\alpha^6\beta\gamma - 3\alpha^5\left(76\beta^2 + 9\right)\delta + \alpha^4\beta\left(172\beta^2 + 9\right)\gamma + 9\alpha^3\delta(19\beta^4 + 10\beta^2 + 3\delta^2) - \alpha^2\beta\gamma\left(121\beta^4 + 62\beta^2 + 99\delta^2\right) + \alpha\beta^2\delta\left(14\beta^4 + 5\beta^2 + 57\gamma^2\right) + 6\beta^7\gamma + 9\beta^5\gamma - 9\beta^3\gamma^3\Big],$$

$$
\begin{aligned}
\frac{c_2}{24} =& - \Big[15\alpha^7\delta - 5\alpha^6\beta\gamma - \alpha^5\left(100\beta^2+9\right)\delta + 3\alpha^4\left(60\beta^3+\beta\right)\gamma + \alpha^3\delta(155\beta^4 \\
& + 54\beta^2 + 9\delta^2) \\
& - \alpha^2\beta\gamma\left(185\beta^4 + 82\beta^2 + 57\delta^2\right) + \alpha\beta^2\delta\left(-50\beta^4 - 17\beta^2 + 99\gamma^2\right) \\
& + 3\beta^3\gamma\left(26\beta^4 + 9\beta^2 - 9\gamma^2\right)\Big], \\
c_3 =& - 8\Big[3\alpha^7\delta - \alpha^6\beta\gamma - 3\alpha^5\left(8\beta^2\delta + \delta\right) + \alpha^4\left(56\beta^3+\beta\right)\gamma + \alpha^3\delta(119\beta^4 + 26\beta^2 \\
& + 3\delta^2) - \alpha^2\beta\gamma\left(285\beta^4 + 62\beta^2 + 27\delta^2\right) + \alpha\beta^2\delta\left(-46\beta^4 - 19\beta^2 + 81\gamma^2\right) \\
& + 9\beta^3\gamma\left(26\beta^4 + 9\beta^2 - 9\gamma^2\right)\Big], \\
c_4 =& \ 24\beta^2\left(\alpha^2+\beta^2\right)^2(3\beta\gamma - \alpha\delta),
\end{aligned}
$$

where $\rho,\ \delta,\ \gamma$ are given in definition (10.3.11)-(10.3.12).

As a final remark, we also notice that the bad Boussinesq equation (10.3.1) has suitable Bäcklund transformations, connecting soliton-like solutions. Some of these ideas will be worked in a forthcoming book.

10.4 Exercises

(10.1) Show that the bad Boussinesq breather (10.3.14) goes to the soliton (10.3.7) at *zero frequency*, namely

$$\lim_{\alpha\to 0} B(t,x) = Q_c(t, x + x_0), \quad c = \sqrt{\beta^2+1}, \ \ x_0 = \log(2).$$

(10.2) Prove that if $B(t,x)$ is a bad Boussinesq breather (10.3.14), then

$$u(t,x) = \mu + (1+3\mu)B\left((1+3\mu)t, \sqrt{1+3\mu}x\right), \quad \mu\in\mathbb{R},$$

is also a *breather solution* of (10.3.1) with nonvanishing boundary value μ at infinity.

(10.3) Prove that *purely periodic* (i.e., without spatial shift) localized bad Boussinesq breathers with nonzero weighted mean cannot exist for any power of the nonlinearity, namely

$$u_{tt} - u_{xx} - \left(u_{xx} + \tfrac{3}{2}u^p\right)_{xx} = 0, \quad p\in\mathbb{N}. \tag{10.4.1}$$

Hint. Use the function $\int_{\mathbb{R}} x \times (10.4.1)dx$.

Chapter 11

Zero mean periodic mKdV breathers

11.1 Introduction

Until this point we have essentially considered only non periodic functions in space as possible breathers. The purpose of this chapter is to bring the attention of the reader to a new class of breathers, which are not only (almost) periodic in time, but also in space. These kinds of solutions are present in several models, like the mKdV equation, the SG, and the NLS equation, among others. However, for the sake of simplicity, and in order to clarify the main ideas and concepts, we will focus our attention to the case of the *focusing* mKdV equation (4.1.1).

Recall that $u = u(x)$ defined on $\mathbb{R}$ is **periodic**, with minimal period $L > 0$, if

$$u(x + L) = u(x), \qquad \text{for all } x \in \mathbb{R}.$$

Note that this condition does not include constant functions for which one can take $L = 0$. This is done to avoid unnecessary complications in some results described below. Usually, the above information is encoded in the notation

$$u = u(x), \quad x \in L\mathbb{T}, \quad \mathbb{T} := \mathbb{R}/\mathbb{Z}.$$

Here $\mathbb{T}$ denotes the one dimensional torus of length one, or the interval $[0, 1]$ with periodic conditions on the boundaries $x = 0, 1$. We will use this notation along this chapter.

Recall the *focusing* mKdV equation studied in the previous chapters, this time posed on the torus of size L

$$\partial_t u + \partial_x(\partial_x^2 u + u^3) = 0, \quad u(t, x) \in \mathbb{R}, \quad x \in L\mathbb{T}. \tag{11.1.1}$$

Now we look for u periodic in x, with minimal period $L > 0$ (to be determined later on). This means that we look for $u(t, x)$ solution of (11.1.1) with the property that

$$u(t, x + L) = u(t, x), \qquad \text{for all } t \in \mathbb{R},\ x \in \mathbb{R}.$$

For the purposes of this chapter, we will only need data in the periodic $H^1(L\mathbb{T})$ space, to be defined in detail below.

11.2 Conservation laws

As in the non periodic case, the periodic mKdV equation (11.1.1) has conserved quantities, actually, infinitely many. Here we recall the most interesting ones from our point of view. One has the mass,

$$M_{\#}[u](t) := \frac{1}{2}\int_{L\mathbb{T}} u^2(t,x)dx = M_{\#}[u](0), \tag{11.2.1}$$

the energy

$$E_{\#}[u](t) := \frac{1}{2}\int_{L\mathbb{T}} u_x^2(t,x)dx - \frac{1}{4}\int_{L\mathbb{T}} u^4(t,x)dx = E_{\#}[u](0), \tag{11.2.2}$$

and a higher order energy or third conserved quantity

$$\begin{aligned} F_{\#}[u](t) := &\frac{1}{2}\int_{L\mathbb{T}} u_{xx}^2(t,x)dx \\ &- \frac{5}{2}\int_{L\mathbb{T}} u^2u_x^2(t,x)dx + \frac{1}{4}\int_{L\mathbb{T}} u^6(t,x)dx = F_{\#}[u](0), \end{aligned} \tag{11.2.3}$$

which are formally preserved if $u(t,\cdot) \in H^2(\mathbb{T})$ is a solution of (11.1.1). Here,

$$\begin{aligned} H^2(\mathbb{T}) := \Big\{ v \in H^2(0,L) \ : \\ v(0) = v(L),\ v_x(0) = v_x(L),\ v_{xx}(0) = v_{xx}(L) \Big\}. \end{aligned}$$

(Note that we work on a bounded interval of $\mathbb{R}$, and this fact will provide nice properties in terms of spectral theory.)

The proof of conservation for the laws $M_{\#}[u](t)$, $E_{\#}[u](t)$, $F_{\#}[u](t)$ is straightforward if we work by density in a space of smooth functions and we note that $u(t,0) = u(t,L)$ and $u_x(t,0) = u_x(t,L)$ for all time imply

$$u_t(t,0) = u_t(t,0), \quad u_{xt}(t,0) = u_{xt}(t,L),$$

and therefore using (11.1.1),

$$u_{xxx}(t,0) = u_{xxx}(t,L).$$

It seems nearly impossible to talk about periodic in space solutions of dispersive models without talking about Jacobi Elliptic Functions. Here, we will concentrate our efforts on the stability of breathers. However, we first recall some standard properties of Jacobi Elliptic Functions.

11.3 Jacobi Elliptic Functions

We present here some basic notions and properties on the Jacobi Elliptic Functions (JEFs hereafter) that we are going to use in the study of the periodic breather. For further reading, see [AbSe81, ByFr71].

The JEF $\operatorname{sn}(\cdot, r)$ with elliptic modulus r is defined as the solution of the following ODE:

$$\begin{aligned}\operatorname{sn}'^2 &= 1-(1-r)\operatorname{sn}^2 + r\operatorname{sn}^4,\\ \operatorname{sn}'' &= -(1+r)\operatorname{sn} + 2r\operatorname{sn}^3.\end{aligned}$$

Additionally, it is related with the following two JEFs $\operatorname{cn}(\cdot, r)$ and $\operatorname{dn}(\cdot, r)$ by the algebraic identities

$$\operatorname{sn}(x,r)^2 + \operatorname{cn}(x,r)^2 = 1, \quad \operatorname{dn}(x,r)^2 + r^2\operatorname{sn}(x,r)^2 = 1,$$

and by the derivatives with respect to x:

$$\begin{aligned}&\frac{d}{dx}\operatorname{sn}(x,r) = \operatorname{cn}(x,k)\operatorname{dn}(x,r), \qquad \frac{d}{dx}\operatorname{cn}(x,r) = -\operatorname{sn}(x,k)\operatorname{dn}(x,r),\\ &\frac{d}{dx}\operatorname{dn}(x,r) = -r^2\operatorname{sn}(x,r)\operatorname{cn}(x,r).\end{aligned}$$

Some special values when $r = 0$ or $r = 1$ are given by:

$$\begin{aligned}&\operatorname{sn}(x,0) = \sin(x), \quad \operatorname{cn}(x,0) = \cos(x),\\ &\operatorname{dn}(x,0) = 1, \quad \operatorname{sn}(x,1) = \tanh(x),\\ &\operatorname{cn}(x,1) = \cosh(x), \quad \operatorname{dn}(x,1) = \operatorname{sech}(x).\end{aligned}$$

The $\operatorname{nd}(\cdot, r)$ is defined as the inverse of the dn JEF, i.e. $\operatorname{nd}(x,r) = \frac{1}{\operatorname{dn}(x,r)}$. It is also characterized to be the solution of the following ODEs (here $\operatorname{nd} \equiv \operatorname{nd}(x,r)$):

$$\begin{aligned}\operatorname{nd}'^2 &= -1 + (2-r)\operatorname{nd}^2 + (r-1)\operatorname{nd}^4,\\ \operatorname{nd}'' &= (2-r)\operatorname{nd} + 2(r-1)\operatorname{nd}^3.\end{aligned}$$

It takes some special values when $r = 0$ or $r = 1$: $\operatorname{nd}(x,0) = 1$, and $\operatorname{nd}(x,1) = \cosh(x)$. Also,

$$\operatorname{nd}(x,0) = 1 \quad \operatorname{nd}(x,1) = \cosh(x),$$

and by the derivatives with respect to x

$$\begin{aligned}&\frac{d}{dx}\operatorname{nd}(x,r) = r^2\operatorname{sd}(x,r)\operatorname{cd}(x,r), \qquad \frac{d}{dx}\operatorname{sd}(x,r) = \operatorname{cn}(x,r)\operatorname{nd}^2(x,r),\\ &\frac{d}{dx}\operatorname{cd}(x,r) = -r^2\operatorname{sn}(x,r)\operatorname{nd}^2(x,r).\end{aligned}$$

11.4 A first class of periodic mKdV breathers

The first class of periodic breather solutions was obtained by Kevrekidis, Khare, Saxena and Herring (KKSH) [KKS04], proposing a well-tailored ansatz, by using JEFs and matching of free parameters.

Definition 11.4.1. Given $\alpha, \beta > 0$, $x_1, x_2 \in \mathbb{R}$ and $k, m \in [0, 1]$, the KKSH breathers are given by the explicit formula [KKS04]

$$\begin{aligned} B &= B(t, x; \alpha, \beta, k, m, x_1, x_2) \\ &:= \partial_x \widetilde{B} := 2\sqrt{2}\partial_x \left[\arctan\left(\frac{\beta}{\alpha}\frac{\operatorname{sn}(\alpha y_1, k)}{\operatorname{nd}(\beta y_2, m)}\right)\right], \end{aligned} \tag{11.4.1}$$

with $\operatorname{sn}(\cdot, k)$ and $\operatorname{nd}(\cdot, m)$ the standard JEFs of elliptic modulus k and m, respectively,

$$y_1 := x + \delta t + x_1, \quad y_2 := x + \gamma t + x_2,$$

and

$$\delta := \alpha^2(1 + k) + 3\beta^2(m - 2), \quad \gamma := 3\alpha^2(1 + k) + \beta^2(m - 2). \tag{11.4.2}$$

Notice that (11.4.2) converges to the mKdV values when $k \to 0$, $m \to 1$. Additionally, in order to be a *spatially* periodic solution of mKdV (11.1.1), the parameters m, k, α and β must satisfy the *commensurability conditions* on the spatial periods

$$\frac{\beta^4}{\alpha^4} = \frac{k}{1 - m}, \qquad K[k] = \frac{\alpha}{2\beta}K[m], \tag{11.4.3}$$

where K denotes the Elliptic function K, which satisfies

$$K[0] = \frac{\pi}{2} \quad \text{and} \quad \lim_{k \to 1^-} K[k] = \infty.$$

Under these assumptions, the spatial period is given by

$$L := \frac{4}{\alpha}K[k] = \frac{2}{\beta}K[m].$$

Note that conditions (11.4.3) formally imply that B (11.4.1) has only four independent parameters. Additionally, if we assume that the ratio β/α stays bounded, we have that k approaches 0 as m is close to 1. Using this information, the standard non periodic mKdV breather (3.2.1)

$$B = 2\sqrt{2}\partial_x \left[\arctan\left(\frac{\beta}{\alpha}\frac{\sin(\alpha y_1)}{\cosh(\beta y_2)}\right)\right],$$

$$y_1 = x + \delta t + x_1, \quad y_2 = x + \gamma t + x_2,$$

and

$$\delta = \alpha^2 - 3\beta^2, \quad \gamma = 3\alpha^2 - \beta^2,$$

can be formally recovered as the limit of very large spatial period $L \to +\infty$, obtained e.g. if $k \to 0$. In that sense, one can think of (11.4.1) as a nontrivial periodic bifurcation at infinity of the aperiodic mKdV breather.

See Fig. 11.1 for a description of a KKHS breather solution. It is important to mention that conditions (11.4.3) impose a particular set of restrictions on k and m. In terms of k, one has $k \in [0, k^*)$, with $k^* \sim 0.058$, while m is decreasing with respect to k, with $m(k = 0) = 1$ and $m(k \sim k^*) \sim 0$. Fig. 11.2 describes the behavior of k and m more clearly.

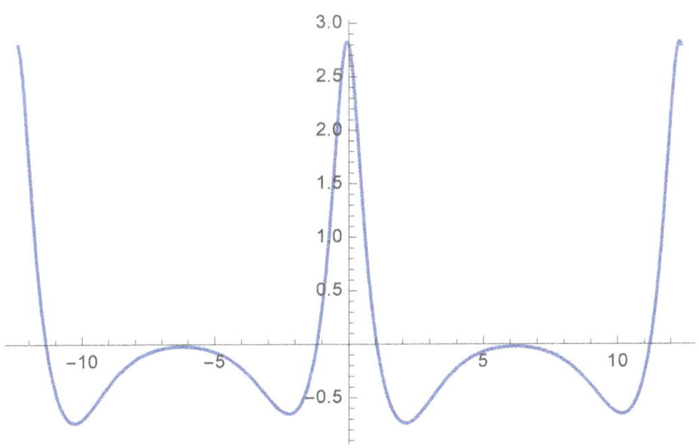

Fig. 11.1 The graph of the KKSH breather at time $t = 0$, for $\beta = 1$, $k = 0.001$, $m = 0.999$ and $L = 12.398$.

In what follows, we will use the convention that

$$m = m(k), \quad \alpha = \alpha(\beta, k), \tag{11.4.4}$$

obtained by solving $m = m(k)$ in (11.4.3) numerically, and then solving for α algebraically.

11.5 Stability analysis of the KKHS breather

An impressive result is the fact that KKHS breathers (11.4.1) do follow a variational characterization.

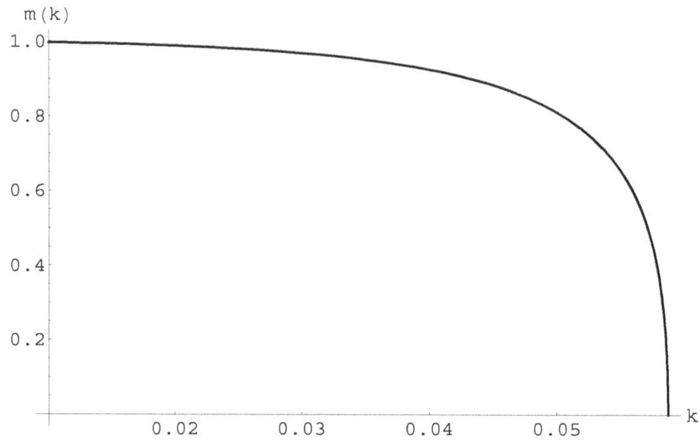

Fig. 11.2 The graph of m as a function of k obtained solving conditions (11.4.3). Although m runs from 0 to 1, not all values of $k \in [0, 1]$ are allowed, being the limiting value of $k \sim 0.05883626$.

11.5.1 *Variational characterization*

Very similar to mKdV posed in the real line, one has the following surprisingly simple result:

Theorem 11.5.1. *Assume* (11.4.3). *Let B be any KKSH breather* (11.4.1). *Define*

$$a_1 := 2(\beta^2(2-m) - \alpha^2(1+k)) = -\frac{1}{2}(\delta + \gamma), \tag{11.5.1}$$

and

$$a_2 := \alpha^4(1 + k^2 - 26k) + 2\alpha^2\beta^2(2-m)(1+k) + \beta^4 m^2. \tag{11.5.2}$$

Then B satisfies the generalized nonlinear elliptic equation

$$B_{4x} + 5BB_x^2 + 5B^2B_{xx} + \frac{3}{2}B^5 - a_1(B_{xx} + B^3) + a_2 B = 0. \tag{11.5.3}$$

We emphasize that the first condition in (11.4.3) is essential for the proof of (11.5.3). However, for the proof we do not need the second one. Note additionally that a_1 and a_2 converge to the corresponding constants for the aperiodic mKdV case when $k \to 0$ and $m \to 1$, see (4.2.1). An important consequence of Theorem 11.5.1 is the following characterization of KKSH breathers as critical points of a suitable conserved functional.

Corollary 11.5.2. *KKSH breathers* (11.4.1) *are critical points of the functional*

$$\mathcal{H}_\#[u] := F_\#[u](t) + a_1 E_\#[u](t) + a_2 M_\#[u](t),$$

defined in the space $H^2(\mathbb{T})$ and preserved along the mKdV periodic flow.

Corollary 11.5.2 must be taken with care. We emphasize that the existence of a suitable elliptic equation **does not imply stability**. Indeed, probably all periodic mKdV breathers are unstable under a suitable type of periodic perturbations, at least for L not so large,[1] see the numerical results obtained by Kevrekidis et al. [KKS04].

In this case, the periodic character of the solution leads to nontrivial interactions between adjacent breathers, which probably play an important role in the instability character of this solution.

As in the previous chapters, the next step is the study of the nonlinear stability of this solution. In fact, we present below compelling evidence that periodic mKdV breathers are *unstable* under a suitable type of periodic perturbations, at least for L not so large. Our theoretical results do agree with the numerical ones obtained by Kevrekidis et al. [KKS03, KKS04], and our numerical computations below. In this case, the periodic character of the solution leads to nontrivial interactions between adjacent breathers, which probably play an important role in the instability character of this solution.

11.6 Spectral analysis of the periodic mKdV breather

11.6.0.1 *Mathematical description*

In this section we give further evidence of the stable-unstable character of the KKSH breather solution (11.4.1) depending on the parameters phase space. First of all, let us notice that, thanks to (11.5.3), the linearized operator for a KKSH breather is given by the expression

$$
\begin{aligned}
\mathcal{L}_{\#}[z] := {} & z_{4x} + (5B^2 - a_1)z_{xx} + 10BB_x z_x \\
& + \left(a_2 + 5B_x^2 + 10BB_{xx} + \frac{15}{2}B^4 - 3a_1 B^2\right) z.
\end{aligned}
\tag{11.6.1}
$$

This operator is defined acting on functions in $H^2(\mathbb{T})$, $\mathbb{T} = (0, L)$, and it reduces to the standard aperiodic mKdV breather operator (7.1.1) as the length of the interval tends to infinity (or $k \to 0$). The constants a_1 and a_2 were introduced in (11.5.1) and (11.5.2), and we assume the convention (11.4.4).

In the following lines, we analyze the spectral stability of the KKSH breather, namely the understanding of the spectrum of $\mathcal{L}_{\#}$ in (11.6.1). First, we prove some useful expression for the mass of a breather. As a second step, we compute

[1] If L is large it seems that the periodic breather is close in a certain topology to the aperiodic breather, which satisfies stability properties in a very general open neighborhood.

numerically the spectrum of $\mathcal{L}_\#$, and conclude that it has the desired spectral properties. Then, we analyze which property is mainly responsible for the KKSH stability.

11.7 Spectral assumptions

Spatial periodic breathers are harder than standard breathers. This affects the spectral properties, that are more complicated to prove than usual. We will make use of a combined theoretical-numerical approach to conclude why periodic breathers may not be stable in some particular regimes. In order to obtain nonlinear stability, additional assumptions will be needed:

Assumption 1: (*Nondegeneracy of the kernel*) For each $k \in (0, k^*)$, $x_1, x_2 \in \mathbb{R}$ and $\beta > 0$, $\ker \mathcal{L}_\#$ is spanned by the two elements $\partial_{x_1} B$ and $\partial_{x_2} B$; and there is a uniform gap between the kernel and the bottom of the positive spectrum.

Assumption 2: (*Unique, simple negative eigenvalue*) For each $k \in (0, k^*)$, $x_1, x_2 \in \mathbb{R}$ and $\beta > 0$, the operator $\mathcal{L}_\#$ has a unique simple, negative eigenvalue $\lambda_1 = \lambda_1(\beta, k, x_1, x_2) < 0$ associated to the unit L^2-norm eigenfunction B_{-1}. Moreover, there is $\lambda_1^0 < 0$ depending on β and k only, such that $\lambda_1 \leq \lambda_1^0$ for all x_1, x_2.

Assumption 3: The following sign condition is satisfied: if $M_\#[B]$ is the mass of the breather solution in terms of β and k, and a_1, a_2 are "variational" parameters given in (11.5.1)-(11.5.2), then

$$\frac{\partial_k a_1 \partial_\beta M_\#[B] - \partial_\beta a_1 \partial_k M_\#[B]}{\partial_k a_1 \partial_\beta a_2 - \partial_k a_2 \partial_\beta a_1} > 0. \tag{11.7.1}$$

(See eqn. (8.11.15) for more details.) This requirement is a sort of generalization of the Weinstein's sign condition for soliton solutions, but it is different from the former because it also considers a certain variation of energy (and not only mass), which has been written above in terms of the mass only. Additionally, this sign condition can be evaluated for non periodic breathers such as mKdV and SG (see Corollary 4.9.1), and it is trivially satisfied, see e.g. (9.8.18) for the corresponding computation in the SG case.

Assumption 3 is needed for ensuring that one can "replace" the first eigenfunction appearing from Assumption 2 above, which is an instability direction, by the breather itself. The advantage of this replacement comes from the fact that perturbative dynamics lying in the L^2 orthogonal space to the breather allows us to control variations of the scalings of the breather, and moreover, their dynamics has formally *quadratic variation* in time, meaning that if the error in our initial data is of order η, then any bad variation of the scalings will be of order at most η^2, and therefore negligible for the stability result.

At the rigorous level, the conservation of mass allows us to control the orthogonal direction to the breather itself in terms of quadratic terms only. The procedure to replace the first eigenfunction by the breather is standard, but needs to ensure that a certain denominator is never zero, see e.g. (9.10.7). Under Assumption 3, such a denominator is never zero. Recall that a similar result has been proved in [AlMu13, eqn. (4.24)], see [DuMe08, Mun14] for more examples in the case of soliton solutions. Adding this new assumption, our method of proof works as in the SG case and we are able to state the following:

Theorem 11.7.1. *Under spectral Assumptions 1, 2 and 3 above, KKSH breathers* (11.4.1) *are stable for L-periodic H^2-perturbations.*

In Fig. 11.3 we describe numerically the meaning of Assumption 3. In particular, it is inferred that for k small enough (depending only on β), Assumption 3 is satisfied. Additionally, note that the condition $k \to 0^+$ is equivalent to taking the spatial period $L \to +\infty$, and formally recovering the standard mKdV aperiodic breather.

Moreover, we believe that the lack or invalidity of Assumption 3 is precisely the source of instability in KKHS breathers, in the sense that if this assumption is not satisfied, then they should be unstable, as it happens in the soliton case in the regime of supercritical nonlinearities (see [Mun14, Lemma 1.6] for example), and just in agreement with the numerical computations performed by Kevrekidis et al. [KKS03], for which (11.7.1) is not satisfied.

11.8 Mass calculations

Across this book we have always computed conserved quantities for breather solutions. These are key elements for understanding the dynamical behavior of breathers. Here, in the periodic case, there is no exception. However, computations are really hard due to the periodic behavior of the solution.

We will express the mass of the KKSH breather (11.4.1) as a function of k and β. And this explicit function will be essential for the study of the nonlinear stability of the solution.

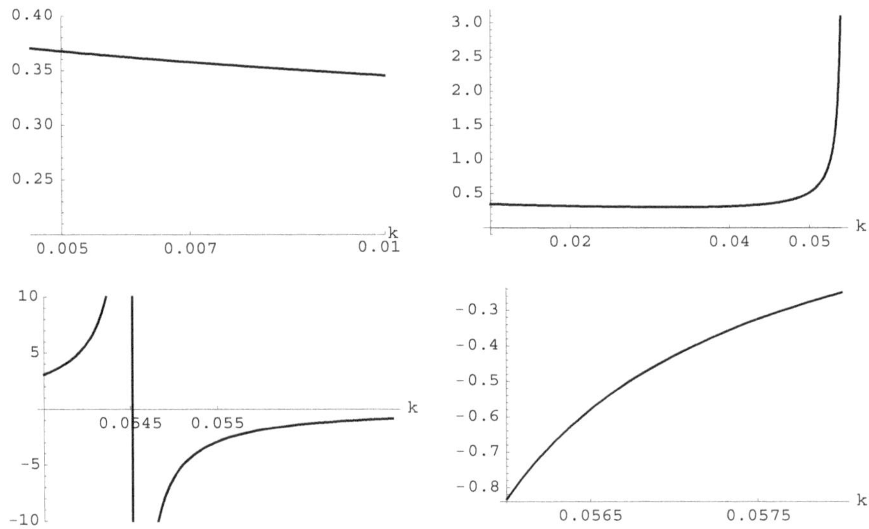

Fig. 11.3 The Weinstein's condition $HG(\beta, k)$ (11.9.4), for the periodic KKSH breather (11.4.1), in the case $\beta = 1$, for $k \in [0.0045, 0.01]$ (above left), for $k \in [0.01, 0.054]$ (above right), $k \in [0.054, 0.056]$ (below left), and $k \in [0.056, 0.058]$ (below right). In order to run our argument for a stability proof [AlMu13], we require $HG(1, k) > 0$, which is only satisfied for $k < k_*$, where $k_* \sim 0.0545$ is the approximate point where $D(1, k)$ vanishes (see Fig. 11.5). Note also that the case $k \sim 0.057$ assumed in [KKS03], that leads to instability, is not included in the stability region described in Theorem 11.9.1, but in the region where $HG(1, k) < 0$.

One has that the mass of the KKSH breather (11.4.1) is [AMP17b]:

$$\begin{aligned} \frac{1}{2}\int_0^L B^2 = {} & 4\beta m \frac{\operatorname{cd}\left(\frac{\beta}{2}L, m\right)\operatorname{sd}\left(\frac{\beta}{2}L, m\right)}{\operatorname{nd}\left(\frac{\beta}{2}L, m\right)} - 4\beta m \frac{\operatorname{cn}\left(\frac{\beta}{2}L, m\right)\operatorname{sn}\left(\frac{\beta}{2}L, m\right)}{\operatorname{dn}\left(\frac{\beta}{2}L, m\right)} \\ & - 8\alpha K(k) + 4\alpha\mathcal{E}\Big(A(2K(k), k), k\Big) + 4\beta\mathcal{E}\left(A\left(\frac{\beta}{2}L, m\right), m\right), \end{aligned} \tag{11.8.1}$$

where $\mathcal{E}$ denotes the *complete elliptic integral of the second kind* [ByFr71], defined as

$$\mathcal{E}(r) := \int_0^{\pi/2} (1 - r\sin^2(s))^{1/2} ds = \int_0^1 (1 - t^2)^{-1/2}(1 - rt^2)^{1/2} dt, \tag{11.8.2}$$

and A is the *Jacobi amplitude*, that can be defined by

$$A(x, r) := \int_0^x dn(s, r) ds. \tag{11.8.3}$$

From (11.8.2) and (11.8.3), one has that some simple exact values for $\mathcal{E}$ are

$$\mathcal{E}(0) = \frac{\pi}{2} \quad \text{and} \quad \mathcal{E}(1) = 1,$$

and for A are

$$A(0,r)=0,\quad A(K(r),r)=\frac{\pi}{2}.$$

Note also that $\mathcal{E}(A(2K(k),k),k) = \mathcal{E}(A(0,k)+\pi,k) = 2\mathcal{E}(k)$. Using (11.4.3), (11.8.1) simplifies as follows:

$$\frac{1}{2}\int_0^L B_0^2 = 4\beta\mathcal{E}(m) + 8\alpha\mathcal{E}(k) - 8\alpha K(k). \tag{11.8.4}$$

We are able to go one step further in the simplification of (11.8.4), using (11.4.3) again. We have

$$\frac{k}{1-m} = \frac{1}{16}\frac{K(m)^4}{K(k)^4}. \tag{11.8.5}$$

Hence, with these relations, (11.8.4) simplifies to the neat expression:

$$M_{\#}[B] = \frac{1}{2}\int_0^L B_0^2 = 4\beta\left(\mathcal{E}(m) + 4\frac{K(k)}{K(m)}(\mathcal{E}(k)-K(k))\right). \tag{11.8.6}$$

When $k\to 0$, we have $m\to 1$, and $M_{\#}[B]\to 4\beta$, which is the value of the mass of the aperiodic mKdV breather solution in the real line (see [AlMu13, p.6, Lemma 2.1]).

For $k\neq 0, m\neq 1$, the dependence of the periodic mass (11.8.6) with respect to the parameter m is computed in the following way: for each value of m, we solve numerically the implicit equation (11.8.5) in k. We then substitute these two pairs of values (m,k) verifying (11.8.5) inside the expression (11.8.6). The resulting plot of (11.8.6) versus m is given in Fig. 11.4.

11.9 Duality stability/instability

Now the main problem is to figure out where our nonlinear stability proof does/does not work. For this purpose, the *discriminant*

$$D = D(\beta,k) := \partial_k a_1 \partial_\beta a_2 - \partial_k a_2 \partial_\beta a_1 \qquad (a_1, a_2 \text{ from } (11.5.1) - (11.5.2)), \tag{11.9.1}$$

is the key element to look at. In order to explain why this element is important, let us notice that from (11.5.3) and (11.6.1), we readily have (compare with Corollary 9.8.5)

$$\mathcal{L}_{\#}\partial_k B = \partial_k a_1 (B_{xx}+B^3) - \partial_k a_2 B,$$
$$\mathcal{L}_{\#}\partial_\beta B = \partial_\beta a_1 (B_{xx}+B^3) - \partial_\beta a_2 B.$$

Therefore, as soon as $D\neq 0$,

$$B_{0,\#} := \frac{1}{D}(\partial_k a_1 \partial_\beta B - \partial_\beta a_1 \partial_k B), \tag{11.9.2}$$

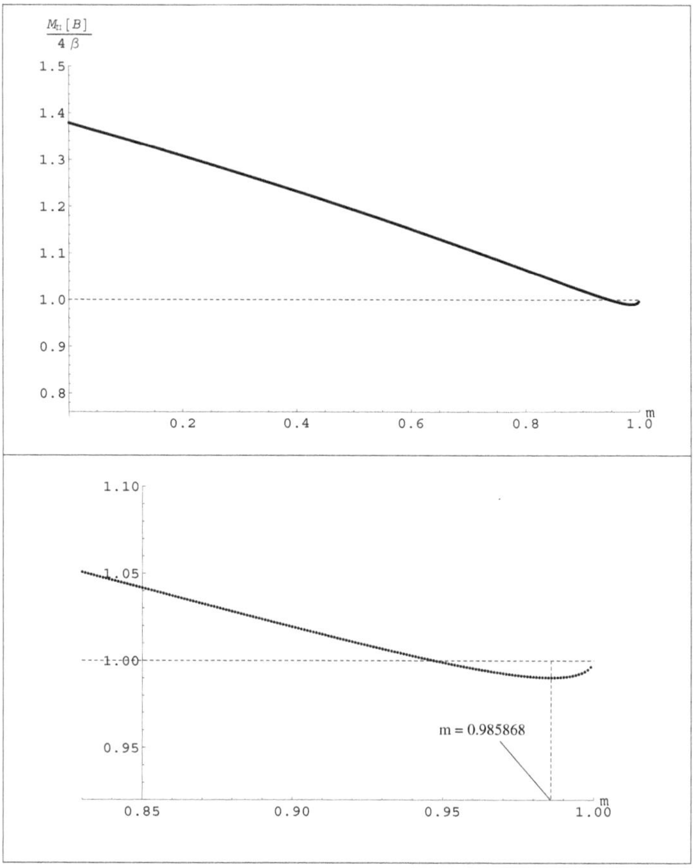

Fig. 11.4 The mass of the periodic breather $M_\#[B]/4\beta$ (11.8.6) as a function of m. Note that the resulting function is **decreasing** except for $m \gtrsim 0.98$ (see zoom figure below), corresponding to $k \lesssim 0.025$, a parameter region very close to the stable, aperiodic mKdV breather.

satisfies the equation

$$\mathcal{L}_\# B_{0,\#} = -B,$$

see also [AlMu13, Corollary 4.5]. Using this fact, we can easily prove that the eigenfunction associated to the negative eigenvalue of $\mathcal{L}_\#$ can be replaced by the breather itself, which has better behavior in terms of error controlling, unlike the first eigenfunction. This simple fact allows us to prove the nonlinear stability result as in the standard approach, without using scaling modulations. Recall that using the first eigenfunction as orthogonality condition does not guarantee a suitable control on the scaling modulation parameter, because the control given by this direction might be not good enough to close the stability estimates. However, the breather can be used as an alternative direction, and all these previous arguments remain valid, exactly as in [AlMu13], provided the Weinstein's sign condition

$$\int_0^L B_{0,\#}B > 0 \qquad \Big(\text{or equivalently} \int_0^L B_{0,\#}\mathcal{L}_\# B_{0,\#} < 0\Big), \tag{11.9.3}$$

do hold. Using (11.9.2), we are lead to the understanding of the quantity

$$\begin{aligned}\int_0^L B_{0,\#}B &= \frac{1}{D}\int_0^L (\partial_k a_1 \partial_\beta B - \partial_\beta a_1 \partial_k B)B \\ &= \frac{1}{D}(\partial_k a_1 \partial_\beta M_\#[B] - \partial_\beta a_1 \partial_k M_\#[B]) =: HG(\beta, k),\end{aligned} \tag{11.9.4}$$

where $M_\#[B]$ was computed in (11.8.6). Recall that a_1 and a_2 are almost explicit from (11.5.1)-(11.5.2). An exact expression for $HG(\beta, k)$ has escaped to us, however, we can graph this new function in some interesting cases. In particular, for the case considered in [KKS03], we assume $\beta = 1$ and we graph $D = D(1, k)$ and $HG(1, k)$, to obtain the results in Fig. 11.5 and Fig. 11.3. We note that condition (11.9.3) holds provided k is small enough. However, the values for which $HG(\beta, k) > 0$ do not coincide with the values for which the standard Weinstein's condition (positive derivative with respect to the scaling), deduced from Fig. 11.4, holds true, and this is totally natural for the case of breathers, as it was explained in [AlMu13, Corollary 2.2].

Now we can present a rigorous statement for Theorem 11.7.1.

Theorem 11.9.1. *Under Assumptions 1, 2 and 3 in page 188, KKHS breathers B* (11.4.1) *are orbitally stable under small, L-periodic H^2 perturbations. More precisely, there are $\eta_0 > 0$ and $K_0 > 0$, only depending on β and k, such that if $0 < \eta < \eta_0$ and if u_0 is L-periodic with*

$$\|u_0 - B(t = 0, \cdot; 0, 0)\|_{H^2(\mathbb{T})} < \eta,$$

then there are real-valued parameters $x_1(t)$ and $x_2(t)$ for which the global H^2, L-periodic solution $u(t)$ of (4.1.1) *with initial data u_0 satisfies*

$$\sup_{t\in\mathbb{R}} \|u(t) - B(t, \cdot; x_1(t), x_2(t))\|_{H^2(\mathbb{T})} < K_0\eta,$$

with similar estimates for the derivatives of the shift parameters x_1, x_2.

Proof. The proof of this result is completely similar to the proof of Theorem 9.9.4, after following the same steps (see also [AlMu13] for a proof in the scalar mKdV case). Assumption 3 is used to ensure that an expression like (9.10.7) has a nonzero denominator. □

Additionally, based on this numerical evidence, we conjecture the following alternative stability theory for KKSH breathers:

Conjecture: *Assume that* $HG(\beta, k) < 0$. *Then* B (11.4.1) *is unstable under small* $H^2(\mathbb{T})$ *perturbations.*

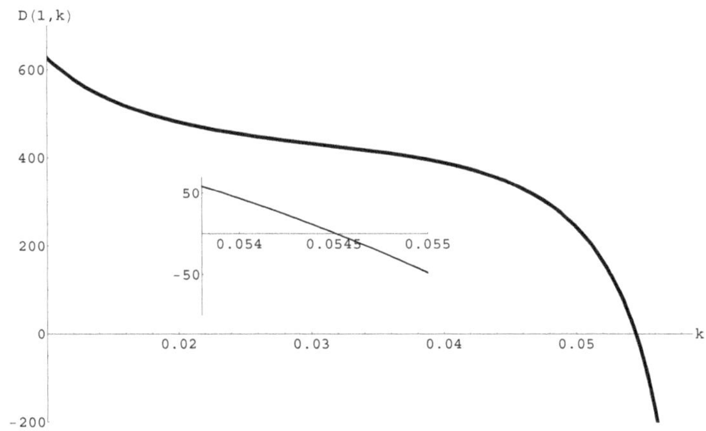

Fig. 11.5 The discriminant function $D(1,k)$ in terms of k, as described in (11.9.1). Note that $D(1,k)$ changes sign (see zoom figure inside) for $k \geq k_* \sim 0.0545$.

11.10 Numerical tests

As it is done in the case of mKdV and SG breathers, one can perform numerical computations to check spectral properties associated to classical breathers. Given the complicated functions that define the KKSH breather (11.4.1) and its related linearized operator, a rigorous description of the spectra of $\mathcal{L}_\#$ has eluded us. We perform then some numerical simulations to get a good understanding of these spectral properties. We could expect, given the instability results in [KKS03], that the KKSH breather (11.4.1) should give rise to a very different spectrum, with respect to the previous aperiodic breathers. However, as we will show later, this is not the case.

In the following lines we explain the main ideas of the numerical method used to compute the eigenvalues of $\mathcal{L}_\#$. The core of the algorithm is the same as in the previous sections, the main difference being the required test functions, which must now be periodic on $[0, L]$. We have used the classical orthonormal basis of $L^2(0, L)$:

$$\left\{\frac{1}{\sqrt{L}}, \sqrt{\frac{2}{L}}\cos\left(\frac{2\pi nx}{L}\right), \sqrt{\frac{2}{L}}\sin\left(\frac{2\pi nx}{L}\right)\right\},$$

where $n \in \{1, \dots, N\}$. For the standard numerical computations, it is enough to take $N = 40$, although when approaching the critical values $k \to k^*$ $(m \to 0)$ or $k \to 0$ $(m \to 1)$, more and more test functions are naturally required.

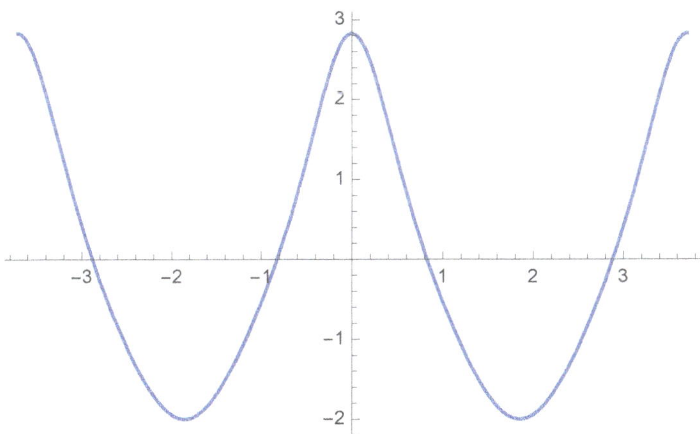

Fig. 11.6 Periodic KKSH breather [KKS03] at time $t = 0$ for $\beta = 1$, $m = 0.5$, $k = 0.057$, here $L \sim 3.71$. Although numerically unstable [KKS03], this breather leads to a linearized operator $\mathcal{L}_{\#}$ that possesses a "standard" one negative eigenvalue and a two dimensional kernel (11.10.1).

First of all, we run a specific computation for the case described in [KKS03], which the authors describe as numerically unstable. In this case, $\beta = 1$, $m = 0.5$, $k = 0.057$, and $L \sim 3.71$ (see Fig. 11.6). Numerically, we have found that this breather possesses the same spectral structure of all our previous breather solutions. To be more precise, running our numerical algorithm with $N = 40$ test functions, we have found the following approximations of the first four eigenvalues of $\mathcal{L}_{\#}$:

$$\{-4.86, \ -1.23 \times 10^{-8}, \ 3.22 \times 10^{-10}, \ 35.35\}. \tag{11.10.1}$$

Clearly the two components of the kernel are recovered with high precision (recall that a second negative eigenvalue but very close to zero is unlikely just by continuity arguments on the coefficients of the original breather), and a distinctive negative eigenvalue appears, far from the kernel itself. This behavior repeats for all cases we have studied.

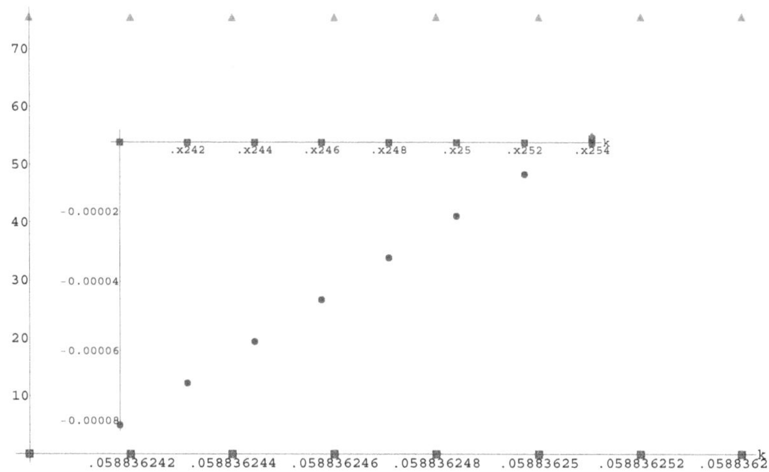

Fig. 11.7 The double zero kernel (purple box and brown diamond) and the first positive eigenvalue (green triangle) of the linearized operator $\mathcal{L}_{\#}$ around a periodic KKSH breather for $N = 40$, $\beta = 1$ and k increasing. Note that the numerical method returns only two eigenvalues very close to zero, as expected from the conjectured linear spectral stability. Inside, the representation of the negative (blue circle) and double zero kernel. We used the notation $.x242 \equiv 0.058836242, \quad x = 058836$.

As a second test, we perform several eigenvalue computations for the same parameter $\beta = 1$, and k moving. For the most difficult case, the one where k approaches the critical value ~ 0.0588, we obtain the results described in Fig. 11.7. Also, in Fig. 11.1, we exactly describe those eigenvalues, that we obtain for different values of k. It is important to mention that we always get one negative eigenvalue and a two dimensional kernel, as well as a clearly defined spectral gap.

In the intermediate case, for pairs (k, m), with $k = 0.01, \ldots, 0.04$, we obtain the results described in Fig. 11.8. Also, in the Table 11.2, we describe the eigenvalues of $\mathcal{L}_{\#}$ that we obtain for different values of k. Once again, we get one negative eigenvalue and a two dimensional kernel.

Finally, for the case, with k small, i.e. $k \approx 10^{-3}$, we obtain in Fig. 11.9 the description of the discrete spectra for the first four eigenvalues for the linearized operator $\mathcal{L}_{\#}$.

Also, in the Table 11.3, we show the explicit numerical eigenvalues which correspond to these small values of k. Once again, we get one negative eigenvalue and a two dimensional kernel.

Table 11.1 The first four eigenvalues of $\mathcal{L}_{\#}$ for $\beta = 1$, $x_1 = 0.1$, $x_2 = 0$, and k varying from $.x240 \equiv 0.058836240$, $x = 058836$ to $.x254$, as in Fig. 11.7. All computations were made with $N = 40$ test functions. The third and fourth columns represent approximate kernel of $\mathcal{L}_{\#}$.

k	1st. eig	2nd	3rd	4th
$.x240$	-0.00008	$-7.750 \cdot 10^{-10}$	$-1.565 \cdot 10^{-9}$	75.490
$.x242$	-0.00006	$-6.449 \cdot 10^{-10}$	$-1.561 \cdot 10^{-9}$	75.502
$.x244$	-0.00005	$-5.952 \cdot 10^{-10}$	$-1.558 \cdot 10^{-9}$	75.514
$.x246$	-0.00004	$-6.333 \cdot 10^{-10}$	$-1.554 \cdot 10^{-9}$	75.528
$.x248$	-0.00003	$-6.830 \cdot 10^{-10}$	$-1.552 \cdot 10^{-9}$	75.544
$.x250$	-0.00002	$-6.160 \cdot 10^{-10}$	$-1.546 \cdot 10^{-9}$	75.563
$.x252$	$-9.189 \cdot 10^{-6}$	$-6.774 \cdot 10^{-10}$	$-1.541 \cdot 10^{-9}$	75.589
$.x254$	$-3.463 \cdot 10^{-7}$	$-1.135 \cdot 10^{-6}$	$-1.869 \cdot 10^{-6}$	75.640

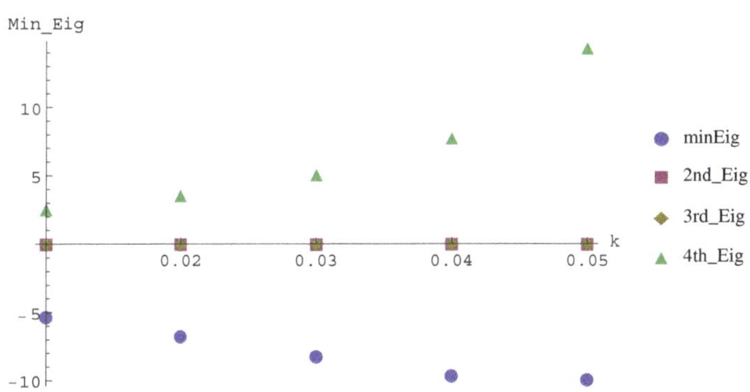

Fig. 11.8 For intermediate values of k: we plot the negative eigenvalue, the double zero kernel and the fourth eigenvalue of the linearized operator $\mathcal{L}_{\#}$ around a periodic KKSH breather, for $\beta = 1$ and k increasing from 0.01 to 0.05, as expressed before in Table 11.2. Note that the numerical method returns only two eigenvalues very close to zero, as expected from the conjectured linear spectral stability. Computations were made with $N = 50$ test functions.

Therefore, we can certainly claim that there is strong evidence that, according to numerical simulations, KKSH breathers (11.4.1) have standard spectrum, in the sense that they have only one negative eigenvalue, and a nondegenerate, two-dimensional kernel.

Table 11.2 The first four eigenvalues of $\mathcal{L}_\#$ for $\beta = 1$, $x_1 = 0.1$, $x_2 = 0$, and k varying, as corresponding to Fig. 11.8. All computations were made with $N = 50$ test functions. The third and fourth columns represent the approximate kernel of $\mathcal{L}_\#$.

k	1st eig.	2nd	3rd	4th
0.01	-5.343	$1.023 \cdot 10^{-6}$	0.0001	2.540
0.02	-6.751	$5.905 \cdot 10^{-10}$	$2.683 \cdot 10^{-6}$	3.561
0.03	-8.216	$-2.651 \cdot 10^{-9}$	$1.163 \cdot 10^{-6}$	5.067
0.04	-9.623	$3.997 \cdot 10^{-9}$	$4.896 \cdot 10^{-8}$	7.756
0.05	-9.922	$-2.173 \cdot 10^{-7}$	$-1.143 \cdot 10^{-8}$	14.329

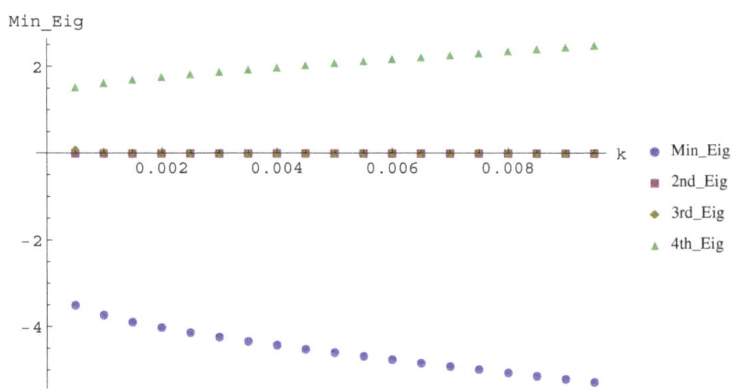

Fig. 11.9 For *k small*: the negative eigenvalue, the double zero kernel and the fourth eigenvalue for the linearized operator $\mathcal{L}_\#$ around a periodic KKSH breather, for $\beta = 1$ and k increasing from 0.0005 to 0.0095, as expressed before in Table 11.3. Note that the numerical method returns only two eigenvalues very close to zero, as expected from the conjectured linear spectral stability. Computations were made with $N = 50$ test functions.

11.11 Nonzero mean periodic mKdV breathers

11.11.1 *Introduction*

KKSH breathers (11.4.1) are periodic in space, but they have zero-mean. This property, classical in mKdV-type breathers, is hard to avoid. Also, non periodic and nonzero mean breather solutions of mKdV were already known, see [Ale12c, Ale12a].

We would like to see if nonzero mean periodic breather solutions may exist. Numerical evidence of the existence of these solutions was given in [Ale12a], since

Table 11.3 The first four eigenvalues for the linearized operator $\mathcal{L}_{\#}$ around a periodic KKSH breather for $\beta = 1$, $x_1 = 0.1$, $x_2 = 0$, and k varying in a sample of points of Fig. 11.9. All computations were made with $N = 50$ test functions. The third and fourth columns (left) and eighth and ninth columns (right) represent respectively the approximate kernel of $\mathcal{L}_{\#}$.

k	1st eig.	2nd $\cdot 10^{-6}$	3rd	4th		k	1st eig.	2nd $\cdot 10^{-5}$	3rd	4th
0.0095	-5.271	1.265	0.0001	2.495		0.0045	-4.505	1.531	0.0017	2.040
0.0085	-5.127	1.962	0.0002	2.405		0.0035	-4.327	2.919	0.0034	1.941
0.0075	-4.971	1.248	0.0142	2.316		0.0025	-4.127	6.234	0.0073	1.833
0.0065	-4.828	5.092	0.0005	2.225		0.0015	-3.883	16.47	0.0198	1.708
0.0055	-4.671	8.609	0.0009	2.134		0.0005	-3.497	79.70	0.0983	1.530

these solutions are connected with the (nonzero) curvature of the planar curve evolving according to a particular law. For more details about this connection, see e.g. [Ale12a].

It turns out that these solutions do exist. More explicitly,

Definition 11.11.1. Given $c_1, c_2, \mu > 0$, p, q nonzero integers, with p, q coprime, such that the following *commensurability* condition is satisfied

$$\frac{2\mu^2 - c_1}{2\mu^2 - c_2} = \frac{p^2}{q^2}, \tag{11.11.1}$$

(note that each c_i is less than $2\mu^2$), we define the breather $B = B(t, x; c_1, c_2, \mu, p, q)$ by the formula

$$B := \mu + 2\sqrt{2}\partial_x \arctan\left(\frac{f(t,x)}{g(t,x)}\right), \tag{11.11.2}$$

where,

$$\rho := \frac{\sqrt{c_1} + \sqrt{c_2}}{\sqrt{c_1} - \sqrt{c_2}},$$

$$f(t,x) := -\sqrt{2}\mu\rho\left(\sqrt{c_1} - \sqrt{c_2} + \sqrt{2\mu^2 - c_2}\tan y_2 - \sqrt{2\mu^2 - c_1}\tan y_1\right),$$

and

$$g(t,x) := 2\mu^2 + \left(\sqrt{2\mu^2 - c_1}\tan y_1 - \sqrt{c_1}\right)\left(\sqrt{2\mu^2 - c_2}\tan y_2 - \sqrt{c_2}\right).$$

Here

$$y_1 = \frac{1}{2}\sqrt{2\mu^2 - c_1}(x - \delta t), \quad y_2 = \frac{1}{2}\sqrt{2\mu^2 - c_2}(x - \gamma t), \tag{11.11.3}$$

and the speeds are

$$\delta := \mu^2 + c_1, \quad \gamma := \mu^2 + c_2.$$

Note that condition (11.11.1) is imposed in order to obtain a truly *spatially* periodic solution, see the formulae for f and g. The spatial period of this breather is given by $L = \frac{2\pi q}{\sqrt{2\mu^2 - c_1}}$.

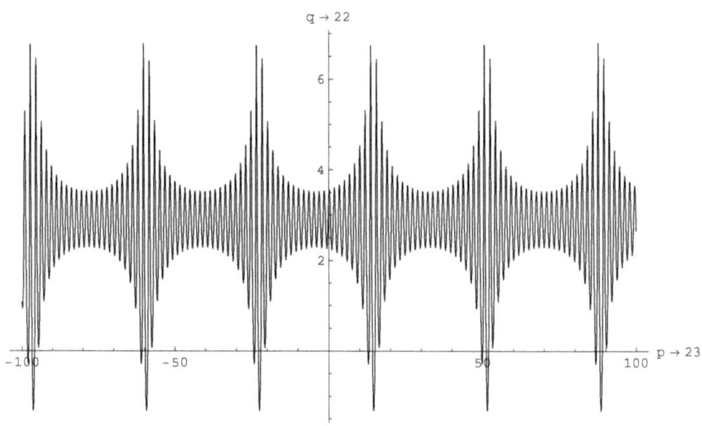

Fig. 11.10 Periodic breather of (4.1.1) with nonzero mean as in Definition 11.11.1. The parameters here are $c_1 = 1.65, c_2 = 2.95, p = 23, q = 22$, and the period L is ~ 35.7.

See also Fig. 11.10-11.11 for some drawings of different breather solutions, depending on the parameters c_1, c_2, μ, p and q.

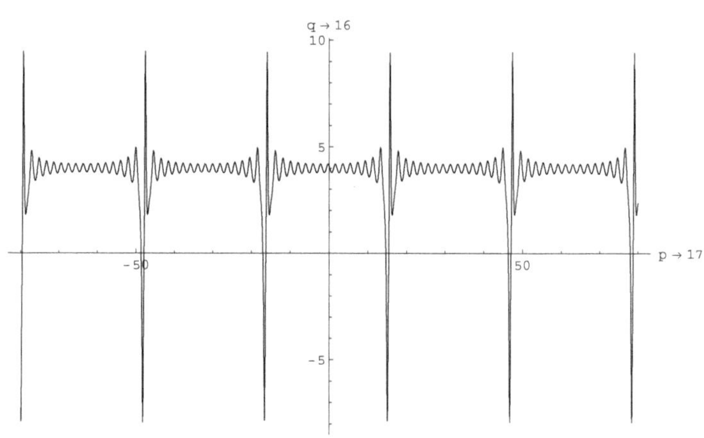

Fig. 11.11 Periodic breather of (4.1.1) with nonzero mean as in Definition 11.11.1, for parameters $c_1 = 20.65,\ c_2 = 21.95,\ p = 17,\ q = 16$.

Recall that KKSH breathers (11.4.1) cannot have any admissible set of parameters k and m; they are constrained by conditions in (11.4.3). Here, the only condition that we need to satisfy is (11.11.1). Therefore, given nonnegative integers p and q, with p and q coprime, and given $c_1 < 2\mu^2$, then there is a unique c_2 solution to (11.11.1).

In that sense, μ being fixed, the breather in (11.11.2) has three different degrees of freedom (in addition to shifts), two of them being discrete. Finally, in terms of p and q, the larger these integers are, the more oscillatory the breather solution is. Note additionally that this breather has not been constructed by using JEFs, but only standard periodic functions, a fact that simplifies many computations.

It is also relevant to mention that the construction of a breather solution with different patterns as the usually required can be very involved. For example, a different class of periodic breather was discovered by Blank et al. in [BCB11]. This particular solution is constructed for the so-called ϕ^4 model, by assuming that the equation is no longer autonomous, but it has suitable, well-chosen periodic coefficients.

11.11.2 *Construction of the breather solution*

Let us describe in a few lines the breather presented in Section 11.11. One uses the Bäcklund Transformation for mKdV (see [AlMu15] for a rigorous setting of the computations below) to construct this solution. Given a solution u_0 of the form $u_0 = \partial_x \widetilde{u}_0$ of mKdV and fixed constants $a_1, a_2 \in \mathbb{R}$, $a_1 \neq a_2$, we can construct a second $u_1 = \partial_x \widetilde{u}_1$ and third solution $u_2 = \partial_x \widetilde{u}_2$ by setting

$$u_1 - u_0 = a_1 \sin\left(\frac{\widetilde{u}_1 + \widetilde{u}_0}{\sqrt{2}}\right), \quad \text{and} \quad u_2 - u_0 = a_2 \sin\left(\frac{\widetilde{u}_2 + \widetilde{u}_0}{\sqrt{2}}\right).$$

These two solutions can be combined to construct a fourth solution u through the *permutability* condition [AlMu15]:

$$\tan\left(\frac{\widetilde{u} - \widetilde{u}_0}{2\sqrt{2}}\right) = -\left(\frac{a_1 + a_2}{a_1 - a_2}\right) \tan\left(\frac{\widetilde{u}_2 - \widetilde{u}_1}{2\sqrt{2}}\right). \tag{11.11.4}$$

Now we will apply these formulae to construct the breather.

Fix $\mu > 0$. Starting with the constant solution $u_0 = \mu$, $\widetilde{u}_0 = \mu x$ and two constants $a_1 = \sqrt{2c_1}$, $a_2 = \sqrt{2c_2}$, $c_1, c_2 > 0$, we have

$$u_1 - \mu = \sqrt{2c_1} \sin\left(\frac{\widetilde{u}_1 + \mu x}{\sqrt{2}}\right),$$

$$u_2 - \mu = \sqrt{2c_2} \sin\left(\frac{\widetilde{u}_2 + \mu x}{\sqrt{2}}\right).$$

Looking for a solution of the form $u_i = \partial_x \widetilde{u}_i$, $i = 1, 2$, we get

$$\widetilde{u}_i(t,x) = -\mu x + 2\sqrt{2}\arctan\left(\frac{1}{2\mu}\left(-\sqrt{2c_i} + \sqrt{4\mu^2 - 2c_i}\tan\left(\frac{\sqrt{4\mu^2 - 2c_i}}{2\sqrt{2}}(x - (\mu^2 + c_i)t)\right)\right)\right). \tag{11.11.5}$$

The factor $(\mu^2 + c_i)t$ in the above solution appears as a constant of integration, and it is chosen in such a form that u_i is actually solution to mKdV.

Calling $y_1 = x + \delta t$, $y_2 = x + \gamma t$ as in (11.11.3), and $\rho = \frac{\sqrt{c_1}+\sqrt{c_2}}{\sqrt{c_1}-\sqrt{c_2}}$, the desired breather (11.11.2) is obtained by using (11.11.4) with $u_0 = \mu$, u_1, u_2 from (11.11.5):

$$B = \mu + \partial_x \widetilde{B},$$

$$\widetilde{B} := 2\sqrt{2}\arctan\left(\frac{-\sqrt{2}\mu\rho(\sqrt{c_1} - \sqrt{c_2} + \sqrt{2\mu^2 - c_2}\tan(y_2) - \sqrt{2\mu^2 - c_1}\tan(y_1))}{2\mu^2 + (\sqrt{2\mu^2 - c_1}\tan(y_1) - \sqrt{c_1})\cdot(\sqrt{2\mu^2 - c_2}\tan(y_2) - \sqrt{c_2})}\right).$$

The fact that this is a solution of mKdV is a tedious, lengthy but straightforward computation.

Note that a simple transformation, actually Lemma 8.1.1, leads to a (localized) *periodic breather solution* of the Gardner equation (1.4.2), since for B as in (11.11.2),

$$\hat{B}(t,x) := B\left(t, x + \frac{\mu^2}{3}t\right) - \frac{\mu}{3},$$

solves (1.4.2) with functions f, g and commensurability condition as in (11.11.2).

11.11.3 *Final comments*

We advance that nonzero mean periodic breather B (11.11.2) satisfies the elliptic ODE

$$\begin{aligned}
&B_{4x} - (c_1 + c_2 - 4\mu^2)(B_{xx} + 3\mu(B-\mu)^2 + (B-\mu)^3) \\
&\quad + (c_1 - 2\mu^2)(c_2 - 2\mu^2)(B-\mu) \\
&\quad + 5(B-\mu)B_x^2 + 5(B-\mu)^2 B_{xx} + \frac{3}{2}(B-\mu)^5 + 5\mu B_x^2 \\
&\quad + \frac{15}{2}\mu(B-\mu)^4 + 10\mu(B-\mu)B_{xx} + 10\mu^2(B-\mu)^3 = 0.
\end{aligned}$$

In particular, B is a critical point of the functional

$$\mathcal{H}_{\mu p}[w](t) := F_{\mu p}[w](t) + (c_1 + c_2 - 4\mu^2)E_{\mu p}[w](t) + (c_1 - 2\mu^2)(c_2 - 2\mu^2)M_{\mu p}[w](t),$$

where $M_{\mu p}$ and $E_{\mu p}$ are defined in the following way:

$$M_{\mu p}[w](t) := \frac{1}{2}\int_{\mathbb{R}}(w-\mu)^2(t,x)dx = M_{\mu p}[w](0),$$

$$E_{\mu p}[w](t) := \frac{1}{2}\int_{\mathbb{R}} w_x^2 - \mu\int_{\mathbb{R}}(w-\mu)^3 - \frac{1}{4}\int_{\mathbb{R}}(w-\mu)^4 = E_{\mu p}[w](0),$$

and

$$\begin{aligned}F_{\mu p}[w](t) := &\frac{1}{2}\int_{\mathbb{R}} w_{xx}^2 dx - 5\mu\int_{\mathbb{R}}(w-\mu)w_x^2 dx + \frac{5}{2}\mu^2\int_{\mathbb{R}}(w-\mu)^4 dx\\ &-\frac{5}{2}\int_{\mathbb{R}}(w-\mu)^2 w_x^2 dx\\ &+\frac{3\mu}{2}\int_{\mathbb{R}}(w-\mu)^5 + \frac{1}{4}\int_{\mathbb{R}}(w-\mu)^6 dx = F_{\mu p}[w](0),\end{aligned}$$

is a third conserved quantity for the mKdV equation. Finally, it is interesting to note that we can obtain an aperiodic mKdV breather with nonzero mean (see [Ale12c, Ale18] substituting in (11.11.2) complex conjugate scalings $\sqrt{c_1} = \beta + i\alpha$, $\sqrt{c_2} = \beta - i\alpha$. For the sake of simplicity, we only show a picture of it in Fig. 11.12 and a geometric application, when considering these breather solutions with complex conjugate scalings as curvatures of a curve. Integrating twice and following the same procedure as in [Ale12a], we can obtain the curve depicted in Fig. 11.13. Note moreover that with the same transformation as in Lemma 8.1.1, we get a (localized) *non periodic* breather solution of the Gardner equation (1.4.2).

11.12 Breathers with periodic background

Recently, it has been possible to construct breather solutions for *defocusing* mKdV (6.4.1), KdV (2.7.1) and BO (11.12.20) equations, but and critical to their existence, raised on a periodic background, and which accompanies its evolution. The key technique was to use Darboux transformations [MaSa91] to explicitly obtain these raised breathers on a periodic background for both models.

Note moreover that the *defocusing* mKdV (6.4.1), KdV (2.7.1) and BO (11.12.20) equations do not admit conventional, regular breather solutions which are localized in space and periodic in time. The proof of nonexistence of breathers

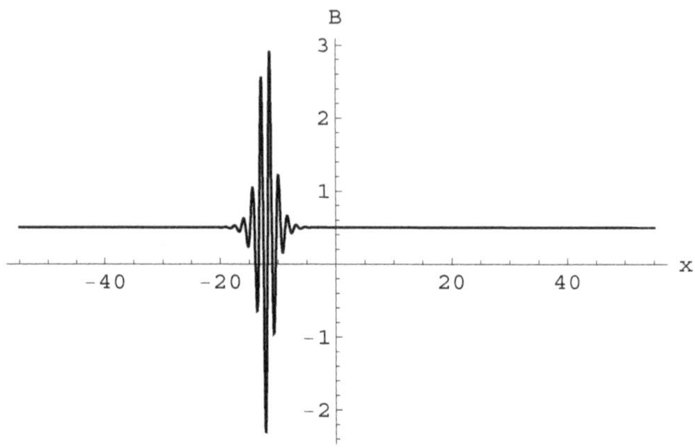

Fig. 11.12 Aperiodic mKdV breather with nonzero mean at $t = 0.2$ with $\beta = 1, \alpha = 4, \mu = 0.5$, obtained from the periodic breather (11.11.2).

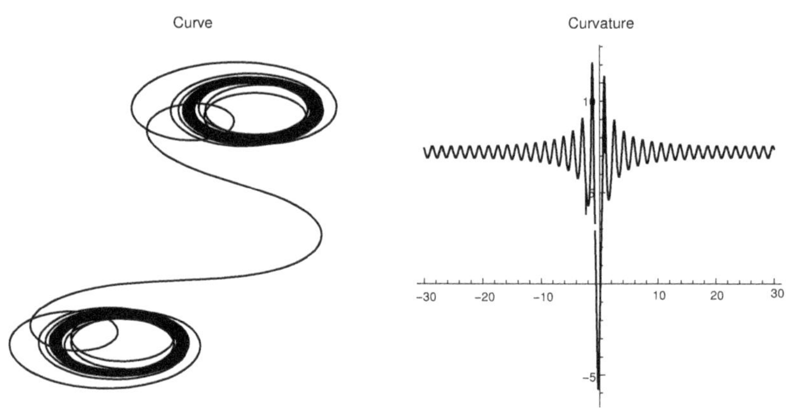

Fig. 11.13 Right: the breather solution with complex scalings considered as curvature. Left: the associated curve obtained from this *breather curvature* at $t = 0$ with parameters $\beta = 11.07, \alpha = 10.09, p = 32, q = 31$.

in these equations was recently given in [MuPo19]-[MuPo19bis]. The breathers presented in [HMP23] and [ChPe24] are obtained as a result of periodic interaction of solitary waves and the traveling periodic wave. Both the KdV and BO equations are the limiting cases of the integrable Intermediate Long-Wave (ILW) equation [Mil90]. Similar breathers are expected to exist in the ILW equation[2] but

[2]see [MPS20] for a proof of their nonexistence.

expressed in the Jacobi theta functions. Only the BO equation (11.12.20) describes such wave interactions in elementary functions. Let us explain a bit more such results.

11.12.1 *The defocusing mKdV case*

The defocusing mKdV equation

$$u_t + (u_{xx} - 2u^3)_x = 0, \quad u(t, x) \in \mathbb{R}, (t, x) \in \mathbb{R} \times \mathbb{R}, \tag{11.12.1}$$

is a well-known canonical model which can be used to describe nonlinear phenomena in the physics of fluids and crystals. Note that (11.12.1) as an integrable model, has a well-known Lax pair or Lax system, written as follows

$$\begin{aligned} \varphi_x &= \begin{pmatrix} i\zeta & u \\ u & -i\zeta \end{pmatrix} \varphi, \\ \varphi_t &= \begin{pmatrix} 4i\zeta^3 + 2i\zeta u^2 & 4\zeta^2 u - 2i\zeta u_x + 2u^3 - u_{xx} \\ 4\zeta^2 u + 2i\zeta u_x + 2u^3 - u_{xx} & -4i\zeta^3 - 2i\zeta u^2 \end{pmatrix} \varphi, \end{aligned} \tag{11.12.2}$$

where ζ is the (t, x)–independent spectral parameter and $\varphi = (p, q)^T$ is the corresponding eigenfunction. Remember that the defocusing mKdV equation (11.12.1) appears as a compatibility condition $\varphi_{xt} = \varphi_{tx}$ of the Lax system (11.12.2) (see [AKNS74] for further details.)

In [MuPe24] breather solutions on a periodic snoidal *sn* JEFs background are obtained. In fact, these breathers describe the periodic interaction of a dark soliton and a traveling periodic wave. As a consequence of the periodicity of such interactions, they named this solution as *dark breathers* of (11.12.1).

Periodic traveling wave solutions to (11.12.1) are written as

$$u(t, x) := \phi_0(x + c_0 t), \quad \phi_0(x) = k sn(x, k), \quad c_0 = 1 + k^2, \tag{11.12.3}$$

with $k \in (0, 1)$ the elliptic modulus and *sn*, the periodic snoidal JEFs. Remember that $\phi_0(x) = 0$ as $k \to 0$ and $\phi_0(x) = \tanh(x)$ as $k \to 1$, this one called as the kink of (11.12.1). The main idea is to use the one-fold Darboux transformation on the snoidal background (11.12.3) to construct breathers of (11.12.1). A few preliminaries are in order: for instance, the following two Jacobi's theta functions will be used

$$\begin{aligned} \theta_1(y) &:= 2 \sum_{n=1}^{\infty} (-1)^{n-1} q^{(n-1/2)^2} \sin((2n-1)y), \\ \theta_4(y) &:= 1 + 2 \sum_{n=1}^{\infty} (-1)^{n-1} q^{n^2} \cos(2ny), \end{aligned} \tag{11.12.4}$$

where $q := e^{-\frac{\pi K'(k)}{K(k)}}$ with $K(k)$ being the complete elliptic integral and $K'(k) = K(k')$ with $k' = \sqrt{1-k^2}$. It is well-known [Law13] that $K(k)$ is a quarter period and $iK'(k)$ is a half period of the Jacobi elliptic function $sn(x,k)$ with the correspondence

$$y = \frac{\pi x}{2K(k)}.$$

In fact, the authors in [MuPe24] used the notation

$$H(x) = \theta_1(y),\ \Theta(x) = \theta_4(y),\ Z(x) = \frac{\Theta'(x)}{\Theta(x)}.$$

Remembering that from [Law13, 2.1.1]

$$sn(x,k) = \frac{H(x)}{\sqrt{k}\Theta(x)},$$

then, the traveling wave ϕ_0 can be written as

$$\phi_0(x) = \frac{H(K)}{\Theta(K)}\frac{H(x)}{\Theta(x)}.$$

Now, using the one-fold Darboux transformation [MaSa91] allows us to obtain a new solution $\hat{u}$ of (11.12.1) from a previous solution u, in the following way:

$$\hat{u} = u - \frac{4i\zeta pq}{p^2 - q^2}, \tag{11.12.5}$$

where $\varphi = (p,q)^T$ is a particular nonzero solution of the linear systems (11.12.2) associated with the potential u for a particular value of the spectral parameter ζ. Note that $\hat{u}$ in (11.12.5) is real if $q = p^*$ is satisfied on the solution $\varphi = (p,q)^T$ of (11.12.2) with real u and real ζ. The main result in [MuPe24] is the following:

Theorem 11.1. *Let $\alpha \in (0,K)$ be a free parameter with $k \in (0,1)$ and define β, γ by*

$$\beta := \left(1 - \frac{2(1+k)^2 sn(\alpha)^2}{(1+ksn(\alpha)^2)^2}\right)\gamma, \quad \gamma := \frac{\Theta(2\alpha)}{\Theta(0)}.$$

Then, the new solution of (11.12.1) *for the dark breather, is*

$$\hat{u} := \frac{H(K)}{\Theta(K)}\frac{H(\xi+2\alpha)e^{-2\eta} + H(\xi-2\alpha)e^{2\eta} + 2\beta H(\xi)}{\Theta(\xi+2\alpha)e^{-2\eta} + \Theta(\xi-2\alpha)e^{2\eta} + 2\gamma\Theta(\xi)}, \tag{11.12.6}$$

where $\xi = x + c_0 t$, with $c_0 = 1 + k^2$ and $\eta = \kappa(x + ct + x_0)$ with

$$\kappa = Z(\alpha) + \frac{ksn(\alpha)cn(\alpha)dn(\alpha)}{1+ksn(\alpha)^2} > 0,$$

$$c := c_0 + \frac{2(1+k)^2(1-ksn(\alpha)^2)sn(\alpha)cn(\alpha)dn(\alpha)}{(Z(\alpha)(1+ksn(\alpha)^2) + ksn(\alpha)cn(\alpha)dn(\alpha))(1+ksn(\alpha)^2)^2} > c_0,$$

for $x_0 \in \mathbb{R}$.

Remark 11.1. The value of α is uniquely defined by the spectral parameter ζ from the characteristic equation:

$$\zeta := \frac{1+k}{2}\frac{1-ksn(\alpha)^2}{1+ksn(\alpha)^2},$$

and hence, if $\alpha \in (0, K)$, then $\zeta \in (\zeta_-, \zeta_+)$ with $\zeta_\pm := \frac{1}{2}(1 \pm k)$.

Remark 11.2. The value $\alpha \in (0, K)$ determines the phase shift of the solitary wave propagating across the snoidal background because

$$\lim_{\eta\to\pm\infty} \hat{u}(t,x) = ksn(\xi \mp 2\alpha).$$

Since the period of $sn(\xi)$ is $4K$, a suitably normalized phase shift can be defined by

$$\Delta := \frac{2\pi(4\alpha)}{4K} = \frac{2\pi\alpha}{K} \in (0, 2\pi).$$

The inverse localization width of the solitary wave is defined by κ and its velocity is defined by c.

Remark 11.3. In the limit $k \to 1$, the family of dark breathers (11.12.6) of Theorem 11.1 generates a two-soliton solution of (11.12.1) as follows

$$\hat{u}(t,x) := \frac{\sinh(\xi+2\alpha)e^{-2\eta} + \sinh(\xi-2\alpha)e^{2\eta} + 2\sinh(\xi)(1-\sinh(2\alpha)^2)\operatorname{sech}(2\alpha)}{\cosh(\xi+2\alpha)e^{-2\eta} + \cosh(\xi-2\alpha)e^{2\eta} + 2\cosh(\xi)\cosh(2\alpha)}, \tag{11.12.7}$$

where $\xi = x+2t$, $\eta = \tanh(2\alpha)(x+2t+4t\operatorname{sech}(2\alpha)^2 + x_0)$, $x_0 \in \mathbb{R}$ and $\alpha \in (0,\infty)$.

11.12.2 *KdV case*

The authors in [HMP23] started with the KdV model

$$v_t + (v_{xx} + 3v^2)_x = 0, \tag{11.12.8}$$

which, as it is well-known since XIX century ([KdV95]), has *cnoidal wave solutions* in the form

$$v(t,x) = \phi_0(x-ct), \quad \phi_0(x) := a + 2d^2k^2cn(d(x+x_0),k)^2, \ k \in (0,1), \ a \in \mathbb{R}, \tag{11.12.9}$$

with $c = 6a + 4d^2(2k^2-1)$ the speed of the background cnoidal wave. In fact, they assumed the simplest case, when $d = 1, a = 0$ to feed the Darboux transformation. On the other hand, as it is well-known in integrability theory [Lam80]-[GGKM67], every smooth solution $v(t,x)$ of (11.12.8) corresponds to the compatibility condition[3] of the stationary Schrödinger equation with potential $v(t,x)$

$$(-\partial_x^2 - v)\psi = \lambda\psi, \tag{11.12.10}$$

[3]Namely, $\psi_{xt} = \psi_{tx}$.

with the time flow determined by

$$\psi_t = (4\lambda - 2v)\psi_x + v_x\psi, \tag{11.12.11}$$

and λ the spectral parameter (independent of (t, x)). In fact, (11.12.10) with the simplest cnoidal wave (11.12.9) is known as the *Lamé equation* [Inc56, p.395]. It can be written as

$$\begin{aligned} &\psi''(\xi) - 2k^2 sn(\xi, k)^2\psi(\xi) + \eta\psi(\xi) = 0, \\ &\eta = \lambda + 2k^2, \ \xi = x - ct. \end{aligned} \tag{11.12.12}$$

It is easy to see, using JEFs basic properties (see [ByFr71] for further details), that (11.12.12) has three particular solutions $\psi = \psi_i, i = 1, 2, 3$ with sorted spectral parameters $\lambda = \lambda_i(k), i = 1, 2, 3, \quad \lambda_1(k) < \lambda_2(k) < \lambda_3(k)$ and written as follows:

$$\begin{aligned} &\lambda_1(k) = -k^2, \quad \psi_1(\xi) := dn(\xi, k), \\ &\lambda_2(k) = 1 - 2k^2, \quad \psi_2(\xi) := cn(\xi, k), \\ &\lambda_3(k) = 1 - k^2, \quad \psi_3(\xi) := sn(\xi, k). \end{aligned}$$

Actually there are only two linearly independent solutions of (11.12.12) for $\lambda \neq \lambda_i(k), i = 1, 2, 3$ and given by

$$\psi_\pm(\xi) = \frac{H(\xi \pm \alpha)}{\Theta(\xi)} e^{\mp\xi Z(\alpha)}, \qquad Z(\alpha) = \frac{\Theta'(\alpha)}{\Theta(\alpha)} = Z(\varphi_\alpha, k), \ \varphi_\alpha = am(\alpha, k), \tag{11.12.13}$$

with $\alpha \in \mathbb{C}$ satisfying the characteristic equation $\eta = \lambda + 2k^2 = k^2 + dn(\alpha, k)^2$, or equivalently $\lambda = 1 - 2^2 + k^2 cn(\alpha, k)^2$.

Here the function $H(x)$ is defined in terms of (11.12.4) as (remember that $q = e^{-\frac{\pi K'(k)}{K(k)}}, \quad K'(k) = K(k'), \ k' = \sqrt{1 - k^2}$)

$$H(x) = \theta_1\left(\frac{\pi x}{2K(k)}\right),$$

and

$$\Theta(x) = \theta_4\left(\frac{\pi x}{2K(k)}\right).$$

After that, it is needed to get the time evolution of the corresponding eigenfunctions of (11.12.10). In fact, assuming $v(t, x) = \phi_0(t, x)$ in the simplest case (11.12.9), and $\psi(t, x) = \psi_\pm(t, x)$ independent solutions of (11.12.10)-(11.12.11), the time dependence of $\psi_\pm(t, x)$ can be found by separation of variables as:

$$\psi_\pm(t, x) = \frac{H(x - ct \pm \alpha)}{\Theta(x - ct)} e^{\mp(x-ct)Z(\alpha)\mp\omega(\alpha)t}, \tag{11.12.14}$$

where $\omega(\alpha)$ has to be determined, by substituting it in (11.12.11), obtaining

$$\omega(\alpha) = (c + 4\lambda - \phi_0(\xi))\left[Z(\alpha) \pm Z(\xi) \mp \frac{H'(\xi \pm \alpha)}{H(\xi \pm \alpha)}\right] \mp \phi_0'(\xi).$$

Since the above equation is true for every $x \in \mathbb{R}$ for the compatibility of (11.12.10)-(11.12.11), taking $c = 4(2k^2 - 1)$ and $x = 0$, $\omega(\alpha)$ reads as follows:

$$\omega(\alpha) = 4(\lambda + k^2 - 1)\left[\frac{\Theta'(\alpha)}{\Theta(\alpha)} - \frac{H'(\alpha)}{H(\alpha)}\right]. \tag{11.12.15}$$

In fact, it is possible to see [HMP23, Prop.4], that $\omega(\alpha) \in \mathbb{R}$ if $\lambda \in (-\infty, \lambda_1(k)) \cup (\lambda_2(k), \lambda_3(k))$ and $\omega(\alpha) \in \mathbb{C}$ if $\lambda \in [\lambda_1(k)), \lambda_2(k)]$. Explicit expressions for the quotients $\frac{\Theta'(\alpha)}{\Theta(\alpha)}$, $\frac{H'(\alpha)}{H(\alpha)}$ can be found in [HMP23, Prop.4] and [GradRyz07, 8.199-3].

From here, and it is possible to apply a Darboux transformation [MaSa91] for the KdV equation (11.12.8). This is an iterative method to obtain solutions, starting with a suitable seed or simplest solution of the corresponding PDE, like constants or traveling wave solutions. Namely for this KdV case, by fixing $\lambda = \lambda_0$ and obtaining a solution ψ_0 of (11.12.10)-(11.12.11) associated with the potential $v(t, x) = \phi_0(x - ct)$, an explicit periodic solution of (11.12.8), a new solution of (11.12.8) can be obtained through the expression

$$\hat{v}(t, x) = \phi_0(x - ct) + 2\partial_x^2 \log \psi_0(t, x). \tag{11.12.16}$$

This new KdV solution is non-singular if $\psi_0(t, x) \neq 0$ everywhere in (t, x) plane. Actually this corresponds to the case $\lambda_0 \in (-\infty, \lambda_1(k))$. The main result for this kind of KdV breathers with background is stated in the following theorem (see [HMP23, Th.1])

Theorem 11.2. *There exists an exact solution $v(t, x)$ to the KdV equation* (11.12.8), *called bright breather, in the form*

$$v(t, x) = 2\left(k^2 - 1 + \frac{E(k)}{K(k)}\right) + 2\partial_x^2 \log(\tau(t, x)), \tag{11.12.17}$$

with

$$\tau(t, x) := \Theta(x - ct + \alpha_b)e^{\kappa_b(x - c_b t + x_0)} + \Theta(x - ct - \alpha_b)e^{-\kappa_b(x - c_b t + x_0)}, \tag{11.12.18}$$

where $x_0 \in \mathbb{R}$ is arbitrary and $\kappa_b > 0$, $c_b > c$, $\alpha_b \in (0, K(k))$ are uniquely defined from $\lambda \in (-\infty, \lambda_1(k))$ by

$$\begin{aligned} \alpha_b &= F(\varphi_\gamma, k), \\ \kappa_b &= \frac{\sqrt{1 - \lambda - k^2}\sqrt{-\lambda - k^2}}{\sqrt{1 - 2k^2 - \lambda}} - Z(\varphi_\gamma, k), \\ c_b &= c + \frac{4\sqrt{1 - \lambda - k^2}\sqrt{-\lambda - k^2}\sqrt{1 - 2k^2 - \lambda}}{\kappa_b}, \end{aligned} \tag{11.12.19}$$

with $\varphi_\gamma \in (0, \pi/2)$ obtained from $\sin(\varphi_\gamma) = \frac{\sqrt{-\lambda - k^2}}{\sqrt{1 - 2k^2 - \lambda}}$.

For plots of these breathers with periodic background, see [HMP23, Fig.6]. Finally note the following:

Remark 11.4. Equations (11.12.16) and (11.12.17) are related once we realize that function τ is just a solution of the system (11.12.10)-(11.12.11) with potential (11.12.9) and the Jacobi cnoidal function cn and the Jacobi theta function Θ are linked, in the following way:

$$k^2 cn(x,k)^2 = k^2 - 1 + \frac{E(k)}{K(k)} + \partial_x^2 \log(\Theta(x)), \quad k \in (0,1).$$

See [HMP23, Prop.5] for further details. Finally note that $E(k) := E\left(\frac{\pi}{2}, k\right)$ with $E(\varphi, k)$ the Elliptic integral of the second kind, defined as

$$E(\varphi, k) := \int_0^{\varphi} \sqrt{1 - k^2 \sin(\alpha)^2} d\alpha,$$

and $K(k) := F\left(\frac{\pi}{2}, k\right)$ with $F(\varphi, k)$ the Elliptic integral of the first kind, defined as

$$F(\varphi, k) := \int_0^{\varphi} \frac{d\alpha}{\sqrt{1 - k^2 \sin(\alpha)^2}}.$$

Remark 11.5. There exists a second family of breathers, called *dark breathers* on a cnoidal wave background, when instead, the function $\tau(t,x)$ is chosen as:

$$\tau(t,x) := \Theta(x - ct + \alpha_d)e^{-\kappa_b(x - c_d t + x_0)} + \Theta(x - ct - \alpha_d)e^{\kappa_b(x - c_d t + x_0)},$$

with uniquely defined $\kappa_b > 0, c_d < c$ and $\alpha_d \in (0, K(k))$. See [HMP23, Fig.7] for further details on these dark breathers.

11.12.3 *The Benjamin-Ono case*

Long internal waves of small amplitude can be described within the integrable evolution equations, being the Benjamin-Ono (BO) equation the model which describes the case of deep fluids [Gri81]. Recently, some authors [ChPe24] showed and contructed breather solutions with periodic background for the BO equation

$$u_t + H(u_x) + 2uu_x = 0, \quad H(f) := \frac{1}{\pi} p.v. \int_{\mathbb{R}} \frac{f(y)}{y - x} dy. \tag{11.12.20}$$

This is a nonlocal integrable model, with explicit traveling periodic wave as follows:

$$u(t,x) = \frac{k \sinh(\phi)}{\cos(k\xi) + \cosh(\phi)}, \tag{11.12.21}$$

where $\xi = x - c_0 t - x_0$, $c_0 = k\ cotanh(\phi)$, $k, \phi > 0$, $x_0 \in \mathbb{R}$. The solitary wave corresponds to the limit $\phi \to 0$, and after the transformation $\xi \to \xi + \pi k^{-1}$, the above periodic wave solution reduces to the (algebraic) solitary wave solution

$$u(t,x) = \frac{2c}{1 + c_0^2 \xi^2}.$$

Moreover, in [ChPe24] authors constructed breathers with a background, namely the new exact *algebro-oscillatory* solutions are (here $\Delta_+ = 1 + \beta^2 + c^2\eta^2$, $\Delta_- = 1 - \beta^2 + c^2\eta^2$)

$$u(t,x) := \frac{2(c + k\eta)\cosh(\phi) + (k\Delta_+ + 2\beta c)\sinh(\phi) + 2c\cos(k\xi)}{\Delta_+ \cosh(\phi) + 2\beta\sinh(\phi) + \Delta_- \cos(k\xi) + 2\beta c\eta\sin(k\xi)}, \qquad (11.12.22)$$

where $\eta = x - ct - \eta_0$, $\eta_0 \in \mathbb{R}$ and $\beta = \frac{2ck}{(c-c_0)^2 - k^2}$, where $c > 0$ is the arbitrary speed of the solitary wave defined in two different regions

- Bright Breather: $c > c_0 + k$,
- Dark Breather: $0 < c < c_0 - k$.

In fact, in [ChPe24] breathers (11.12.22) are generalized to multi-breathers, in the following result

Theorem 11.3. *The BO equation* (11.12.20) *admits the following general multi-breather solution on the traveling periodic wave* (11.12.21) *in the form*

$$u(t,x) := i\partial_x \log\left(\frac{det(F')}{det(F)}\right), \qquad (11.12.23)$$

with

$$F = \begin{pmatrix} 1 + e^{ik\xi+\phi} & \frac{2k}{k+c_0-c_1} & \cdots & \frac{2k}{k+c_0-c_N} \\ \frac{2c_1}{k+c_1-c_0} & -ic_1\eta_1 - 1 - \frac{2kc_1}{(c_0-c_1)^2-k^2} & \cdots & \frac{2c_1}{c_1-c_N} \\ \vdots & \vdots & \ddots & \vdots \\ \frac{2c_N}{k+c_N-c_0} & \frac{2c_N}{c_N-c_1} & \cdots & -ic_N\eta_N - 1 - \frac{2kc_N}{(c_0-c_N)^2-k^2} \end{pmatrix}$$

and

$$F' = \begin{pmatrix} 1 + e^{ik\xi-\phi} & \frac{2k}{k+c_0-c_1} & \cdots & \frac{2k}{k+c_0-c_N} \\ \frac{2c_1}{k+c_1-c_0} & -ic_1\eta_1 + 1 - \frac{2kc_1}{(c_0-c_1)^2-k^2} & \cdots & \frac{2c_1}{c_1-c_N} \\ \vdots & \vdots & \ddots & \vdots \\ \frac{2c_N}{k+c_N-c_0} & \frac{2c_N}{c_N-c_1} & \cdots & -ic_N\eta_N + 1 - \frac{2kc_N}{(c_0-c_N)^2-k^2} \end{pmatrix},$$

where for $1 \le j \le N$, $\eta_j = x - c_j t - x_j$ *with arbitrary* $x_j \in \mathbb{R}$ *and arbitrary distinct* $c_j > 0$ *satisfying* $|c_j - c_0| > k$. *Actually* (11.12.23) *is real-valued and bounded for every* $(t,x) \in \mathbb{R} \times \mathbb{R}$.

Both (11.12.22) and (11.12.23) can be obtained following the multi-periodic solutions constructed with the bilinear method [SaIs79] and also by using Bäcklund-Darboux transformation of the BO equation (11.12.20) (see [Nak79] for further details).

11.13 Exercises

(11.1) Prove that low order conservation laws (11.2.1)-(11.2.2)-(11.2.3), are actually conserved.

(11.2) Prove by hand Corollary 11.5.2.

(11.3) Is the aperiodic mKdV breather with nonzero mean from Fig.11.12 equal to the mKdV breather with NVBC in [Ale12c]? Is the non periodic Gardner breather equal to the classical Gardner breather (3.2.2)?

(11.4) Using the corresponding Miura transform (see exercise (6.3) for details), map the new defocusing mKdV solution (11.12.6) to a (periodic) KdV solution.

Chapter 12

Hierarchies of breathers

12.1 The KdV/mKdV and Gardner hierarchies

In this chapter we are concerned with the *focusing* Gardner hierarchy[1]

$$u_t = -\frac{\partial}{\partial x}\left(-i\frac{\partial}{\partial x} + 2(\mu + u)\right)\mathcal{L}_n[iu_x + (\mu + u)^2],$$
$$u(t,x) \in \mathbb{R}, \qquad \mu \in \mathbb{R}^+, \qquad n \in \mathbb{N}, \tag{12.1.1}$$

which we will call it hereafter as (2n+1)th-order *focusing* Gardner equations. Here $\mathcal{L}_n$ are the *Lenard* operators defined recursively by

$$\frac{\partial}{\partial x}\mathcal{L}_{n+1}[v] = \left(\frac{\partial^3}{\partial x^3} + 4v\frac{\partial}{\partial x} + 2v_x\right)\mathcal{L}_n[v], \quad n \in \mathbb{N}. \tag{12.1.2}$$

For instance, starting with $\mathcal{L}_0[v] = \frac{1}{2}$, we get

$$\begin{aligned}
\mathcal{L}_1[v] &= v,\\
\mathcal{L}_2[v] &= v_{xx} + 3v^2,\\
\mathcal{L}_3[v] &= v_{4x} + 10vv_{xx} + 5v_x^2 + 10v^3,
\end{aligned} \tag{12.1.3}$$

and

$$\begin{aligned}
\mathcal{L}_4[v] &= v_{6x} + 14vv_{4x} + 28v_xv_{3x} + 21v_{xx}^2 + 70v^2v_{xx} + 70vv_x^2 + 35v^4,\\
\mathcal{L}_5[v] &= v_{8x} + 18vv_{6x} + 54v_xv_{5x} + 114v_{2x}v_{4x} + 126v^2v_{4x} + 69v_{3x}^2\\
&\quad + 504vv_xv_{3x} + 420v^3v_{xx} + 390vv_{2x}^2 + 630v^2v_x^2\\
&\quad + 462v_x^2v_{2x} + 966v_x^2v_{3x} + 1260v^5v_x^2 + 126v^5.
\end{aligned} \tag{12.1.4}$$

This recursion relation (12.1.1) generates the whole *focusing* Gardner hierarchy.

Using the Lenard operators $\mathcal{L}_n$ (12.1.2)-(12.1.3)-(12.1.4) and a suitable spatial translation to remove linear terms, a first few members of the Gardner hierarchy

---

[1] Note that here and for the sake of simplicity, we slightly abuse the notation, using u for both mKdV and Gardner equations of the corresponding hierarchy.

are (see Section 12.8.1 for 9th and 11th order Gardner equations): the *focusing* Gardner equation

$$u_t + (u_{xx} + 6\mu u^2 + 2u^3)_x = 0, \qquad (12.1.5)$$

the 5th-order Gardner equation

$$\begin{aligned} &u_t + u_{5x} + 10\mu^2 u_{3x} + 20\mu u u_{3x} + 10u^2 u_{3x} + 120\mu^3 u u_x \\ &+ 180\mu^2 u^2 u_x + 120\mu u^3 u_x + 10u_x^3 + 30u^4 u_x + 40\mu u_x u_{xx} + 40 u u_x u_{xx} = 0, \end{aligned} \qquad (12.1.6)$$

and *the 7h-order Gardner* equation

$$\begin{aligned} &u_t + u_{7x} + 14\mu^2 u_{5x} + 28\mu u u_{5x} + 14u^2 u_{5x} + 70\mu^4 u_{3x} + 280\mu^3 u u_{3x} \\ &+ 420\mu^2 u^2 u_{3x} + 280\mu u^3 u_{3x} + 70u^4 u_{3x} + 840\mu^5 u u_x + 2100\mu^4 u^2 u_x + 2800\mu^3 u^3 u_x \\ &+ 420\mu^2 u_x^3 + 2100\mu^2 u^4 u_x + 840\mu u u_x^3 + 840\mu u^5 u_x + 420 u^2 u_x^3 + 140 u^6 u_x \\ &+ 84\mu u_{4x} u_x + 84 u u_x u_{4x} + 140\mu u_{xx} u_{3x} + 140 u u_{xx} u_{3x} + 126 u_x^2 u_{3x} + 560\mu^3 u_x u_{xx} \\ &+ 1680\mu^2 u u_x u_{xx} + 1680\mu u^2 u_x u_{xx} + 182 u_x u_{xx}^2 + 560 u^3 u_x u_{xx} = 0. \end{aligned} \qquad (12.1.7)$$

Note, moreover that selecting $\mu = 0$ in (12.1.1), we get the *focusing* mKdV hierarchy of equations, defined by

$$u_t = -\frac{\partial}{\partial x}\left(-i\frac{\partial}{\partial x} + 2u\right)\mathcal{L}_n[iu_x + u^2], \quad n \in \mathbb{N}. \qquad (12.1.8)$$

This recursion relation (12.1.8) generates the whole mKdV hierarchy, recovering the one depicted in [Olv77]-[Mat80]. Many previous works have shown explicit solutions of hierarchies. For the mKdV hierarchy, Matsuno [Mat80] proved the existence and built explicitly the N-soliton solution.

12.2 Breathers of the hierarchy

With respect to breather solutions for any member of the hierarchies already exposed, these are defined as localized in space and periodic in time functions (up to symmetries of the equation). For instance, in the Gardner case, they are *shortly* defined as follows:

Definition 12.2.1 (Gardner breather). Let $\alpha, \beta \in \mathbb{R}\backslash\{0\}$, $\mu \in \mathbb{R}^+\backslash\{0\}$ such that $\Delta = \alpha^2 + \beta^2 - 4\mu^2 > 0$. A breather solution B_μ of the *classical* Gardner equation (12.1.5) is given by the formula

$$B_\mu \equiv B_{\alpha,\beta,\mu}(t, x; x_1, x_2) := 2\partial_x \left[\arctan\left(\frac{\frac{\beta\sqrt{\alpha^2+\beta^2}}{\alpha\sqrt{\Delta}}\sin(\alpha y_1) - h_1(t,x)}{\cosh(\beta y_2) - h_2(t,x)}\right)\right],$$

where h_1, h_2 are precise trigonometric and hyperbolic functions.

The main purpose of this chapter is to show that the breather solutions for the whole Gardner and mKdV hierarchies (12.1.1)-(12.1.8) share the same functional profile for the whole hierarchy, up to corresponding speed parameters, which will depend on the level of the hierarchy considered (see (12.2.1) for a detailed definition).

Theorem 12.2.2 (Existence of Gardner hierarchy breathers). *For all $n \in \mathbb{N}$ there exists a breather solution of the corresponding* $(2n+1)$*-th order Gardner equation of the Gardner hierarchy* (12.1.1). *More precisely*

(1) Structure: *All these breather solutions have the same functional structure, namely given $\alpha, \beta \in \mathbb{R}\backslash\{0\}$ and $\mu \in \mathbb{R}^+\backslash\{0\}$ such that $\Delta = \alpha^2 + \beta^2 - 4\mu^2 > 0$, and $x_1, x_2 \in \mathbb{R}$ we have*

$$B_\mu := 2\partial_x \left[\arctan \left(\frac{\frac{\beta\sqrt{\alpha^2+\beta^2}}{\alpha\sqrt{\Delta}}\sin(\alpha y_1) - \frac{2\mu\beta[\cosh(\beta y_2)+\sinh(\beta y_2)]}{\Delta}}{\cosh(\beta y_2) - \frac{2\mu\beta[\alpha\cos(\alpha y_1)-\beta\sin(\alpha y_1)]}{\alpha\sqrt{\alpha^2+\beta^2}\sqrt{\Delta}}} \right) \right], \tag{12.2.1}$$

with

$$y_1 = x + \delta_{2n+1,\mu} t + x_1, \quad y_2 = x + \gamma_{2n+1,\mu} t + x_2.$$

(2) Velocities: *The velocities $\gamma_{2n+1,\mu}$ and $\delta_{2n+1,\mu}$ are the only parameters depending on the level of the hierarchy considered, in the following explicit form:*

$$\gamma_{2n+1,\mu} := -\frac{1}{\beta}\,\mathrm{Re}\left[\sum_{p=1}^{n} a_{p,n}(\beta + i\alpha)^{2p+1}\mu^{2(n-p)}\right],$$

$$\delta_{2n+1,\mu} := -\frac{1}{\alpha}\,\mathrm{Im}\left[\sum_{p=1}^{n} a_{p,n}(\beta + i\alpha)^{2p+1}\mu^{2(n-p)}\right],$$

where

$$a_{p,n} \equiv \textit{coefficient of the term } \mu^{2(n-p)}u_{(2p+1)x}$$
in the (2n+1)th-order Gardner equation (12.1.1).

(3) Smoothness: *Each breather solution of the Gardner hierarchy above presented is smooth in time and space, and belong to the Schwartz class in space.*

(4) Convergence to mKdV: *For all $n \in \mathbb{N}$, these breather solutions of the Gardner hierarchy reduce to breather solutions of the mKdV hierarchy* (12.1.8) *as $\mu \to 0$, namely*

$$B \equiv B_{\alpha,\beta,n}(t, x; x_1, x_2) := 2\partial_x\left[\arctan\left(\frac{\beta}{\alpha}\frac{\sin(\alpha y_1)}{\cosh(\beta y_2)}\right)\right], \tag{12.2.2}$$

with y_1 and y_2 given by

$$y_1 = x + \delta_{2n+1}t + x_1, \qquad y_2 = x + \gamma_{2n+1}t + x_2,$$

and with velocities

$$\delta_{2n+1} := -\frac{1}{\alpha}\,\mathrm{Im}\left[(\beta + i\alpha)^{2n+1}\right], \qquad \gamma_{2n+1} := -\frac{1}{\beta}\,\mathrm{Re}\left[(\beta + i\alpha)^{2n+1}\right].$$

This is an unexpected and surprising *universality property*. The previous result defines and characterizes regular breather and N-soliton solutions of the Gardner and mKdV hierarchies of equations. In the periodic mKdV setting, it describes *periodic* breather solutions of the corresponding 5th and 7th order mKdV equations.

12.3 Formulae for 1-solitons

As a consequence of the complete integrability of (12.1.1), and following Matsuno [Mat80, (3.11)] and the ISM, it is possible to see that the Gardner hierarchy (12.1.1) has explicit 1-soliton solutions. For the sake of completeness, since we have not found a formal statement of this result in the literature, we include it here:

Definition 12.3.1. The higher order 1-soliton solution $Q_c \equiv Q_{c,n}$ of any equation member of the Gardner hierarchy (12.1.1), i.e. of any $(2n+1)$-th order Gardner equation is given by

$$\begin{aligned} Q_\mu(t,x) &:= Q_{\mu,c}(x - v_{\mu,c}t), \\ Q_{\mu,c}(z) &:= \frac{c^2}{2\mu + \sqrt{4\mu^2 + c^2}\cosh(cz)}, \qquad c > 0, \end{aligned} \tag{12.3.1}$$

with

$$v_{\mu,c} := \sum_{p=1}^{n} a_{p,n} c^{2p} \mu^{2(n-p)}, \qquad \forall n \in \mathbb{N},$$

and $a_{p,n}$ as in the Definition (12.2.1) above.

Moreover, it is easy to see that any 1-soliton solution $Q_{\mu,c}$ (12.3.1) of any equation member of the Gardner hierarchy (12.1.1) satisfies the *same* nonlinear stationary elliptic equation:

Lemma 12.3.2. *Any soliton solution* $Q_{\mu,c} \equiv Q_{c,n}$ (12.3.1) *of the* $(2n+1)$*-th order Gardner equation* (12.1.1) *satisfy the* same *second order stationary elliptic equation, independently of the index* $n \in \mathbb{N}$,

$$Q_\mu'' - c\,Q_\mu + 6\mu Q_\mu^2 + 2Q_\mu^3 = 0, \quad Q_\mu > 0, \quad Q_\mu \in H^1(\mathbb{R}). \tag{12.3.2}$$

Proof. By substituting directly Q_μ (12.3.1) in (12.3.2), for all $n \in \mathbb{N}$. □

Finally note that selecting the parameter $\mu = 0$ in (12.3.1) we reduce to the mKdV limit, and we get from the 1-soliton solution of the Gardner hierarchy (12.3.1), the 1-soliton solution of the mKdV hierarchy, as it was depicted by Matsuno [Mat80]:

Definition 12.3.3 (1-soliton solution of the mKdV hierarchy). The 1-soliton solution of the (2n+1)th-order mKdV equation (12.1.8) is given by

$$Q_\mu(t,x) := Q_{\mu,c}(x - v_n t), \quad Q_{\mu,c}(z) := \operatorname{sech}(cz), \qquad c > 0,$$

with $v_n := c^{2n}$, for all $n \in \mathbb{N}$.

Aside from these higher order 1-soliton solution (12.3.1), we are also able to obtain *breather* solutions of the Gardner hierarchy (12.1.1).

12.4 Higher order breathers. Proof of Theorem 12.2.2

Note that item (4) in Theorem 12.2.2 follows directly from proving items (1), (2) and (3). On the other hand, item (3) is a direct consequence of item (1). Given the speeds in item (2), we are only left to prove item (1) in Theorem 12.2.2.

First of all, we present the following identity valid for any solution of the Gardner hierarchy (12.1.1), and that will be a key tool in the proof of item (1) in Theorem 12.2.2:

Lemma 12.4.1. *Let*

$$u(t,x) = i\partial_x \log\left(\frac{f_\mu - ig_\mu}{f_\mu + ig_\mu}\right) = 2\partial_x \arctan\left(\frac{g_\mu}{f_\mu}\right)$$

be any solution of the Gardner hierarchy (12.1.1) *for any $t \in \mathbb{R}$ with the particular form stated above. Then u satisfies*

$$u^2 - \frac{\partial^2}{\partial x^2}\log\left(f_\mu^2 + g_\mu^2\right) + 2\mu u = 0. \tag{12.4.1}$$

Proof. We select a solution u of (12.1.1) in the form

$$u(t,x) := \phi_x, \quad \phi(t,x) := i\log\left(\frac{\mathbf{g}(t,x)}{\mathbf{f}(t,x)}\right),$$

$$\mathbf{f} := f_\mu + ig_\mu, \quad \mathbf{g} = f_\mu - ig_\mu = \mathbf{f}^*.$$

First note that here f_μ and g_μ are not necessarily the same functions introduced in (12.2.1) but generic ones for this ansatz. Note moreover, that the left part of the

Gardner hierachy (12.1.1) can be rewritten as a finite sum of terms like $\prod_{l=0}^{L}(u_{lx})^{d_l}$, with $u_{lx} = \partial^l u/\partial x^l$ and d_l, $l = 0, \ldots, L$ are nonnegative integers. Then, substituting the above expression in (12.1.1), and using Hirota's D-operators[2] (D_t, D_x), we arrive to the following conditions on $\mathbf{g}$ and $\mathbf{f}$:

$$D_x^2(\mathbf{gf}) - 2i\mu D_x(\mathbf{gf}) = 0, \tag{12.4.2}$$

and

$$D_t(\mathbf{gf}) + \tilde{a}_{n,\mu} D_x^{2n+1}(\mathbf{gf}) + \sum_{j=0}^{n-1} \tilde{b}_{n,\mu} D_t D_x^{2(n-j)}(\mathbf{gf}) = 0,$$

where $\tilde{a}_{n,\mu}$, $\tilde{b}_{n,\mu}$ are coefficients depending on μ and associated to each member of (12.1.1). Then, dividing by $\mathbf{gf}$ the first equation (12.4.2), and taking into account the following identity

$$\frac{D_x^2(\mathbf{gf})}{\mathbf{gf}} = \partial_x^2 \log(\mathbf{gf}) + \left(\partial_x \log\left(\frac{\mathbf{g}}{\mathbf{f}}\right)\right)^2,$$

we obtain:

$$\begin{aligned}\frac{D_x^2(\mathbf{gf})}{\mathbf{gf}} - 2i\mu\frac{D_x(\mathbf{gf})}{\mathbf{gf}} &= \partial_x^2 \log(\mathbf{gf}) + \left(\partial_x \log\left(\frac{\mathbf{g}}{\mathbf{f}}\right)\right)^2 - 2i\mu\partial_x \log\left(\frac{\mathbf{g}}{\mathbf{f}}\right)\\ &= \partial_x^2 \log(\mathbf{gf}) + \left(\frac{u}{i}\right)^2 - 2\mu u = 0.\end{aligned}$$

Hence,

$$u^2 = \frac{\partial^2}{\partial x^2}\log(\mathbf{g}\cdot\mathbf{f}) - 2\mu u.$$

□

Proof of Theorem 12.2.2. (1) Structure: Let $\alpha, \beta \in \mathbb{R}\backslash\{0\}$ and $\mu \in \mathbb{R}^+\backslash\{0\}$ such that $\Delta = \alpha^2 + \beta^2 - 4\mu^2 > 0$, and $x_1, x_2 \in \mathbb{R}$. We check now that B_μ in (12.2.1) satisfies, at $t = 0$, the identity (12.4.1) valid solutions of the Gardner hierarchy (12.1.1). For ease of notation let us use the following expression for B_μ (12.2.1):

$$B_\mu = 2\partial_x \arctan\left[\frac{a_1 \sin(y_1) - a_2 e^{y_2}}{\cosh(y_2) - a_3(\alpha\cos(y_1) - \beta\sin(y_1))}\right] = \frac{H(t,x)}{N(t,x)}, \tag{12.4.3}$$

with

$$\begin{aligned}H(t,x) \equiv H &= 2 - a_3\left(\alpha^2 a_1 + a_2\left(\beta^2 - \alpha^2\right)e^{y_2}\sin(y_1) - 2\alpha a_2\beta e^{y_2}\cos(y_1)\right)\\ &\quad + \beta\sinh(y_2)\left(a_2 e^{y_2} - a_1\sin(y_1)\right) + \cosh(y_2)\left(\alpha a_1\cos(y_1) - a_2\beta e^{y_2}\right),\\ N(t,x) \equiv N &= F_\mu^2 + G_\mu^2 = \left(a_2 e^{y_2} - a_1\sin(y_1)\right)^2 + (\cosh(y_2) + a_3\beta\sin(y_1)\\ &\quad - a_3\alpha\cos(y_1))^2,\end{aligned} \tag{12.4.4}$$

[2] e.g. $D_t f \cdot g = (\partial_{t'} - \partial_t) f(t') g(t)|_{t'=t} = f_t g - f g_t$.

where

$$a_1 = \frac{\beta\sqrt{\alpha^2+\beta^2}}{\alpha\sqrt{\Delta}}, \quad a_2 = \frac{2\beta\mu}{\Delta}, \quad a_3 = \frac{2\beta\mu}{\alpha\sqrt{\Delta}\sqrt{\alpha^2+\beta^2}}. \tag{12.4.5}$$

We will use the following notation for the sake of simplicity:

$$\begin{aligned}
N_x :=\ & 2\alpha\left(a_1{}^2 - \alpha^2 a_3{}^2 + a_3{}^2\beta^2\right)\sin(y_1)\cos(y_1) - 2\alpha a_1 a_2 e^{y_2}\cos(y_1) \\
& - 2a_1a_2\beta e^{y_2}\sin(y_1) + 2a_2{}^2\beta e^{2y_2} + 2\alpha^2 a_3{}^2\beta\,(\sin(y_1))^2 \\
& - 2\alpha^2 a_3{}^2\beta\,(\cos(y_1))^2 + 2\alpha^2 a_3\sin(y_1)\cosh(y_2) - 2\alpha a_3\beta\cos(y_1)\sinh(y_2) \\
& + 2\alpha a_3\beta\cos(y_1)\cosh(y_2) + 2a_3\beta^2\sin(y_1)\sinh(y_2) \\
& + 2\beta\sinh(y_2)\cosh(y_2), \\
N_{xx} :=\ & 8\alpha^3 a_3{}^2\beta\sin(y_1)\cos(y_1) - 4\alpha a_1a_2\beta e^{y_2}\cos(y_1) + 4a_2{}^2\beta^2 e^{2y_2} \\
& + 2\alpha^2\left(a_1{}^2 - \alpha^2 a_3{}^2 + a_3{}^2\beta^2\right)(\cos(y_1))^2 \\
& - 2\alpha^2\left(a_1{}^2 - \alpha^2 a_3{}^2 + a_3{}^2\beta^2\right)(\sin(y_1))^2 \\
& - 2a_3\beta\left(\alpha^2 - \beta^2\right)\sin(y_1)\cosh(y_2) + 4\alpha a_3\beta^2\cos(y_1)\sinh(y_2) \\
& + 4\alpha^2 a_3\beta\sin(y_1)\sinh(y_2) + 2\beta^2(\cosh(y_2))^2 + 2\beta^2(\sinh(y_2))^2, \\
& + 2a_1a_2\left(\alpha^2 - \beta^2\right)e^{y_2}\sin(y_1) + 2\alpha a_3\left(\alpha^2 - \beta^2\right)\cos(y_1)\cosh(y_2), \\
N_{3x} :=\ & -8\alpha^3\left(a_1{}^2 - \alpha^2 a_3{}^2 + a_3{}^2\beta^2\right)\sin(y_1)\cos(y_1) + 8a_2{}^2\beta^3 e^{2y_2} \\
& + 2a_1a_2\alpha\left(\alpha^2 - 3\beta^2\right)e^{y_2}\cos(y_1) + 8\alpha^4 a_3{}^2\beta\,(\cos(y_1))^2 \\
& - 8\alpha^4 a_3{}^2\beta(\sin(y_1))^2 \\
& + 8\beta^3\sinh(y_2)\cosh(y_2) - 2\alpha^2 a_3\left(\alpha^2 - 3\beta^2\right)\sin(y_1)\cosh(y_2) \\
& - 2\beta^2 a_3\left(3\alpha^2 - \beta^2\right)\sin(y_1)\sinh(y_2) + 2a_1a_2\beta\left(3\alpha^2 - \beta^2\right)e^{y_2}\sin(y_1) \\
& - 2\alpha a_3\beta\left(\alpha^2 - 3\beta^2\right)\cos(y_1)\cosh(y_2) \\
& + 2a_3\alpha\beta\left(3\alpha^2 - \beta^2\right)\cos(y_1)\sinh(y_2) \\
N_{4x} :=\ & -32\alpha^5 a_3{}^2\beta\cos(y_1)\sin(y_1) + 8\alpha a_1a_2\beta\left(\alpha^2 - \beta^2\right)e^{y_2}\cos(y_1) \\
& - 2a_1a_2\left(\alpha^4 + \beta^4 - 6\alpha^2\beta^2\right)e^{y_2}\sin(y_1) + 16a_2{}^2\beta^4 e^{2y_2} + 8\beta^4(\sinh(y_2))^2 \\
& + 8\alpha^4\left(a_1{}^2 - \alpha^2 a_3{}^2 + a_3{}^2\beta^2\right)(\sin(y_1))^2 + 8\beta^4(\cosh(y_2))^2 \\
& - 2\alpha a_3\left(\alpha^4 + \beta^4 - 6\alpha^2\beta^2\right)\cos(y_1)\cosh(y_2) \\
& - 8\alpha a_3\beta^2\left(\alpha^2 - \beta^2\right)\cos(y_1)\sinh(y_2) \\
& - 8\alpha^4\left(a_1{}^2 - \alpha^2 a_3{}^2 + a_3{}^2\beta^2\right)(\cos(y_1))^2 \\
& + 2a_3\beta\left(\alpha^4 + \beta^4 - 6\alpha^2\beta^2\right)\sin(y_1)\cosh(y_2) \\
& - 8\alpha^2 a_3\left(\alpha^2 - \beta^2\right)\sin(y_1)\beta\sinh(y_2),
\end{aligned}$$

and

$$H_x := 2\left(\alpha^2+\beta^2\right)\left(a_2a_3\alpha\, e^{y_2}\cos(y_1) - a_1\cosh(y_2)\sin(y_1) - a_2a_3\beta\, e^{y_2}\sin(y_1)\right),$$

$$\begin{aligned} H_{xx} := & -2\left(\alpha^2+\beta^2\right)\left(\alpha\, a_1\cosh(y_2)\cos(y_1) + a_1\beta\,\sinh(y_2)\sin(y_1)\right.\\ &\left. + a_2a_3\left(\alpha^2+\beta^2\right)e^{y_2}\sin(y_1)\right),\end{aligned}$$

$$\begin{aligned} H_{3x} := & \, 2\,a_1\left(\alpha^4-\beta^4\right)\cosh(y_2)\sin(y_1) - 4\,a_1\alpha\beta\left(\alpha^2+\beta^2\right)\sinh(y_2)\cos(y_1)\\ & - 2\,a_2a_3\beta\left(\alpha^2+\beta^2\right)^2 e^{y_2}\sin(y_1) - 2\,a_2a_3\alpha\left(\alpha^2+\beta^2\right)^2 e^{y_2}\cos(y_1),\end{aligned}$$

$$\begin{aligned} H_{4x} := & \, 2\,a_1\alpha\left(\alpha^4-2\,\alpha^2\beta^2-3\,\beta^4\right)\cosh(y_2)\cos(y_1)\\ & - 4\,a_2a_3\alpha\beta\left(\alpha^2+\beta^2\right)^2 e^{y_2}\cos(y_1)\\ & + 2\,a_1\beta\left(3\,\alpha^4+2\,\alpha^2\beta^2-\beta^4\right)\sinh(y_2)\sin(y_1)\\ & + 2\,a_2a_3\left(\alpha^6+\alpha^4\beta^2-\alpha^2\beta^4-\beta^6\right)e^{y_2}\sin(y_1).\end{aligned}$$

Then, having in mind notation $N, N_x, N_{xx}, N_{3x}, N_{4x}$ and $H, H_x, H_{xx}, H_{3x}, H_{4x}$, (12.4.1) simplifies as

$$u^2 - \frac{\partial^2}{\partial x^2}\log\left(f_\mu^2+g_\mu^2\right) + 2\mu u = \frac{1}{N^2}\Big(H^2+N_x^2-N_{xx}N+2\mu HN\Big).$$

Finally, substituting explicitly $H's$ and $N's$ terms, we verify, using the symbolic software *Mathematica*, simple trigonometric and hyperbolic identities and some rearrangements, that

$$H^2+N_x^2-N_{xx}N+2\mu HN = 0,$$

and we conclude. Now, in order to get a complete dynamical description of B_μ (12.2.1) at any t, we have to obtain the velocities $\gamma_{2n+1,\mu}$ and $\delta_{2n+1,\mu}$.

(2) Velocities: Now, we determine the velocities $\gamma_{2n+1,\mu}$ and $\delta_{2n+1,\mu}$. Using a matching method with ansatz B_μ (12.2.1) and free parameters $\gamma_{2n+1,\mu}$ and $\delta_{2n+1,\mu}$, we proceed substituting directly B_μ in the 5th and 7th order Gardner equations (12.1.6)-(12.1.7), and upon lengthy algebraic manipulations we find that B_μ is indeed a breather solution of the 5th and 7th order Gardner equations (12.1.6)-(12.1.7) respectively, provided that

$$\begin{aligned}\delta_{5,\mu} := & -\alpha^4+10\alpha^2\beta^2-5\beta^4-10\mu^2(3\beta^2-\alpha^2)\\ = & -\frac{1}{\alpha}\,\mathrm{Im}\left[\sum_{p=1}^{2} a_{p,2}(\beta+i\alpha)^{2p+1}\mu^{2(2-p)}\right],\\ \gamma_{5,\mu} := & -\beta^4+10\alpha^2\beta^2-5\alpha^4-10\mu^2(\beta^2-3\alpha^2)\\ = & -\frac{1}{\beta}\,\mathrm{Re}\left[\sum_{p=1}^{2} a_{p,2}(\beta+i\alpha)^{2p+1}\mu^{2(2-p)}\right],\end{aligned}$$

with $a_{1,2} = 10,\ a_{2,2} = 1$, and

$$\begin{aligned}\delta_{7,\mu} &:= \alpha^6 - 21\alpha^4\beta^2 + 35\alpha^2\beta^4 - 7\beta^6 + 14\mu^2(-\alpha^4 + 10\alpha^2\beta^2 - 5\beta^4)\\ &\quad - 70\mu^4(3\beta^2 - \alpha^2)\\ &= -\frac{1}{\alpha}\,\mathrm{Im}\left[\sum_{p=1}^{3} a_{p,3}(\beta + i\alpha)^{2p+1}\mu^{2(3-p)}\right],\end{aligned}$$

$$\begin{aligned}\gamma_{7,\mu} &:= -\beta^6 + 21\alpha^2\beta^4 - 35\alpha^4\beta^2 + 7\alpha^6 + 14\mu^2(-\beta^4 + 10\alpha^2\beta^2 - 5\alpha^4)\\ &\quad - 70\mu^4(\beta^2 - 3\alpha^2)\\ &= -\frac{1}{\beta}\,\mathrm{Re}\left[\sum_{p=1}^{3} a_{p,3}(\beta + i\alpha)^{2p+1}\mu^{2(3-p)}\right],\end{aligned}$$

with $a_{1,3} = 70,\ a_{2,3} = 14,\ a_{3,3} = 1$. By induction for any order $2n + 1,\ n \in \mathbb{N}$, we get that

$$\begin{aligned}\gamma_{2n+1,\mu} &:= -\frac{1}{\beta}\,\mathrm{Re}\left[\sum_{p=1}^{n} a_{p,n}(\beta + i\alpha)^{2p+1}\mu^{2(n-p)}\right],\\ \delta_{2n+1,\mu} &:= -\frac{1}{\alpha}\,\mathrm{Im}\left[\sum_{p=1}^{n} a_{p,n}(\beta + i\alpha)^{2p+1}\mu^{2(n-p)}\right],\end{aligned} \tag{12.4.6}$$

where $a_{p,n}$ is the coefficient of the term $\mu^{2(n-p)}u_{(2p+1)x}$ in the (2n+1)th-order Gardner equation. □

Remark 12.4.2. See Section 12.8 for a few additional examples of these higher order mKdV and Gardner breather solutions in cases 9th and 11th order.

12.5 Universality

Now we describe a characterization of each new Gardner breather as solution of a nonlinear fourth order ODE. Note that this ODE does not depend on the order of the hierarchy. Furthermore, any breather solution of the hierarchy is a critical point of the same Lyapunov functional.

Theorem 12.5.1 (Universal nonlinear ODE and variational characterization). *Any breather solution B_μ of the* (2n+1)*-th order Gardner equation* (12.1.1) *satisfies, for any $n \in \mathbb{N}$, the* same *fourth order elliptic equation*

$$\begin{aligned}&B_{\mu,4x} + 2(\alpha^2 - \beta^2)(B_{\mu,xx} + 6\mu B_\mu^2 + 2B_\mu^3) + (\alpha^2 + \beta^2)^2 B_\mu\\ &\quad + 10B_\mu^2 B_{\mu,xx} + 10B_\mu B_{\mu,x}^2 + 6B_\mu^5 + 10\mu B_{\mu,x}^2\\ &\quad + 20\mu B_\mu B_{\mu,xx} + 40\mu^2 B_\mu^3 + 30\mu B_\mu^4 = 0.\end{aligned} \tag{12.5.1}$$

Moreover, for any $n \in \mathbb{N}$, the breather solution B_μ of the Gardner hierarchy (12.1.1) *is a critical point of a universal Lyapunov functional $\mathcal{H}_\mu$, which is written as linear combination of three conserved laws in the following way*

$$\mathcal{H}_\mu[u(t)] := F_\mu[u](t) + 2(\beta^2 - \alpha^2)E_\mu[u](t) + (\alpha^2 + \beta^2)^2 M[u](t). \tag{12.5.2}$$

The conserved functionals in (12.5.2) *can be explicitly found in* (12.6.1), (12.6.2) *and* (12.6.3).

As a direct corollary of the main result Theorem 12.2.2, we get, when $\mu = 0$ in (12.5.1), the fourth-order elliptic ODE satisfied by all breather solutions of the mKdV hierarchy.

Corollary 12.5.2 (Universality for mKdV hierarchy breathers). *Any breather solution B of the* $(2n + 1)$*-th order mKdV equation* (12.1.8) *satisfies for any $n \in \mathbb{N}$ the* same *fourth order elliptic equation*

$$\begin{aligned} &B_{4x} + 2(\alpha^2 - \beta^2)(B_{xx} + 2B^3) + (\alpha^2 + \beta^2)^2 B \\ &\quad + 10B^2 B_{xx} + 10BB_x^2 + 6B^5 = 0. \end{aligned}$$

Moreover, mKdV breathers of the hierarchy are classical critical points of an associated functional, just as in 12.5.2, but with $\mu = 0$.

This corollary complements and completes the main result in [Ale24, Th.2.4], which essentially considered the 5th, 7th and 9th mKdV equations only.

12.6 Conserved quantities

The Gardner hierarchy (12.1.1), as being a completely integrable scheme of equations, has infinitely many conserved quantities. Some standard conservation laws at the H^1-level are the *mass*

$$M[u](t) := \frac{1}{2}\int_{\mathbb{R}} u^2(t,x)dx = M[u](0), \tag{12.6.1}$$

the *energy*

$$E_\mu[u](t) := \int_{\mathbb{R}} \left(\frac{1}{2}u_x^2 - 2\mu u^3 - \frac{1}{2}u^4\right)(t,x)dx = E_\mu[u](0), \tag{12.6.2}$$

and the *higher order energy*, defined respectively in $H^2(\mathbb{R})$,

$$\begin{aligned} F_\mu[u](t) := \int_{\mathbb{R}} \Bigg(&\frac{1}{2}u_{xx}^2 - 10\mu u u_x^2 + 10\mu^2 u^4 \\ &- 5u^2 u_x^2 + 6\mu u^5 + u^6\Bigg)(t,x)dx = F_\mu[u](0). \end{aligned} \tag{12.6.3}$$

Now, considering $M[u]$, $E_\mu[u]$ and $F_\mu[u]$, we define the following Lyapunov functional, as in (12.5.2):

$$\mathcal{H}_\mu[u(t)] := F_\mu[u](t) + 2(\beta^2 - \alpha^2)E_\mu[u](t) + (\alpha^2 + \beta^2)^2 M[u](t).$$

Therefore, $\mathcal{H}_\mu[u]$ is a real-valued conserved quantity, well-defined for H^2-solutions of the Gardner hierarchy (12.1.1), provided they exist.

12.7 Proof of Theorem 12.5.1

We first prove (12.5.1): we recast the l.h.s. of (12.5.1) as follows:

$$\begin{aligned}
G_\mu[B_\mu] :=& \ \partial_x^2(B_{\mu,xx} + 6\mu B_\mu^2 + 2B_\mu^3) + 2\left(\alpha^2 - \beta^2\right)(B_{\mu,xx} + 6\mu B_\mu^2 + 2B_\mu^3) \\
&+ 8\mu B_\mu B_{\mu,xx} - 2\mu B_{\mu,x}^2 + 4B_\mu^2 B_{\mu,xx} - 2B_\mu B_{\mu,x}^2 + \left(\alpha^2 + \beta^2\right)^2 B_\mu \\
&+ 40\mu^2 B_\mu^3 + 30\mu B_\mu^4 + 6B_\mu^5 \\
=& \ \partial_x^2(B_{\mu,xx} + 6\mu B_\mu^2 + 2B_\mu^3) - 2(\mu + B_\mu)(B_{\mu,x}^2 + 4\mu B_\mu^3 + B_\mu^4) \\
&+ (\alpha^2 + \beta^2)^2 B_\mu + (4B_\mu^2 + 8\mu B_\mu + 2(\alpha^2 - \beta^2))(B_{\mu,xx} + 6\mu B_\mu^2 + 2B_\mu^3) \\
&- 2(\mu + B_\mu)(B_{\mu,x}^2 + 4\mu B_\mu^3 + B_\mu^4) + (\alpha^2 + \beta^2)^2 B_\mu.
\end{aligned} \tag{12.7.1}$$

Now we compute explicitly the last line in (12.7.1). For simplicity, we use the same notation (12.4.3) as in the proof of Theorem 12.2.2 in Section 12.4:

$$B_\mu(t,x) = \frac{H(t,x)}{N(t,x)},$$

with H, N as in (12.4.4) and a_i, $i = 1,2,3$ already defined in (12.4.5). We first compute the term

$$\begin{aligned}
B_{xx} + 6\mu B^2 + 2B^3 =& \frac{1}{N^3}\Big(2H^3 + 6H^2\mu N + H\left(2N_x^2 - N_{xx}N\right) \\
&+ N(H_{xx}N - 2H_xN_x)\Big),
\end{aligned}$$

and we get

$$\partial_x^2(B_{xx} + 6\mu B^2 + 2B^3) = \frac{M_1}{N^5},$$

with

$$\begin{aligned}
M_1 :=& \Big(6H^3(4N_x^2 - N_{xx}N) - 6H^2N(6H_xN_x - N(H_{xx} - 2\mu N_{xx}) - 6\mu N_x^2) \\
&+ H(2N^2(6H_x^2 - 24H_x\mu N_x + 4N_xN_{3x} + 3N_{xx}^2) + N^3(12H_{xx}\mu - N_{4x}) \\
&+ 24N_x^4 - 36N_x^2N_{xx}N) \\
&+ N(2N^2\left(6H_x^2\mu - 2H_xN_{3x} - 3H_{xx}N_{xx} - 2H_{3x}N_x\right) \\
&+ 12N_xN(2H_xN_{xx} + H_{xx}N_x) - 24H_xN_x^3 + H_{4x}N^3)\Big).
\end{aligned} \tag{12.7.2}$$

Moreover, we have that

$$(4B^2 + 8\mu B + 2(\alpha^2 - \beta^2))(B_{xx} + 6\mu B^2 + 2B^3) = \frac{M_2}{N^5},$$

with

$$M_2 := 2(2H^2 + 4\mu HN + N^2(\alpha^2 - \beta^2))(2H^3 + 6H^2\mu N + H(2N_x^2 - N_{xx}N) + N(H_{xx}N - 2H_xN_x)), \tag{12.7.3}$$

and

$$-2(B + \mu)(B_x^2 + 4\mu B^3 + B^4) + (\alpha^2 + \beta^2)^2 B = \frac{M_3}{N^5},$$

with

$$M_3 := HN^4\left(\alpha^2 + \beta^2\right)^2 - 2(H + \mu N)\left(H^4 + 4\mu H^3 N + (H_xN - HN_x)^2\right). \tag{12.7.4}$$

Hence, we get the following simplification of (12.5.1):

$$G_\mu[B_\mu] = \frac{M_1 + M_2 + M_3}{N^5}, \tag{12.7.5}$$

with M_1, M_2, M_3 in (12.7.2), (12.7.3) and (12.7.4) respectively. In fact, we verify, using the symbolic software *Mathematica*, that after substituting $H's$ and $N's$ terms explicitly in (12.7.5), basic trigonometric and hyperbolic identities and lengthy rearrangements, we get

$$M_1 + M_2 + M_3 = 0,$$

and we conclude the proof of (12.5.1).

Now, we prove that B_μ is actually a critical point of $\mathcal{H}_\mu$. We will show that

$$\mathcal{H}_\mu[B_\mu + z] - \mathcal{H}_\mu[B_\mu] = \frac{1}{2}\mathcal{Q}[z] + \mathcal{N}[z],$$

with $\mathcal{Q}$ being the quadratic form defined in (12.7.6), and $\mathcal{N}[z]$ satisfying $|\mathcal{N}[z]| \leq K\|z\|^3_{H^2(\mathbb{R})}$. We evaluate and expand the Lyapunov functional H_μ in terms of a perturbation of the breather B_μ, with $z \in H^2(\mathbb{R})$. A direct computation with integration by parts yields

$$\begin{aligned}
F_\mu[B_\mu + z] = \ & F_\mu[B_\mu] \\
& + \int [B_{\mu,4x} + 10\mu B_{\mu,x}^2 + 20\mu B_\mu B_{\mu,xx} + 10 B_\mu B_{\mu,x}^2 + 10 B_\mu^2 B_{\mu,x} + 40\mu^2 B_\mu^3 \\
& \qquad + 30\mu B_\mu^4 + 6B_\mu^5]z \\
& + \frac{1}{2}\int \Big(z_{4x} + (20\mu B_\mu + 10B_\mu^2)z_{xx} - 20(\mu B_{\mu,x} + B_\mu B_{\mu,x})z_x \\
& \qquad + (-10B_{\mu,x}^2 + 120\mu^2 B_\mu^2 + 120\mu B_\mu^3 + 30B_\mu^4)z\Big)z \\
& + \int \left(-10\mu z z_x^2 - 10B_\mu z z_x - 10B_{\mu,x}z_x z^2 + 40\mu^2 B_\mu z^3 + 60\mu B_\mu^2 z^3 + 20B_\mu^3 z^3\right).
\end{aligned}$$

Similarly,

$$\begin{aligned} E_\mu[B_\mu + z] = E_\mu[B_\mu] &- \int (B_{\mu,xx} + 6\mu B_\mu^2 + 2B_\mu^3)z \\ &- \frac{1}{2}\int (z_{xx} + (12\mu B_\mu + 6B_\mu^2)z)z \\ &- \int \left(2\mu z^3 + 2B_\mu z^3 + \frac{1}{2}z^4\right), \end{aligned}$$

and

$$M[B_\mu + z] = M[B_\mu] + \int B_\mu z + \frac{1}{2}\int z^2.$$

Collecting all, we get that

$$\mathcal{H}_\mu[B_\mu + z] = \mathcal{H}_\mu[B_\mu] + \int_{\mathbb{R}} G_\mu(B_\mu)z + \frac{1}{2}Q_\mu[z] + \mathcal{N}_\mu[z],$$

where the quadratic form

$$Q_\mu[z] := \int z\mathcal{L}_\mu[z], \tag{12.7.6}$$

is associated to the linearized operator $\mathcal{L}_\mu$ given by

$$\begin{aligned} \mathcal{L}_\mu := \partial_x^4 &+ (20\mu B_\mu + 10B_\mu^2 - 2(\beta^2 - \alpha^2))\partial_x^2 \\ &- 20(\mu B_{\mu,x} + B_\mu B_{\mu,x})\partial_x \\ &+ \Big(-10B_{\mu,x}^2 + 120\mu^2 B_\mu^2 + 120\mu B_\mu^3 + 30B_\mu^4 \\ &\quad -2(\beta^2 - \alpha^2)(12\mu B_\mu + 6B_\mu^2) + (\alpha^2 + \beta^2)^2\Big). \end{aligned} \tag{12.7.7}$$

Gathering all higher order terms (with respect to z) in $N_\mu[z]$ we get

$$\begin{aligned} \mathcal{N}_\mu[z] := \int \Big(&- 10\mu z z_x^2 - 10B_\mu z z_x - 10B_{\mu,x}z_x z^2 + 40\mu^2 B_\mu z^3 + 60\mu B_\mu^2 z^3 \\ &+ 20B_\mu^3 z^3\Big) - 2(\beta^2 - \alpha^2)\int \left(2\mu z^3 + 2B_\mu z^3 + \frac{1}{2}z^4\right). \end{aligned}$$

Part (i) above guarantees $\int_{\mathbb{R}} G_\mu(B_\mu)z = 0$, and then we have $\mathcal{H}_\mu'[B_\mu] = 0$. Moreover, from direct estimates, one has $\mathcal{N}_\mu[z] = O\left(\|z\|_{H^2(\mathbb{R})}^3\right)$, and we conclude.

12.8 Additional higher order mKdV and Gardner equations

For the sake of completeness we list below higher order members of the mKdV and Gardner hierarchies (12.1.8)-(12.1.1). We start with the 9th order mKdV equation which is written as follows

$$
\begin{aligned}
u_t + \partial_x\Big(&u_{8x} + 18u^2u_{6x} + 108uu_xu_{5x} + 228uu_{2x}u_{4x} + 210u_x^2u_{4x} + 126u^4u_{4x}\\
&+ 138u(u_{3x})^2 + 756u_xu_{2x}u_{3x} + 1008u^3u_xu_{3x} + 182u_{2x}^3 + 756u^3u_{2x}^2\\
&+ 3108u^2u_x^2u_{2x} + 420u^6u_{2x} + 798uu_x^4 + 1260u^5(u_x)^2 + 70u^9\Big) = 0.
\end{aligned}
\tag{12.8.1}
$$

The 11th order mKdV equation is written as follows

$$
\begin{aligned}
u_t &+ \partial_x\Big(u_{10x} + 22u^2u_{8x} + 198u^4u_{6x} + 924u^6u_{4x} + 506uu_{4x}^2 + 3036u^3u_{3x}^2\\
&+ 2310u^8u_{xx} + 8316u^5u_{xx}^2 + 9372u^2u_{xx}^3 + 9240u^7u_x^2 + 26796u^3u_x^4\\
&+ 176uu_xu_{7x} + 484uu_{xx}u_{6x} + 462u_x^2u_{6x} + 836uu_{3x}u_{5x} + 2376u^3u_xu_{5x}\\
&+ 5016u^3u_{xx}u_{4x} + 2706u_{xx}^2u_{4x} + 11220u^2u_x^2u_{4x} + 3498u_{xx}u_{3x}^2 + 11088u^5u_xu_{3x}\\
&+ 54516u^4u_x^2u_{xx} + 44748uu_x^2u_{xx}^2 + 13398u_x^4u_{xx} + 2376u_xu_{xx}u_{5x}\\
&+ 21120uu_x^3u_{3x} + 3696u_xu_{3x}u_{4x} + 39336u^2u_xu_{xx}u_{3x} + 252u^{11}\Big) = 0.
\end{aligned}
\tag{12.8.2}
$$

We finally present the 13th order mKdV equation

$$
\begin{aligned}
u_t &+ \partial_x\Big(u_{12x} + 1846\,uu_{5x}^2 + 191620u^4u_{xx}^3 + 924u^{13} + 30888u^5u_xu_{5x}\\
&+ 1733160u^3u_{xx}^2u_x^2 + 823680u^3u_x^3u_{3x} + 1398540u^2u_x^4u_{xx}\\
&+ 648648u^6u_x^2u_{xx} + 96096u^7u_xu_{3x} + 3172uu_{4x}u_{6x} + 1976uu_{3x}u_{7x}\\
&+ 884uu_{xx}u_{8x} + 260uu_xu_{9x} + 324324uu_x^2u_{3x}^2 + 78936uu_x^3u_{5x}\\
&+ 511368uu_x^2u_{xx}u_{4x} + 919776uu_xu_{xx}^2u_{3x} + 566280u_x^3u_{3x}u_{xx}\\
&+ 17160u_xu_{4x}u_{5x} + 5720u_xu_{xx}u_{7x} + 12012u_xu_{3x}u_{6x} + 33176\,u_{xx}u_{3x}u_{5x}\\
&+ 21736u^3u_{3x}u_{5x} + 12584u^3u_{xx}u_{6x} + 4576u^3u_xu_{7x} + 29172u^2u_x^2u_{6x}\\
&+ 231660u^4u_x^2u_{4x} + 65208u^5u_{xx}u_{4x} + 78078u_x^4u_{4x} + 403260u_x^2u_{xx}^3\\
&+ 8866u_{xx}^2u_{6x} + 20306u_{xx}u_{4x}^2 + 858u_x^2u_{8x} + 26598u_{3x}^2u_{4x}\\
&+ 468468u^5u_x^4 + 60060u^9u_x^2 + 157300uu_x^6 + 13156u^3u_{4x}^2 + 286u^4u_{8x}\\
&+ 39468u^5u_{3x}^2 + 1716u^6u_{6x} + 12012u^{10}u_{xx} + 72072u^7u_{xx}^2 + 6006u^8u_{4x}\\
&+ 108966uu_{xx}^4 + 26u^2u_{10x} + 156156u^2u_{xx}^2u_{4x} + 197340u^2u_{xx}u_{3x}^2\\
&+ 219648u^2u_xu_{3x}u_{4x} + 144144u^2u_xu_{xx}u_{5x} + 806520u^4u_xu_{xx}u_{3x}\Big) = 0.
\end{aligned}
\tag{12.8.3}
$$

Now, associated with these additional higher order mKdV equations, we provide the corresponding breather solutions, as in (12.2.2):

Definition 12.8.1 (9th-11th-mKdV breathers). Let $\alpha, \beta > 0$ and $x_1, x_2 \in \mathbb{R}$. The real-valued breather solution associated to the 9th-11th-mKdV equations (12.8.1)-

(12.8.2) are given explicitly by the formula

$$B \equiv B_{\alpha,\beta}(t,x;x_1,x_2) := 2\partial_x\left[\arctan\left(\frac{\beta}{\alpha}\frac{\sin(\alpha y_1)}{\cosh(\beta y_2)}\right)\right],$$

with y_1 and y_2

$$y_1 = x + \delta_i t + x_1, \quad y_2 = x + \gamma_i t + x_2,$$

and with velocities (δ_i, γ_i), $i = 9, 11$

$$\begin{aligned}\delta_9 &:= -\alpha^8 + 36\alpha^6\beta^2 - 126\alpha^4\beta^4 + 84\alpha^3\beta^6 - 9\beta^8,\\ \gamma_9 &:= -\beta^8 + 36\alpha^2\beta^6 - 126\alpha^4\beta^4 + 84\alpha^6\beta^2 - 9\alpha^8,\end{aligned}$$

and

$$\begin{aligned}\delta_{11} &= \alpha^{10} - 55\alpha^8\beta^2 + 330\alpha^6\beta^4 - 462\alpha^4\beta^6 + 165\alpha^2\beta^8 - 11\beta^{10},\\ \gamma_{11} &= 11\alpha^{10} - 165\alpha^8\beta^2 + 462\alpha^6\beta^4 - 330\alpha^4\beta^6 + 55\alpha^2\beta^8 - \beta^{10}.\end{aligned}$$

Finally, and directly from definition of the Gardner hierarchy (12.1.1) with $n = 4$ or alternatively, from the 9th and 11th order mKdV equations (12.8.1)-(12.8.2) with the transformation $u \to \mu + u$, we present the 9th order Gardner equation:

$$\begin{aligned}u_t + \partial_x\Big(&u_{8x} + 18(\mu+u)^2u_{6x} + 108(\mu+u)u_xu_{5x} + 228(\mu+u)u_{2x}u_{4x} + 210u_x^2u_{4x}\\ &+ 126(\mu+u)^4u_{4x} + 138(\mu+u)u_{3x}^2 + 756u_xu_{2x}u_{3x} + 1008u^3u_xu_{3x} + 182u_{2x}^3\\ &+ 756(\mu+u)^3u_{2x}^2 + 3108(\mu+u)^2(u_x)^2u_{2x} + 420(\mu+u)^6u_{2x} + 798(\mu+u)u_x^4\\ &+ 1260(\mu+u)^5u_x^2 + 70(\mu+u)^9\Big) = 0,\end{aligned} \tag{12.8.4}$$

and the 11th order Gardner equation

$$\begin{aligned}u_t + \partial_x\Big(&u_{10x} + 22(\mu+u)^2u_{8x} + 198(\mu+u)^4u_{6x} + 924(\mu+u)^6u_{4x} + 506(\mu+u)u_{4x}^2\\ &+ 3036(\mu+u)^3u_{3x}^2 + 2310(\mu+u)^8u_{xx} + 8316(\mu+u)^5u_{xx}^2 + 9372(\mu+u)^2u_{xx}^3\\ &+ 9240(\mu+u)^7\,(u_x)^2 + 26796(\mu+u)^3u_x^4 + 176(\mu+u)u_xu_{7x} + 484(\mu+u)u_{xx}u_{6x}\\ &+ 462u_x^2u_{6x} + 836(\mu+u)u_{3x}u_{5x} + 2376(\mu+u)^3u_xu_{5x} + 5016(\mu+u)^3u_{xx}u_{4x}\\ &+ 2706u_{xx}^2u_{4x} + 11220(\mu+u)^2u_x^2u_{4x} + 3498u_{xx}u_{3x}^2 + 11088(\mu+u)^5u_xu_{3x}\\ &+ 54516(\mu+u)^4u_x^2u_{xx} + 44748(\mu+u)u_x^2u_{xx}^2 + 13398u_x^4u_{xx} + 2376u_xu_{xx}u_{5x}\\ &+ 21120(\mu+u)u_x^3u_{3x} + 3696u_xu_{3x}u_{4x} + 39336(\mu+u)^2u_xu_{xx}u_{3x}\\ &+ 252(\mu+u)^{11}\Big) = 0.\end{aligned} \tag{12.8.5}$$

The corresponding breather solutions of the 9th and 11th order Gardner equations (12.8.4)-(12.8.5) are the following:

Definition 12.8.2 (9th-11th-Gardner breathers). Let α, β, μ as in definition 12.2.1 and $x_1, x_2 \in \mathbb{R}$. Then the real-valued breather solution associated with the 9th and 11th-Gardner equations (12.8.4)-(12.8.5) is given explicitly by the same formula as in (12.2.1) but respectively with velocities (δ_9, γ_9) and $(\delta_{11}, \gamma_{11})$:

$$\begin{aligned}
\delta_9 :=\ & \alpha(\alpha^8 - 18\alpha^6\left(2\beta^2 + \mu^2\right) + 126\alpha^4\left(\beta^4 + 3\beta^2\mu^2 + \mu^4\right) \\
& - 42\alpha^2\left(2\beta^6 + 15\beta^4\mu^2 + 30\beta^2\mu^4 + 10\mu^6\right) + 9(\beta^8 + 14\beta^6\mu^2 + 70\beta^4\mu^4 \\
& + 140\beta^2\mu^6 + 70\mu^8)), \\
\gamma_9 :=\ & -\beta(9\alpha^8 - 42\alpha^6\left(2\beta^2 + 3\mu^2\right) + 126\alpha^4\left(\beta^4 + 5\beta^2\mu^2 + 5\mu^4\right) \\
& - 18\alpha^2\left(2\beta^6 + 21\beta^4\mu^2 + 70\beta^2\mu^4 + 70\mu^6\right) + \beta^8 + 18\beta^6\mu^2 + 126\beta^4\mu^4 \\
& + 420\beta^2\mu^6 + 630\mu^8),
\end{aligned}$$

and

$$\begin{aligned}
\delta_{11} :=\ & \alpha(\alpha^{10} - 11\alpha^8\left(5\beta^2 + 2\mu^2\right) + 66\alpha^6(5\beta^4 + 12\beta^2\mu^2 + 3\mu^4) \\
& - 462\alpha^4\left(\beta^6 + 6\beta^4\mu^2 + 9\beta^2\mu^4 + 2\mu^6\right) + 33\alpha^2(5\beta^8 + 56\beta^6\mu^2 + 210\beta^4\mu^4 \\
& + 280\beta^2\mu^6 + 70\mu^8) - 11(\beta^{10} + 18\beta^8\mu^2 + 126\beta^6\mu^4 + 420\beta^4\mu^6 + 630\beta^2\mu^8 \\
& + 252\mu^{10})), \\
\gamma_{11} :=\ & -\beta(-11\alpha^{10} + 33\alpha^8\left(5\beta^2 + 6\mu^2\right) - 462\alpha^6(\beta^4 + 4\beta^2\mu^2 + 3\mu^4) \\
& + 66\alpha^4\left(5\beta^6 + 42\beta^4\mu^2 + 105\beta^2\mu^4 + 70\mu^6\right) - 11\alpha^2(5\beta^8 + 72\beta^6\mu^2 + 378\beta^4\mu^4 \\
& + 840\beta^2\mu^6 + 630\mu^8) + \beta^{10} + 22\beta^8\mu^2 + 198\beta^6\mu^4 + 924\beta^4\mu^6 + 2310\beta^2\mu^8 \\
& + 2772\mu^{10}).
\end{aligned}$$

12.9 Standard and new mKdV periodic breathers

A family of periodic breathers (named KKSH breathers) for the *classical* mKdV equation was found by Kevrekidis et al. by using elliptic functions and a matching of free parameters (see [KKS04, KKS03, AMP17b] for further reading).

We consider now, from (12.1.6) and (12.1.7) when $\mu = 0$, the periodic case of the 5th-order mKdV:

$$u_t + (u_{4x} + f_5(u))_x = 0, \qquad f_5(u) := 10uu_x^2 + 10u^2u_{xx} + 6u^5, \tag{12.9.1}$$

and the 7th-order mKdV

$$\begin{aligned}
& u_t + (u_{6x} + f_7(u))_x = 0, \\
& f_7(u) := 14u^2u_{4x} + 56uu_xu_{3x} + 42uu_{xx}^2 + 70u_x^2u_{xx} \\
& \quad + 70u^4u_{xx} + 140u^3u_x^2 + 20u^7.
\end{aligned} \tag{12.9.2}$$

For the higher order mKdV equations (12.9.1) and (12.9.2), an equivalent expression of periodic breathers is available, by following a similar matching of parameters. Namely, we consider here the 5th and 7th mKdV equations (12.9.1) (12.9.2) where now

$$u : \mathbb{R}_t \times \mathbb{T}_x \mapsto \mathbb{R}_x,$$

is periodic in space, and $\mathbb{T}_x = \mathbb{T} = \mathbb{R}/L\mathbb{Z} = (0, L)$ denotes a torus with period L, to be fixed later. Higher order periodic cases can also be described but for the sake of simplicity, we will keep our discussion with these 5th and 7th orders. We refer the reader to [AbSe81, ByFr71] for a more detailed account on the JEFs sn and nd presented below.

Lemma 12.9.1 (Periodic breathers of 5th and 7th mKdV equations). *Given* $\alpha, \beta > 0$, $x_1, x_2 \in \mathbb{R}$ *and* $k, m \in [0, 1]$, *the following is satisfied.*

(1) Periodic breather solutions of the 5th and 7th mKdV equations (12.9.1) (12.9.2), *are given by the explicit formula (see [KKS04] and [AMP17b] for checking and comparison reasons)* .

$$\begin{aligned} B &= B(t, x; \alpha, \beta, k, m, x_1, x_2) \\ &:= \partial_x \tilde{B} := 2\partial_x \left[\arctan\left(\frac{\beta}{\alpha} \frac{\text{sn}(\alpha y_1, k)}{\text{nd}(\beta y_2, m)} \right) \right], \end{aligned} \tag{12.9.3}$$

with $\text{sn}(\cdot, k)$ *and* $\text{nd}(\cdot, m)$ *the standard JEFs of elliptic modulus* k *and* m, *respectively, but now*

$$y_1 := x + \delta_{i,m} t + x_1, \quad y_2 := x + \gamma_{i,m} t + x_2, \quad i = 5, 7.$$

(2) The velocities $(\delta_{5,m}, \gamma_{5,m})$ *in the 5th order case and* $(\delta_{7,m}, \gamma_{7,m})$ *in the 7th order case are given by, respectively:*

$$\begin{aligned} \delta_{5,m} := &- \alpha^4(k^2 - 26k + 1) \\ &+ 10\alpha^2\beta^2(1 + k)(2 - m) - 5\beta^4(m^2 - 16m + 16), \\ \gamma_{5,m} := &- \beta^4(m^2 + 24m - 24) \\ &+ 10\alpha^2\beta^2(1 + k)(2 - m) - 5\alpha^4(k^2 + 14k + 1), \end{aligned}$$

and

$$\begin{aligned}\delta_{7,m} := \ &\alpha^6(k^3 + 135k^2 + 135k + 1)\\ &+ 21\alpha^4\beta^2(-2 + k^2(m-2) + m + 2k(7m-6))\\ &+ 7\alpha^2\beta^4(1+k)(5m^2 - 24m + 24)\\ &+ 7\beta^6(m^3 - 2m^2 + 48m - 48),\end{aligned}$$

$$\begin{aligned}\gamma_{7,m} := \ &-\beta^6(-m^3 - 254m^2 - 2256m + 2512)\\ &+ 7\alpha^2\beta^4(1+k)(3m^2 + 88m - 88)\\ &+ 7\alpha^4\beta^2(5(k^2+1)(m-2) + k(70m + 292))\\ &+ 7\alpha^6(k^3 + 135k^2 + 135k + 1).\end{aligned}$$

(3) *Additionally, in order to be a periodic solution of 5th and 7th-mKdV equations (and also for the classical non periodic mKdV), the parameters m, k, α and β must satisfy the following commensurability conditions on the spatial periods as in* (11.4.3)

$$\frac{\beta^4}{\alpha^4} = \frac{k}{1-m}, \qquad K(k) = \frac{\alpha}{2\beta}K(m),$$

where K is the complete elliptic integral of the first kind, defined as [ByFr71]

$$\begin{aligned}K(r) :=& \int_0^{\pi/2} (1 - r\sin^2(s))^{-1/2} ds\\ =& \int_0^1 ((1-t^2)(1-rt^2))^{-1/2} dt,\end{aligned}$$

and which satisfies

$$K(0) = \frac{\pi}{2} \quad \text{and} \quad \lim_{k\to 1^-} K(k) = \infty.$$

(4) *The spatial period of the breather is given by*

$$L := \frac{4}{\alpha}K(k) = \frac{2}{\beta}K(m).$$

Note that conditions (11.4.3) formally imply that the periodic breather B (12.9.3) has only four independent parameters (e.g. β, k and translations x_1, x_2). Additionally, if we assume that the ratio β/α stays bounded, we have that k approaches 0 as m is close to 1. Using this information, the standard non periodic 5th and 7th-mKdV breathers can be formally recovered as the limit of very large spatial period $L \to +\infty$, obtained e.g. if $k \to 0$.

Note that these breathers can be written using only two parametric variables, say β and k, and have a characteristic period $L = L(\beta, k)$, with $L \to +\infty$ as $k \to 0$.

Moreover, compare the periodic higher order velocities $(\delta_{i,m}, \gamma_{i,m})$, $i = 5, 7$ above, with the equivalent periodic ones in the simpler classical mKdV case [AMP17b, Def.1.1]:

$$\delta := \alpha^2(1 + k) + 3\beta^2(m - 2), \quad \text{and} \quad \gamma := 3\alpha^2(1 + k) + \beta^2(m - 2),$$

and with velocities (δ_i, γ_i), $i = 5, 7$, ((12.4.6) when $\mu = 0$) in the non periodic case:

$$\delta_5 := -\alpha^4 + 10\alpha^2\beta^2 - 5\beta^4, \quad \gamma_5 := -\beta^4 + 10\alpha^2\beta^2 - 5\alpha^4,$$

and

$$\begin{aligned} \delta_7 &:= \alpha^6 - 21\alpha^4\beta^2 + 35\alpha^2\beta^4 - 7\beta^6, \\ \gamma_7 &:= -\beta^6 + 21\alpha^2\beta^4 - 35\alpha^4\beta^2 + 7\alpha^6. \end{aligned}$$

12.10 Exercises

(12.1) Obtain the corresponding 13th order Gardner equation from the 13th order mKdV equation (12.8.3).

(12.2) Construct the 13th-Gardner breather solution of the above 13th order Gardner equation.

Bibliography

[AbCl91] M. Ablowitz and P. Clarkson, *Solitons, nonlinear evolution equations and inverse scattering*, London Mathematical Society Lecture Note Series, 149. Cambridge University Press, Cambridge, 1991.

[AKNS74] M.J. Ablowitz, D.J. Kaup, A.C. Newell, and H. Segur, *The inverse scattering transform-Fourier analysis for nonlinear problems*, Stud. Appl. Math. 53, 249-315 (1974).

[AbSe81] M.J. Ablowitz, and H. Segur, *Solitons and the inverse scattering transform*. SIAM Studies in Applied Mathematics, 4. Society for Industrial and Applied Mathematics (SIAM), Philadelphia, Pa., 1981. x+425 pp. ISBN: 0-89871-174-6.

[AAT09] N. Akhmediev, A. Ankiewicz, and M. Taki, *Waves that appear from nowhere and disappear without a trace*, Phys. Lett. A 373 (2009) 675–678.

[AKF10] N. Akhmediev, B. Kibler, J. Fatome, C. Finot, G. Millot, F. Dias and G. Genty, *The Peregrine soliton in nonlinear fibre optics*, Nature Physics **6**, 790 (2010).

[AkKo86] N. Akhmediev, and V.I. Korneev, *Modulation instability and periodic solutions of the nonlinear Schrödinger equation*. Theor. Math. Phys. 69, 1089–1093 (1986).

[ASA09] N. Akhmediev, J. M. Soto-Crespo, and A. Ankiewicz, *Extreme waves that appear from nowhere: On the nature of rogue waves*, Physics Letters A 373 (2009) 2137–2145.

[ACM24] F. Alegría, G. Chen, C. Muñoz, F. Poblete and B. Tardy, *On uniqueness of KP soliton structures*, J. Differential Equations 445 (2025), Paper No. 113569, 54 pp.

[Ale12a] M.A. Alejo, *Geometric breathers of the mKdV*, Acta App. Math., **121** no. 1, p.137–155 (2012).

[Ale12b] M.A. Alejo, *Well-posedness and stability results for the Gardner equation.* Nonlinear Differential Equations and Applications NoDEA **19**, no. 4 (2012), 503–520.

[Ale12c] M.A. Alejo, *Focusing mKdV breather solutions with nonvanishing boundary conditions by the Inverse Scattering Method*, Jour. NonLin. Math. Phys. **19**, no. 1 (2012) 1250009.

[Ale12d] M.A. Alejo, *On the ill-posedness of the Gardner equation*, Journal of Mathematical Analysis and Applications **396**, Issue 1 (2012), 256–260.

[Ale18] M.A. Alejo, *Nonlinear stability of Gardner breathers.* J. Differential Equations 264, no. 2, 1192–1230, (2018).

[Ale24] M.A. Alejo, *Higher order mKdV breathers: nonlinear stability,* Proyecciones, vol. 43, no. 2, pp. 495-520, (2024).

[AaCa18] M.A. Alejo and E. Cardoso, *On the ill-posedness of the 5th-order Gardner equation*, São Paulo J. Math. Sci. (2019).

[Ale23] M.A. Alejo, *Dynamics of breathers in the Gardner hierarchy: universality of the variational characterization*, Proyecciones (Antofagasta, On line), vol. 42, no. 4, pp. 833-860, Jul. (2023).

[Ale24] M.A. Alejo, *Higher order mKdV breathers: nonlinear stability*, Proyecciones (Antofagasta) **43,** no. 2, 495-520, (2024).

[ACC22] M.A. Alejo, M. Cavalcante, and A. Corcho, *Stability of mKdV breathers on the half-line*. Partial Differ. Equ. Appl. 3, no. 6, Paper No. 70, (2022).

[ACK22] M.A. Alejo, A. Corcho, H. Kalisch and D. Pilod *Breather Solutions to the Cubic Whitham Equation,* arXiv:2201.12074v1.

[AFM18] M.A. Alejo, L. Fanelli and C. Muñoz, *The Akhmediev breather is unstable*, São Paulo Journal Math. Sci. vol. 13, pp. 391–401 (2019).

[AFM21] M.A. Alejo, L. Fanelli, and C. Muñoz, *Stability and instability of breathers in the* $U(1)$ *Sasa-Satsuma and Nonlinear Schrödinger models*, Nonlinearity 34 (2021), no. 5, 3429–3484.

[AlMu12] M.A. Alejo and C. Muñoz, *On the nonlinear stability of mKdV breathers*, J. Phys. A: Math. Theor. **45** 432001 (2012).

[AlMu13] M.A. Alejo, and C. Muñoz, *Nonlinear stability of mKdV breathers*, Comm. Math. Phys. (2013), **324**, Issue 1, pp. 233–262.

[AlMu15] M.A. Alejo, and C. Muñoz, *Dynamics of complex-valued modified KdV solitons with applications to the stability of breathers*, Anal. and PDE. 8 (2015), no. 3, 629–674.

[AMP17] M.A. Alejo, C. Muñoz, and J.M. Palacios, *On the Variational Structure of Breather Solutions I: Sine-Gordon case*, J. Math. Anal. Appl.

Vol.453/2 (2017) pp. 1111–1138. DOI 10.1016/j.jmaa.2017.04.056.

[AMP17b] M.A. Alejo, C. Muñoz, and J. M. Palacios, *On the variational structure of breather solutions II: Periodic mKdV equation.* Electron. J. Differential Equations (2017), Paper No. 56, 26 pp.

[AMP20] M.A. Alejo, C. Muñoz, and J.M. Palacios, *On asymptotic stability of the Sine-Gordon kink in the energy space*, arXiv:2003.09358 (2020). Comm. Math. Phys. **402**, 581-636, (2023).

[AMV13] M.A. Alejo, C. Muñoz, and L. Vega, *The Gardner equation and the L^2-stability of the N-soliton solution of the Korteweg-de Vries equation* Trans. of the AMS 365 (2013), 1, 195–212.

[Aub97] S. Aubry, *Breathers in nonlinear lattices: Existence, linear stability and quantization*, Physica D no. 103, 201–250 (1997).

[BEM71] A. Barone, F. Esposito, C.J. Magee, A.C. Scott. *Theory and applications of the Sine-Gordon equation,* La Rivista del Nuovo Cimento (1971-1977), (1971), 1(2): 227-267.

[BeZa78] V. Belinskii, and V. Zakharov, *Integration of the Einstein Equations by Means of the Inverse Scattering Problem Technique and Construction of Exact Soliton Solutions*, Sov. Phys. JETP. 48 (6): 985–994 (1978).

[Ben72] T.B. Benjamin, *The stability of solitary waves*, Proc. Roy. Soc. London A **328**, 153–183 (1972).

[BiMy76] I. Bialynicki-Birula and J. Mycielski, *Nonlinear wave mechanics.* Ann. Physics, 100, 1-2, p.62-93, (1976).

[BCB11] C. Blank, M. Chirilus-Bruckner, V. Lescarret and G. Schneider. *Breather solutions in periodic media*, Comm. Math. Phys. **302** (3) p. 815–841 (2011).

[BPS14] J.L. Bona, G. Ponce, J.-C. Saut, and C. Sparber, *Dispersive blow-up for nonlinear Schrödinger equations revisited*, J. Math. Pures Appl. 102 (2014), no. 4, 782–811.

[BSS87] J.L. Bona, P. Souganidis and W. Strauss, *Stability and instability of solitary waves of Korteweg-de Vries type*, Proc. Roy. Soc. London **411**, 395–412 (1987).

[BoKa00] H. Kalisch and J.L. Bona, *Models for internal waves in deep water*, Discrete & Continuous Dynamical Systems, **6**-1, (2000).

[Bou99] J. Bourgain, *Global solutions of nonlinear Schrödinger equations*, American Mathematical Society Colloquium Publications, 46. American Mathematical Society, Providence, RI, 1999. viii+182 pp.

[39] J. Boussinesq, *Théorie des ondes et des remous qui se propagent le long d'un canal rectangulaire horizontal, en communiquant au liquide contenu dans ce canal des vitesses sensiblement pareilles de la surface*

au fond, J. Math. Pure Appl. (2) 17 (1872), 55–108.

[ByFr71] P.F. Byrd and M.D. Friedman, *Handbook of Elliptic Integrals for Engineers and Scientists*, 2nd ed., Springer-Verlag, New York and Heidelberg, 1971.

[BCN80] H. Brezis, J.M. Coron, L. Nirenberg, *Free vibrations for a nonlinear wave equation and a theorem of P. Rabinowitz*. Commun. Pure Appl. Math. 33(5), 667-684 (1980).

[ByFr71] P.F. Byrd and M.D. Friedman, *Handbook of Elliptic Integrals for Engineers and Scientists*, 2nd ed., Springer-Verlag, New York and Heidelberg, 1971.

[CaGa18] R. Carles and I. Gallagher, *Universal dynamics for the defocusing logarithmic Schrödinger equation*. Duke Math. J. 167, p.1761-1801, (2018).

[Car18] J.D. Carter, *Bidirectional Whitham equations as models of waves on shallow water*, Wave Motion 82, 51 (2018).

[Car22] J.D. Carter, H. Kalisch, C. Kharif and M. Abid, *The Cubic Vortical Whitham Equation*, Wave Motion, **110** (2022).

[CaMu19] M. Cavalcante and C. Muñoz, *Stability of KdV solitons on the half-line*, Rev. Mat. Iberoam. **35** (2019), no. 6, 1809–1834.

[Caz03] T. Cazenave, *Semilinear Schrödinger equations*, Courant Lecture Notes in Mathematics, 10. New York University, Courant Institute of Mathematical Sciences, New York; American Mathematical Society, Providence, RI, 2003. xiv+323 pp. ISBN: 0-8218-3399-5.

[CaLi82] T. Cazenave, and P.-L. Lions, *Orbital stability of standing waves for some nonlinear Schrödinger equations*, Comm. Math. Phys. 85 (1982), no. 4, 549–561.

[CNH12] A. Chabchoub, S. Neumann, N. Hoffmann and N. Akhmediev, *Spectral properties of the Peregrine soliton observed in a water wave tank*, Journal of Geophysical Research: Oceans **117**, (2012), 054104.

[CLL20] G. Chen, J. Liu, and B. Lu, *Long-time asymptotics and stability for the Sine-Gordon equation*. arXiv: 2009.04260 (Submitted).

[ChLi21] G. Chen, and J. Liu, *Soliton resolution for the focusing modified KdV equation*, Ann. Inst. H. Poincaré C Anal. Non Linéaire 38 (2021), no. 6, 2005–2071.

[ChLi22] G. Chen, and J. Liu, *Long-time asymptotics of the modified KdV equation in weighted Sobolev spaces*. Forum Math. Sigma 10 (2022), No. e66, 52 pags.

[CLQY21] X.Y. Cheng, D. Li, C.Y. Quan and W. Yang. *On a parabolic Sine-Gordon model*, Numer. Math. Theory Methods Appl, 14 (2021), no.

4, 1068-1084.

[ChPe24] J. Chen and D.E. Pelinovsky, *Bright and dark breathers of the Benjamin-Ono equation on the traveling periodic background*, Wave Motion **126**, 103263, (2024).

[CKS03] J. Colliander, M. Keel, G. Staffilani, H. Takaoka, and T. Tao, *Sharp global well-posedness for KdV and modified KdV on* $\mathbb{R}$ *and* $\mathbb{T}$, J. Amer. Math. Soc. **16** (2003), no. 3, 705–749 (electronic).

[Cor82] J.M. Coron, *Periodic solutions of a nonlinear wave equation without assumption of monotonicity.* Math. Ann. 262(2), 273-285 (1983)

[CoFr20] R. Côte and X. Friederich, *On smoothness and uniqueness of multi-solitons of the non-linear Schrödinger equations,* Comm. Partial Diff. Eqs, **46** (12), 2325-2385, (2021).

[CoSe23] R. Côte, and A. Semenov, *Improved uniqueness of multi-breathers of the modified Korteweg–de Vries equation.* Vietnam J. Math. 51 (2023), no. 4, 935–949.

[CGK05] W. Craig, P. Guyenne, and H. Kalisch, *Hamiltonian long-wave expansions for free surfaces and interfaces,* Comm. Pure and App. Math. **58**, 1587 (2005).

[JKW14] J. Cuevas-Maraver, P. G. Kevrekidis, and F. Williams *The Sine-Gordon model and its applications. From pendula and Josephson junctions to gravity and high-energy physics.* Edited by Jesús Cuevas-Maraver, Panayotis G. Kevrekidis and Floyd Williams. Nonlinear Systems and Complexity, 10. Springer, Cham, 2014. xiv+263 pp. ISBN: 978-3-319-06721-6.

[DHB75] R.F. Dashen, B. Hasslacher and A. Neveu, *Particle spectrum in model field theories from semiclassical functional integral techniques*, Physical Review D, **11**-12, p.3424, (1975).

[DaPe06] T. Dauxois, and M. Peyrard, *Physics of solitons*, Cambridge University Press, Cambridge, 2010. xii+ 422pp.

[Den93] J. Denzler, *Nonpersistence of breather families for the perturbed Sine-Gordon equation*, Comm. Math. Phys. **158**, 397–430 (1993).

[DyTr99] K. Dysthe and K. Trulsen, *Note on breather type solutions of the NLS as models for freak-waves*, Phys. Scripta **82**, 48 (1999).

[DuMe08] T. Duyckaerts, and F. Merle , *Dynamic of threshold solutions for energy-critical NLS*, GAFA **18** (2008), 1787–1840.

[EhKa09] M. Ehrnström and H. Kalisch, *Traveling waves for the Whitham equation*, Differential and Integral Equations, **22**-11/12, p.1193–1210, (2009).

[EGW12] M. Ehrnström, M. Groves and E. Wahlén, *On the existence and sta-*

bility of solitary-wave solutions to a class of evolution equations of Whitham type, Nonlinearity , **25**-10, p.2903, (2012).

[EhKa13] M. Ehrnström and H. Kalisch, *Global bifurcation for the Whitham equation*, Mathematical Modelling of Natural Phenomena, **8**-5, p.13–30, (2013).

[Eme21] L. Emerald, *Rigorous derivation from the water waves equations of some full dispersion shallow water models*, SIAM Journal on Mathematical Analysis, **53**-4, p.3772–3800, (2021).

[EFM85] N. Ercolani, M. G. Forest and D. W. McLaughlin, *Modulational stability of two-phase Sine-Gordon wave trains*, Studies in Applied Math (2), 91-101 (1985).

[FFM24] J. Faya, P. Figueroa, C. Muñoz and F. Poblete, *Uniqueness of quasi-monochromatic breathers for the generalized Korteweg-de Vries and Zakharov-Kuznetsov models,* arXiv:2404.09100 (2024).

[Fer20] G. Ferriere, *The focusing logarithmic Schrödinger equation: analysis of breathers and nonlinear superposition,* DCDS-A, 40-11, p.6247-6274, (2020).

[GGKM67] C.S. Gardner, J.M. Greene, M.D. Kruskal and R.M. Miura, *Method for solving the Korteweg-de Vries equation*, Phys. Rev. Lett. 19 1095-7, (1967).

[GKM68] C.S. Gardner, M.D. Kruskal and R. Miura, *Korteweg-de Vries equation and generalizations. II. Existence of conservation laws and constants of motion*, J. Math. Phys. **9**, no. 8, 1204–1209 (1968).

[GaPe15] T. Gallay, and D.E. Pelinovsky, *Orbital stability in the cubic defocusing NLS equation. Part I: Cnoidal periodic waves,* J. Diff. Equations **258** (2015), 3607–3638.

[GaXu23] M. Gao and J. Xu, *Convergence to steady states of parabolic Sine-Gordon,* Dynamics of Partial Differential Equations, **20** no.3, (2023), 227-248.

[GNN81] B. Gidas, Wei Ming Ni, and L. Nirenberg. *Symmetry of positive solutions of nonlinear elliptic equations in* $\mathbb{R}^n$. In Mathematical analysis and applications, Part A, volume 7 of Adv. in Math. Suppl. Stud., pages 369-402. Academic Press, New York-London, (1981).

[GilTru98] D. Gilbarg; N.S. Trudinger. *Elliptic partial differential equations of second order.* Reprint of the 1998 edition. Classics in Mathematics. Springer-Verlag, Berlin, (2001). xiv+517 pp. ISBN: 3-540-41160-7

[GoPe91] R. Goldstein, D.M. Petrich, *The Korteweg-de Vries hierarchy as dynamics of closed curves in the plane*, Phys. Rev. Lett. **67** (1991) 3203–3206.

[GradRyz07] I.S. Gradshteyn and I.M. Ryzhik, *Table of Integrals, Series, and Products*, Academic Press 2007.

[Gri81] R. Grimshaw, *Evolution equations for long nonlinear internal waves in stratified shear flows,* Stud. Appl. Math. **65** (1981) 159-188.

[GFZ11] J.F. Gomes, G.S. França and A.H. Zimerman, *Nonvanishing boundary condition for the mKdV hierarchy and the Gardner equation* (2012) J. Phys. A: Math. Theor. 45 015207.

[AGV13] C. Gorria, M.A. Alejo and L. Vega, *Discrete conservation laws and the convergence of long time simulations of the mKdV equation*, J. Comp. Phys. 235 274–285 (2013).

[Gre91] L. Greenberg, *An oscillation method for fourth order, self-adjoint, two-point boundary value problems with nonlinear eigenvalues*, SIAM J. Math. Anal. 22 (1991), no. 4, 1021–1042.

[GSS87] M. Grillakis, J. Shatah, and W. Strauss, *Stability theory of solitary waves in the presence of symmetry. I.* J. Funct. Anal. 74 (1987), no. 1, 160–197.

[GSP10] R. Grimshaw, A. Slunyaev and E. Pelinovsky, *Generation of solitons and breathers in the extended Korteweg-de Vries equation with positive cubic nonlinearity.* Chaos 20 (2010), n.1, 01310201–01310210.

[GLN10] P. Guerrero, J.L. Lopez and J. Nieto, *Global H^1 solvability of the 3D logarithmic Schrödinger equation,* Nonlinear Analysis: Real World Applications 11-1, p.79-87, (2010).

[GGK12] D.R. Gulevich, M.B. Gaifullin and F.V. Kusmartsev, *Controlled dynamics of Sine-Gordon breather in long Josephson junctions*, Eur. Phys. J. B **85**: 24 (2012).

[HaOy24] M. Hayashi and T. Ozawa, *The Cauchy Problem for the Logarithmic Schrödinger Equation Revisited,* Annales de l´ Institute Henri Poincaré, (2024).

[HPW82] D.B. Henry, J.F. Perez and W.F. Wreszinski, *Stability Theory for Solitary-Wave Solutions of Scalar Field Equations*, Comm. Math. Phys. **85**, 351–361 (1982).

[HeRe81] E. Hernandez and B. Remaud, *General properties of gausson-conserving descriptions of quantal damped motion.* Physica A: Statistical Mechanics and its Applications, 105-1, p.130-146, (1981).

[Hir19] A. Hirsch and W. Reichel, *Real-valued, time-periodic localized weak solutions for a semilinear wave equation with periodic potentials*, Nonlinearity, **32**-4, p. 1408, (2019).

[Hir72] R. Hirota, *Exact solution of the* modified *Korteweg-de Vries equation for multiple collisions of solitons*, J. Phys. Soc. Japan, **33**, no. 5, 1456–

1458 (1972).

[HMP23] M.A. Hoefer, A. Mucalica and D.E Pelinovsky , *KdV breathers on a cnoidal wave background,* J. Phys. A: Math. Theor. **56**, p.185701 (2023).

[HoWa13] A. Hoffman, and C.E. Wayne, *Orbital stability of localized structures via Bäcklund transformations*, Differential Integral Equations 26:3-4 (2013), 303–320.

[HPZ11] J. Holmer, G. Perelman, and M. Zworski, *Effective dynamics of double solitons for perturbed mKdV*, Comm. Math. Phys. 305, 363–425 (2011).

[Hol06] J. Holmer, *The initial-boundary value problem for the Korteweg-de Vries equation*, CPDE, **31** (2006), 1151–1190.

[Inc56] E.L. Ince, *Ordinary Differential Equations*, New York: Dover Publications, 1956.

[IKT08] A. Ionescu, C. Kenig, D. Tataru, *Global well-posedness of the KP-I initial-value problem in the energy space*, Invent. Math. 173 (2) (2008) 265–304.

[JoPe15] E.R. Johnson, and D. Pelinovsky, *Orbital stability of periodic waves in the class of reduced Ostrovsky equations*, Jour. Diff. Eqs **261**, 3268-3304 (2016).

[Ka69] T. Kakutani, *Weakly nonlinear Hydromagnetic waves in a cold collision free plasma*, J.Phys.Soc.Japan, **26** 5 (1969).

[KaNg10] H. Kalisch and N. T. Nguyen, *On the stability of internal waves,* Journal of Physics A 43, 495205 (2010).

[KhAb18] C. Kharif and M. Abid, *Nonlinear water waves in shallow water in the presence of constant vorticity: A Whitham approach,* European Journal of Mechanics-B/Fluids 72, 12 (2018).

[Kap07] T. Kapitula, *On the stability of N-solitons in integrable systems*, Nonlinearity, 20 (2007) 879–907.

[Kat59] T. Kato, *Growth properties of solutions of the reduced wave equation with a variable coefficient.* Comm. Pure Appl. Math. 12, 403-425 (1959).

[KEB00] W. Krolikowski, D. Edmundson, and O. Bang. *Unified model for partially coherent solitons in logarithmically nonlinear media.* Phys. Rev. E, 61, p.3122-3126, (2000).

[KPV93] C.E. Kenig, G. Ponce, and L. Vega, *Well-posedness and scattering results for the generalized Korteweg-de Vries equation via contraction principle*, Comm. Pure Appl. Math. 46 (1993), 527–620.

[KPV01] C.E. Kenig, G. Ponce and L. Vega, *On the ill-posedness of some canonical dispersive equations*, Duke Math. J. **106**, no. 3, 617–633 (2001).

[KKS04] P.G. Kevrekidis, A. Khare, A. Saxena and G. Herring, *On some classes of mKdV periodic solutions*, Journal of Physics A: Mathematical and General, **37**, 10959-10965 (2004).

[KKS03] P.G. Kevrekidis, A. Khare and A. Saxena, *Breather lattice and its stabilization for the modified Korteweg-de Vries equation*, Phys. Rev. E, **68**, 0477011-0477014 (2003).

[KFF10] B. Kibler, J. Fatome, C. Finot, G. Millot, F. Dias, G. Genty, N. Akhmediev, and J.M. Dudley, *The Peregrine soliton in nonlinear fibre optics*, Nature Physics 6, 790–795 (2010) doi:10.1038/nphys1740.

[KFF12] B. Kibler, J. Fatome, C. Finot, G. Millot, G. Genty, B. Wetzel, N. Akhmediev, F. Dias, and J.M. Dudley, *Observation of Kuznetsov-Ma soliton dynamics in optical fibre*, Sci. Rep. 2, 463; DOI:10.1038/srep00463 (2012).

[KiOl92] S. Kichenassamy and P.J. Olver, *Existence and non-existence of solitary wave solutions to higher order model evolution equations,* SIAM J. Math. Anal. **3** (1992), 1141-1166.

[KlHa17] C. Klein, and M. Haragus, *Numerical study of the stability of the Peregrine breather*, Annals of Mathematical Sciences and Applications Vol. 2 (2017) No 2, 217–239.

[KLP18] C. Klein, F. Linares, D. Pilod, and J.-C. Saut, *On Whitham and related equations,* Studies in Applied Mathematics 140, 133 (2018).

[KS15] C. Klein, and J.-C. Saut, *IST versus PDE: a comparative study.* Hamiltonian partial differential equations and applications, 383–449, Fields Inst. Commun., 75, Fields Inst. Res. Math. Sci., Toronto, ON, (2015).

[KSW22] C. Klein, J.-C. Saut and Y. Wang, *On the modified fractional Korteweg-de Vries and related equations*, Nonlinearity, **35**, p.1170–1212, (2022).

[KoSa95] T.S. Komatsu and S.I. Sasa, *Kink soliton characterizing traffic congestion*, Phys. Rev. E 52, 5574 (1995);

[KoBu81] C.G. Koop and G. Butler, *An investigation of internal solitary waves in a two-fluid system,* Journal of Fluid Mechanics 112, 225 (1981).

[KdV95] D. J. Korteweg and G. de Vries, *On the change of form of long waves advancing in a rectangular canal, and on a new type of long stationary waves*, Philos. Mag. **39** (1895), 422–443.

[KoKo74] A.M. Kosevich and A.S. Kovalev, *Selflocalization of vibrations in a one-dimensional anharmonic chain*, Sov. Phys. JETP, **67**, 1793, (1974).

[KMM17] M. Kowalczyk, Y. Martel, and C. Muñoz, *Nonexistence of small, odd breathers for a class of nonlinear wave equations*, Letters in Mathematical Physics, (2017) Vol. 107, Issue 5, 921–931.

[KMM21] M. Kowalczyk, Y. Martel, C. Muñoz, and H. Van Den Bosch, *A sufficient condition for asymptotic stability of kinks in general (1+1)-scalar field models*. Ann. PDE 7 (2021), no. 1, Paper No. 10, 98 pp.

[KSB98] J.D. Kress, A. Saxena, A.R. Bishop and R.L. Martin, Phys. Rev B **58**, no.10 6161 (1998).

[Kur78] R. Kurth, *A global particular solution of the initial-value problem of stellar dynamics*, Quat. Appl. Mathematics **325**, 1978.

[Kuz77] E. Kuznetsov, *Solitons in a parametrically unstable plasma*, Sov. Phys. Dokl. 22, 507–508 (1977).

[Lam80] G.L. Lamb, *Elements of Soliton Theory*, Pure Appl. Math., Wiley, New York, 1980.

[Lam71] G.L. Lamb, Jr. *Analytical descriptions of ultrashort optical pulse propagation in a resonant medium,* Rev. Modern Phys, 43 (1971), 99–124.

[Lan13] D. Lannes, *The water-wave problem: mathematical analysis and asymptotics*, American Mathematical Soc. **188**, 2013.

[Law13] D.F. Lawden, *Elliptic Functions and Applications,* Springer New York, NY, (2013).

[Lax68] P.D. Lax, *Integrals of nonlinear equations of evolution and solitary waves*, Comm. Pure Appl. Math. **21**, 467–490 (1968).

[LQT22] D. Li, C.Y. Quan, T. Tang and W. Yang. *On symmetry breaking of Allen-Cahn,* CSIAM Trans. Appl. Math. 3 (2022), no 2, 221-243.

[LQT21] D. Li, C.Y. Quan, T, Tang and W. Yang. *Sharp convergence to steady states of Allen-Cahn,* (2021), arXiv:2106.08534.

[LiPo14] F. Linares and G. Ponce, *Introduction to nonlinear dispersive equations*, Second edition. Universitext. Springer, New York, 2015. xiv+301 pp.

[Ma79] Y.C. Ma, *The perturbed plane-wave solutions of the cubic Schrödinger equation*, Stud. Appl. Math. 60, 43–58 (1979).

[MaSa93] J.H. Maddocks and R.L. Sachs, *On the stability of KdV multi-solitons*, Comm. Pure Appl. Math. 46, 867–901 (1993).

[Man21] R. Mandel, *A uniqueness result for the Sine-Gordon breather*, Part. Diff. Eqns. Appl. Springer, vol. 2(2), pp. 1–8 (2021).

[MaSc21] R. Mandel and D. Schneider, *Variational methods for breather solutions of nonlinear wave equations*, Nonlinearity, **34**, 6, (2021).

[MaSm90] T.R. Marchant and N.F. Smyth, *The extended Korteweg-de Vries equa-*

tion and the resonant flow of a fluid over topography, J. Fluid Mech. **221** (1990), 263-288.

[Mar05] Y. Martel, *Asymptotic N-soliton-like solutions of the subcritical and critical generalized Korteweg-de Vries equations,* American Journal of Mathematics, 127 (2005) no. 5, 1103-1140.

[MaMe01] Y. Martel, and F. Merle, *Asymptotic stability of solitons for subcritical gKdV equations*, Arch. Ration. Mech. Anal. **157**, no. 3, 219–254 (2001).

[MaMe06] Y. Martel, F. Merle, *Multi solitary waves for nonlinear Schrödinger equations*, AN 23 (2006), 849-864.

[MMT02] Y. Martel, F. Merle and T.-P. Tsai, *Stability and asymptotic stability in the energy space of the sum of N solitons for subcritical gKdV equations*, Comm. Math. Phys. **231** 347–373 (2002).

[MaMe09] Y. Martel and F. Merle, *Stability of two soliton collision for nonintegrable gKdV equations*, Comm. Math. Phys. **286**, 39–79 (2009).

[MaMe05] Y. Martel and F. Merle, *Asymptotic stability of solitons of the subcritical gKdV equations revisited*, Nonlinearity **18**, 55–80 (2005).

[Mat80] Y. Matsuno, *Bilinearization of Nonlinear Evolution Equations: Higher Order mKdV*, Jour. Phys. Soc. Japan, **49**, n.2 (1980).

[MaSa91] V.B. Matveev, and M.A. Salle, *Darboux transformations and solitons.* Springer Ser. Nonlinear Dynam., Springer-Verlag, Berlin, 1991.

[MeVe03] F. Merle and L. Vega, *L^2 stability of solitons for KdV equation*, Int. Math. Res. Not., no. 13, 735–753 (2003).

[Mil90]] T. Miloh, *On periodic and solitary wavelike solutions of the intermediate long-wave equation,* J. Fluid Mech. 211, 617-627 (1990).

[Miu68] R.M. Miura, C. Gardner, and M. Kruskal, *Korteweg-de Vries equation and generalizations. II. Existence of conservation laws and constants of motion*, J. Math. Phys. 9, no. 8 (1968), 1204-1209. crdfxv

[MiPe12] T. Mizumachi, and D. Pelinovsky, *Bäcklund transformation and L^2-stability of NLS solitons*, Int. Math. Res. Not. IMRN (2012), no. 9, 2034–2067.

[MKD15] D. Moldabayev, H. Kalisch, and D. Dutykh, *The Whitham equation as a model for surface water waves,* Physica D 309, 99 (2015).

[Mun14] C. Muñoz, *Stability of integrable and nonintegrable structures*, Adv. Differential Equations Volume 19, Number 9/10 (2014), 947–996.

[Mu17] C. Muñoz, *Instability in nonlinear Schrodinger breathers*, Proyecciones Vol. 36, no. 4 (2017) 653–683.

[Mun16] C. Muñoz, *The Gardner equation and the stability of the multi-kink solutions of the mKdV equation.* DCDS 36 (7), 3811–3843 (2016).

[Mun17] C. Muñoz, *Instability in nonlinear Schrödinger breathers*, Proyecciones 36 (2017), no. 4, 653–683.

[MuPa19] C. Muñoz, and J.M. Palacios, *Nonlinear stability of 2-solitons of the Sine-Gordon equation in the energy space*, Ann. IHP C Analyse Nonlinéaire. 36 (2019), no. 4, 977–1034.

[MuPe24] A. Mucalica and D. Pelinovsky, *Dark breathers on a snoidal wave background in the defocusing mKdV equation*, Lett. Math. Phys. **114**, 100 (2024).

[MuPo19] C. Muñoz and G. Ponce, *Breathers and the dynamics of solutions to the KdV type equations*, Comm. Math. Phys. **367** (2019), no. 2, 581–598.

[MuPo19bis] C. Muñoz and G. Ponce, *On the asymptotic behavior of solutions to the Benjamin-Ono equation*, arXiv:1810.02329, Proc. Amer. Math. Soc. 147 (2019), no. 12, 5303-5312.

[MPS20] C. Muñoz, G. Ponce and J.-C. Saut *On the long time behavior of solutions to the Intermediate Long Wave equation*, SIAM J. Math. Anal. 53 (2021), no. 1, 1029-1048.

[Naga99] T. Nagatani, *Jamming transitions and the modified Korteweg-de Vries equation in a two-lane traffic flow*, Physica A 265, 297 (1999).

[Nak79] A. Nakamura, *Bäcklund transform and conservation laws of the Benjamin-Ono equation,* J. Phys. Soc. Japan **47**, 1335-1340 (1979).

[NSW92] K. Nakayama, H. Segur, M. Wadati, *Integrability, the motion of curves*, Phys. Rev. Lett. **69** (1992) 2603–2606.

[NaWa93] K. Nakayama, M. Wadati, *Motion of curves in the plane*, J. Phys. Soc. Jpn. **62** (1993) 473–479.

[NeLo06] A. Neves and O. Lopes, *Orbital stability of double solitons for the Benjamin-Ono equation*, Commun. Math. Phys. **262**, 757–791 (2006).

[OhWa82] K. Ohkuma and M. Wadati, *Multi-pole solutions of the modified Korteweg-de Vries equation*, Journ. Phys. Soc. Japan **51** no.6 (1982) 2029–2035.

[Olv84] P.J. Olver, *Hamiltonian perturbation theory and water waves*, Contemp. Math. 28 (1984), 231-249.

[Olv77] P.J. Olver, *Evolution equations possessing infinitely many symmetries*, J. Math. Phys. 18, 1212 (1977).

[Ono92] H. Ono, *Soliton Fission in Anharmonic Lattices with Reflectionless Inhomogeneity*, J. Phys. Soc. Jpn. 61, 4336 (1992).

[PeWe94] R.L. Pego and M.I. Weinstein, *Asymptotic stability of solitary waves*, Commun. Math. Phys. **164**, 305–349 (1994).

[PeYa05] D. Pelinovsky, and J. Yang, *Stability analysis of embedded solitons in the generalized third-order nonlinear Schrödinger equation*, CHAOS

15, 037115 (2005).

[PeGr97] D. Pelinovsky and R. Grimshaw, *Structural transformation of eigenvalues for a perturbed algebraic soliton potential,* Phys. Lett. A 229, no.3, 165–172 (1997).

[PeSh14] D. Pelinovsky and Y. Shimabukuro, *Orbital stability of Dirac solitons*, Lett. Math. Phys. **104**, 21–41 (2014).

[Per83] D.H. Peregrine, *Water waves, nonlinear Schrödinger equations and their solutions*, J. Austral. Math. Soc. Ser. B 25, 16–43 (1983).

[Per67] D.H. Peregrine, *Long waves on a beach.* J. Fluid Mechanics 27 (1967), 815–827.

[PeSky62] J.K. Perring, T.H.R. Skyrme. *A model unified field equation,* Nuclear Phys. 31, (1962), 550-555.

[PluRei16] M. Plum and W. Reichel, *A breather construction for a semilinear curl-curl wave equation with radially symmetric coefficients.* J. Elliptic Parabol. Equ. 2(1-2), 371-387 (2016).

[Rab78] P.H. Rabinowitz, *Free vibrations for a semilinear wave equation.* Commun. Pure Appl. Math. 31(1),31-68 (1978).

[RaRe18] T. Ramming and G. Rein, *Oscillating solutions of the Vlasov-Poisson system: A numerical investigation*, Phys. D **365** 72-79 (2018).

[SaWa21] J.C. Saut and Y. Wang, *Long time behavior of the fractional Korteweg-de Vries equation with cubic nonlinearity*, Discrete & Continuous Dynamical Systems, **41**-3, p.1133, (2021).

[SaSa91] N. Sasa, and J. Satsuma, *New-type of soliton solutions for a higher-order nonlinear Schrödinger equation*, J. Phys. Soc. Japan Vol. 60, No. 2, Feb. (1991), pp. 409–417.

[SaYa74] J. Satsuma, and N. Yajima, *Initial Value Problems of One-Dimensional Self-Modulation of Nonlinear Waves in Dispersive Media*, Supplement of the Progress of Theoretical Physics, No. 55,(1974), 284–306.

[SaIs79] J. Satsuma and Y. Ishimori, *Periodic wave and rational soliton solutions of the Benjamin-Ono equation,* J. Phys. Soc. Japan **46**, 681-687 (1979).

[Schi95] W.K. Schief, *An infinite hierarchy of symmetries associated with hyperbolic surfaces,* Nonlinearity 8 (1995) 1-9.

[Sem22] A. Semenov, *Orbital stability of a sum of solitons and breathers of the modified Korteweg–de Vries equation.* Nonlinearity 35 (2022), no. 8, 4211–4249.

[Sem23] A. Semenov, *On the uniqueness of multi-breathers of the modified Korteweg-de Vries equation.* Rev. Mat. Iberoam. 39 (2023), no. 4, 1247–1322.

[Sch86] P.C. Schuur, *Asymptotic analysis of soliton problems. An inverse scattering approach*, Lecture Notes in Mathematics, 1232. Springer-Verlag, Berlin, (1986). viii+180 pp.

[Seg87] H. Segur, and M. D. Kruskal, *Nonexistence of small-amplitude breather solutions in ϕ^4 theory*, Phys. Rev. Letters Vol. 58, (1987), No. 8, p. 747.

[Schei20] D. Scheider, *Breather solutions of the cubic Klein-Gordon equation.* Nonlinearity 33(12), 7140-7166 (2020).

[SHB10] V.N. Serkin, A. Hasegawa, and T.L. Belyaeva, *Solitary waves in nonautonomous nonlinear and dispersive systems: nonautonomous solitons*, Jour. Mod. Optics **57**, n.14, (2010) p. 1456-1472.

[ShGe10] V.I. Shrira, and Y.V. Geogjaev, *What make the Peregrine soliton so special as a prototype of freak waves?* J. Eng. Math. 67, 11–22 (2010).

[SoWe99] A. Soffer, M.I. Weinstein, *Resonances, radiation damping and instability in Hamiltonian nonlinear wave equations*, Invent. Math. **136**, no. 1, 9–74 (1999).

[StOs98] T. Stanton and L. Ostrovsky, *Observations of highly nonlinear internal solitons over the continental shelf,* Geophysical Research Letters 25, 2695 (1998).

[Swa79] C.A. Swanson, *Semilinear second-order elliptic oscillation.* Canad. Math. Bull. 22(2), 139-157 (1979).

[196] M. Tajiri and Y. Murakami, *On Breather solutions of the Boussinesq equation*, Jour. Phys. Soc. Japan, **58**, n.10 (1989), 3585-3590.

[TKDV00] K. Trulsen, I. Kliakhandler, K. Dysthe and M. Velarde, *On weakly nonlinear modulation of waves on deep water*, Physics of Fluids, **12**-10, p. 2432–2437, (2000).

[Vac06] T. Vachaspati, *Kinks and domain walls*, Cambridge University Press, New York, 2006, An introduction to classical and quantum solitons.

[Wad73] M. Wadati, *The modified Korteweg-de Vries equation*, J. Phys. Soc. Japan 34 (5), 1973, 1289-1296.

[Wei85] M.I. Weinstein, *Modulational stability of ground states of nonlinear Schrödinger equations*, SIAM J. Math. Anal. **16** (1985), no. 3, 472–491.

[Wei86] M.I. Weinstein, *Lyapunov stability of ground states of nonlinear dispersive evolution equations*, Comm. Pure Appl. Math. **39**, (1986) 51–68.

[Wei87] M.I. Weinstein, *Existence and dynamic stability of solitary wave solutions of equations arising in long wave propagation*. Comm. Partial Differential Equations 12 (1987), no. 10, 1133–1173.

[ZaDy06] V.E. Zakharov, A.I. Dyachenko, and A.O. Prokofiev, *Freak waves as nonlinear stage of Stokes wave modulation instability*, Eur. J. Mech. B - Fluids 5, 677–692 (2006).

[Wex99] C. Wexler, A.T. Dorsey, Contour dynamics, waves, solitons in the quantum hall effect, Phys. Rev. B **60** (1999) 10971–10983.

[Whi65] G.B. Whitham, *A general approach to linear and non-linear dispersive waves using a Lagrangian,* Journal of Fluid Mechanics 22, 273 (1965).

[Whi74] G.B. Whitham, Linear, Nonlinear Waves, Wiley, New York, 1974.

[Yan10] J. Yang, *Nonlinear waves in integrable and nonintegrable systems*, SIAM Mathematical Modeling and Computation, 2010.

[ZaGa71] N.J. Zabusky and C.J. Galvin, *Shallow-water waves, the Korteweg-de Vries equation and solitons,* J. Fluid Mech. **47** (1971), 811–824.

[ZaSh72] V.E. Zakharov, and A.B. Shabat, *Exact theory of two-dimensional self-focusing and one-dimensional self-modulation of waves in nonlinear media*, JETP, 34 (1): 62–69 (1972).

[Zhi87] P. Zhidkov, *The Cauchy problem for a nonlinear Schrödinger equation*, Dubna, 1987.

Index

www.ingramcontent.com/pod-product-compliance
Ingram Content Group UK Ltd.
Pitfield, Milton Keynes, MK11 3LW, UK
UKHW061933100926
8556IPUK00007B/899

9 789819 832309